T0100764

HUMANS

HUMANS

PERSPECTIVES ON OUR EVOLUTION FROM WORLD EXPERTS

Edited by Sergio Almécija

Columbia University Press
New York

Columbia University Press
Publishers Since 1893
New York   Chichester, West Sussex
cup.columbia.edu

Copyright © 2023 Columbia University Press
All rights reserved

Library of Congress Cataloging-in-Publication Data
Names: Almécija, Sergio, editor.
Title: Humans : reflections about our present, past & future from the
world experts / a project by Sergio Almécija, Senior Research Scientist,
Division of Anthropology, American Museum of Natural History.
Description: First Edition. | New York : Columbia University Press, 2023. |
Includes index. Identifiers: LCCN 2022030723 | ISBN 9780231201209 (Hardback) |
ISBN 9780231201216 (Trade Paperback) | ISBN 9780231553988 (eBook)
Subjects: LCSH: Human beings—Origin. | Human evolution.
Classification: LCC GN281 .H855 2023 | DDC 599.93/8—dc23/eng/20221021
LC record available at https://lccn.loc.gov/2022030723

Columbia University Press books are printed on permanent and
durable acid-free paper.
Printed in the United States of America

Cover design: Philip Pascuzzo
Cover illustration: Christopher Smith

COVER ILLUSTRATION The three-panel figure represents—from left to right—humans' *past* (a fossil cranium, KNM-ER 3733, attributed to *Homo erectus*), *present* (a modern, gritty face looking at the reader), and *future* (a dissolving posthuman face).

To all past, present, and future students of humans, their nature, and their evolution, in or outside academia. May your minds stay curious and creative.

And remember:

"It might be useful to view the study of human evolution as a game, rather than as a science."

—SHERWOOD "SHERRY" WASHBURN

CONTENTS

PART I: PRELUDE: ANSWERS FROM THE EXPERTS LOOKING AT THE ANTECEDENTS TO HUMAN EVOLUTION

PART II: BEGINNINGS: ANSWERS FROM EXPERTS FOCUSING ON THE EARLY STAGES OF THE HUMAN CAREER

PART III: BECOMING HUMAN: ANSWERS FROM THOSE WITH VAST EXPERTISE ON MORE RECENT ASPECTS OF HUMAN EVOLUTION

PART IV: NOW: ANSWERS FROM EXPERTS WHO FOCUS LARGELY ON THE PRESENT (BIOLOGY. ETHNOLOGY, PRIMATOLOGY, ETC.)

PART V: OUTRO: VIEW FROM THOSE WITH EXPERIENCE IN OTHER FIELDS (E.G., MEDICINE, PSYCHOLOGY, MATHEMATICS, AND PHILOSOPHY)

PREFACE
What Is This Book?

I cannot teach anybody anything, I can only make them think.

—SOCRATES

This book is the result of a "life-hack experiment," something that might or might not have worked at all but that was worth trying. This book started years before I came up with the idea of making it. It was 2015, and as a new assistant professor of anthropology, I was honored to oversee educating young undergraduate students about human evolution. I was concerned about finding the best venue for such a mission. I didn't want to deceive the students with missing, imprecise, or wrong information. God forbid. What could I do?

Ideally, there would be a great textbook out there that my students and I could follow. The truth is that some great colleagues have authored many excellent and exciting books. However, even those books become dated after a couple of years because of continuous new discoveries (e.g., fossils, artifacts, and ancient molecules). As the media would say, "this new discovery shakes up the human evolutionary tree." Different studies produce results that are used to support opposite hypotheses about crucial aspects of human history, such as: Why do humans look the way they do today? What did the apes from which humans evolved look like? and When did all this happen?

The answers to these questions are essential for understanding fundamental aspects of our daily lives, such as our behavior, physiology, diet, and sex. Because of this, I usually prefer to *discuss* different aspects of humans and their evolution rather than "teach it." However, I *had* to teach it. That was my job! Thus, clever me, I decided, "why not just read as many scientific articles on each topic as possible the day before each lecture and present a summary of *all* of the available information the following day to my students?" I wanted to be thorough and objective; there are many exciting, well-articulated ideas out there and, likely, as many points of view as folks working in the field.

I decided to expose the fact that most topics in human evolution are still being heavily debated in the scientific community. How did this go with my students? I believe I inspired or at least entertained some of them. For others, it was a disaster. At the end of class, some students looked completely confused, lost in thought, and frustrated. I remember once a student shaking her head side to side and proclaiming: "We know nothing!" I would proudly answer: "You are exactly right." There is extensive evidence clearly indicating that humans have evolved. Simply put, we have changed over time, both physically and mentally. However, I intended to remark that there is still a ton of missing information—our evolutionary history book is still missing some pages.

Socrates would have said that by knowing how much one does not know, at least one actually *knows more* than before, when one was ignorant of this fact. The reaction of some students indicated that I failed at communicating this.

For this reason, I spent the last years wondering how I could summarize all the essential perspectives while making them attractive to a larger audience and without misinterpreting or misrepresenting them. Even more, is it possible to talk about the big picture of human evolution and its role in fundamental aspects of human life such as politics, economics, and demographics? If so, what are the take-home messages? In the end, I decided the best way to accomplish my mission was to *not* write my own book on human evolution. Instead, a much better idea came into focus: enlist some of the world's experts and let them speak their mind using their own words. "If you want to go far, go together," the old say goes.

To make the story short, I wrote down a dream list with dozens of wise experts from different knowledge areas in human evolution (anthropology, paleontology, genetics, behavior, etc.). Then I put together enough courage to reach out to colleagues whom I know, explain my goal, and ask them to reply to some of my questions. When I began to receive some feedback and answers from my test group, I knew I was onto something. I decided to move beyond my comfort zone, reaching out to more colleagues, even from other areas (such as psychology, neuroscience, and philosophy). The mere thought of picking their brains filled me with joy and excitement. I would be able to satisfy my curiosity by interviewing some of my heroes. At a higher level, I was about to curate, in a single book, the digested knowledge of these humans, the results of their lifelong work, and present it to the world.

If you are reading these lines, it means the plan worked out. You are now about to access essential pieces of the big puzzle of our very existence as *humans*. Buckle up for intense time traveling and enjoy the journey!

PS: It took me almost three years (2019–2021) to compile all the contributions to this book. Hence, some are from pre-COVID-19 days, and others are more recent. As you will gather, the pandemic affected some of the answers.

HUMANS

INTRODUCTION
User Manual

THE MISSION

Humans seeks to serve as a vessel for education through entertainment:

- This book provides an encyclopedia of stories (and theories) about *us*. This book tells as many stories as the number of people featured in it. These are well-informed and important stories told by people like us about our collective history.
- This book seeks to make a small change in our relationship with the academic world. This book shows that researchers, professors, and book celebrities are normal people. People like us who decided to act, inform themselves, do research, come up with new ideas, and write about them.
- This book demonstrates that better comprehending human evolution (our own history in deep time) is relevant for everyone. Having an informed view of our natural origins can only help us find better answers to our many whys and hows.
- This book provides a new way to learn about human evolution. It compiles and summarizes essential knowledge—and wisdom—about humans. This book presents information gathered during decades of combined work and study by more than one hundred world experts. This book is also an homage to all of them and to their work.

HOW TO READ THIS BOOK

Your choice. The ideas, concepts, and references in the book naturally follow the ideas presented by the experts featured here. There is no reason to feel overwhelmed by the amount of information that it captures; not many people would read an encyclopedia from A to Z. Browse freely, pick and read whatever feels interesting at the time, and get inspired!

This book scales with each reading. You will probably learn something new every time you read it (I still do). Some of the stories are simple yet so profound that your mind will incorporate different layers of information each time.

The "Recommended Readings" section at the end of the book represents an incredible stand-alone resource of "game-changing" pieces curated by the experts featured in the book. I am confident you'll find some jewels in there.

THE QUESTIONS

The questions posed to the experts were designed to be simple yet summon years of accumulated knowledge and experience in the form of personal stories. Another goal was to provide a venue for the experts to express some ideas to a wider audience, beyond those who focus on jargon-based research articles. Many colleagues (some of them in the book) were my first test subjects and suggested people to include on the final list.

I sent the questions to dozens of world experts and asked them to answer at least five of them. For conciseness and simplicity, in the main body of the book each participant's answers will follow the icon and short description that summarizes the question.

Introductory Questions

I conceived the following three questions as "ice breakers" intended to introduce the experts and their interests, in their own terms.

 YOUR BEGINNINGS

Can you identify and describe the moment in your life when you decided to do the work that you do?

This question provides an opportunity for the experts to present the catalytic moment or collection of moments that led to their current life and work. It is not surprising that many were hugely inspired by small events during their childhood or key people in their surroundings (such as a particularly involved and caring teacher or a visit to a museum!).

 GAME CHANGER

Which discovery, research study, or book would you highlight as a "game changer" in the way we look at our own evolution? How did it influence your career or life?

The intention of this question is to highlight works or discoveries that were important enough to influence the perspective and motivation of the experts. I believe that these could be rediscovered after their exposure here and illuminate the nature of evolution for many others as well.

 AMAZING FACT

What fact about humans or human evolution amazes you the most?

This straightforward and simple question was aimed at dragging a particularly fascinating idea out of the experts' minds. My intention also was to highlight the idea that many experts do not believe that there are many facts but instead see evolutionary scenarios and hypotheses to test.

Core Questions

These questions were designed to constitute the heart of the intellectual feedback from the experts participating in the book. If they answered some of the "Introductory Questions," the formal presentations are now over. What do they really have to say?

 TIME TRAVEL

If you had a one-shot round trip in a time machine, to which specific time period—past or future—would you go and why?

This was my way of asking, "What is the most important time period of humanity for you and why?" For some experts, it represented too much of a "sci-fi" approach. For others, it offered an excellent opportunity to pin down moments in time that are key for understanding the major transitions in our evolution. This question also provided the opportunity to formulate predictions about where we could be headed—in other words, a chance for time travel in mental space!

 DRIVING FACTORS

What were the main driving factors that set an ape apart and onto the road leading to the human lineage?

This question is simply asking why a particular group evolved into humans and which circumstances sparked the beginnings of the human career. As you will see, this is a significant event in our evolution, and there is no consensus even about how to approach the question.

 FUTURE EVOLUTION

What will be shaping human evolution in the future? What will humans look like in 100, 100,000, or 1 million years?

In other words, "Are humans still evolving?" If so, which aspects are changing and at what rate? The first part of the question refers to "why" we are evolving. Some experts consider that evolution is too contingent (i.e., subject to chance) to make any predictions. Others had a clear image of what to expect in the future. It may be fair to argue that the future cannot be predicted with precision. However, some trends can perhaps be observed today (and in the past) to which we should pay attention.

 EVOLUTIONARY LESSONS

What can human evolutionary studies teach us about our past that can be helpful for our present or future?

This question is especially important to me and captures one of the main reasons I decided to compile this book. I followed it up by asking, "If you think human evolutionary studies cannot teach us anything helpful for our present or future, please explain why it is important to study human evolution."

With this combination of questions, I was able to gather thoughtful perspectives. Responses included the intrinsic values of studying human evolution and history. The experts also provided reasons it is essential to find solutions to contemporary problems in health, society, and politics, to name a few.

Philosophical Questions

The following three questions were designed to bring up a discussion about the unique physical and metaphysical aspects of what makes a human, as well as their interplay. Furthermore, I aimed to solicit practical advice on how to stay motivated and focused on working on questions relevant to humans and their evolution.

 SPECIAL?

Are humans "special" when compared to other animals? Can the study of our closest living relatives (apes and other primates) help us elucidate this question?

With this question, I intended to provoke some itch among the different experts and readers of the book. Although there is a broad spectrum of opinions, we could easily point toward two opposite extremes. On the one hand, some think that humans are unique among all the creatures that have ever existed. On the other hand, others have a less "anthropocentric" view of our role in the universe.

 RELIGION AND SPIRITUALITY

How is human evolution compatible with having a spiritual or religious view of the world?

The answers to this question constituted my biggest positive surprise in this project. I consider this to be one of the most challenging questions in the book. Initially, because of its sensitivity, I expected that most of the scientists would ignore this question and focus on others. However, this question produced the most thoughtful and illuminating responses and—I believe—is one of the most valuable discussions in the book.

 ADVICE

What advice would you give to anyone interested in working on questions related to understanding human nature, our past, or our future?

This question was not explicitly framed to benefit students. I was thinking of *anyone*—irrespective of their age, profession, and nation—willing to participate in the work of learning. I was seeking advice for those who want to learn more and to provide something new and meaningful in the area.

"Bonus" Question

Finally, I presented this last question:

 INSPIRING PEOPLE

To which person—living or dead—would you like to ask these questions?

This question had a twofold mission: First, I wanted to be directed to interesting characters and their ideas. Second, and in more practical terms, I wanted to identify other possible participants who might have escaped my attention. Based on the results, I am glad I included this one!

AN ILLUSTRATED GUIDE TO HUMAN EVOLUTION

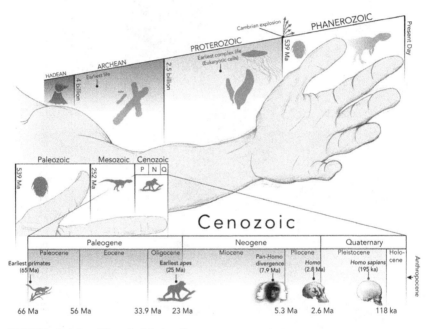

FIGURE 1. Evolution of life on Earth in deep time.

A human arm serves as the temporal scale for the advent of selected key events (top panel). At the beginning of the current Phanerozoic eon, during the radiation events encompassed in what is commonly called the "Cambrian explosion" (the process lasted several millions of years), almost all major animal groups originated. The Phanerozoic is shown in more detail using the hand as a scale (middle panel). The bottom panel highlights details of the Cenozoic era, which started with the end of the dinosaur's dominance in favor of mammals (we still have the birds). Humans of modern aspect (*Homo sapiens*) did not appear until the Late Pleistocene, the very end of the index fingertip. For some, we are still living in the Holocene epoch, which began after the last glacial period. For others, we are currently living in the Anthropocene, a geological epoch that started when humans began to dramatically affect the ecosystems. There are alternative, unofficial proposals about the Anthropocene starting date. The idea is to capture when humans first influenced the geological record (we have been affecting the global ecosystems for much longer). Thus, some propose that the Anthropocene began with the "Trinity Test" (the first detonation of a nuclear device in 1945). For simplicity, in the figure only the mean value for each age estimate (rounded up to one decimal) is provided (the full ranges sometimes encompass millions of years). Abbreviations: Ma = megaannum (or million years ago), ka = kiloannum (or thousand years ago).

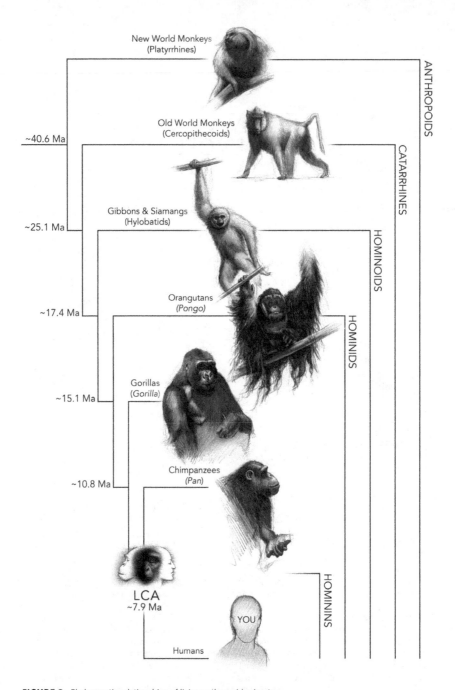

FIGURE 2. Phylogenetic relationships of living anthropoid primates.
You and I are more closely related to a chimpanzee (including bonobos) than to gorillas. We share a last common ancestor (LCA) that lived approximately 7.9 million years ago (with the estimated range from 9.3 to 6.5 million years ago [Ma], based on the latest molecular clock estimates). The mean age estimate of other ancestral nodes is also shown (rounded to one decimal). The superfamily Hominoidea (the "hominoids") includes living and fossil apes and humans. There was a time when the family Hominidae (the "hominids") was widely used to include all species of the human lineage (modern humans and fossil relatives). Given that we now know that humans are closer to some great apes (chimpanzees and gorillas, the African apes) than those apes are to other great apes (orangutans), the term hominid is commonly used to refer to the great ape and human family. Gibbons and siamangs (family Hylobatidae, or "hylobatids") are colloquially called "lesser apes," based on the size differences with the former. Hence, members of the human lineage are widely referred to as "hominins" (tribe Hominini). However, as you will see in the answers given by various experts, some still rely on the classic nomenclature. Branch lengths are not proportional to time.

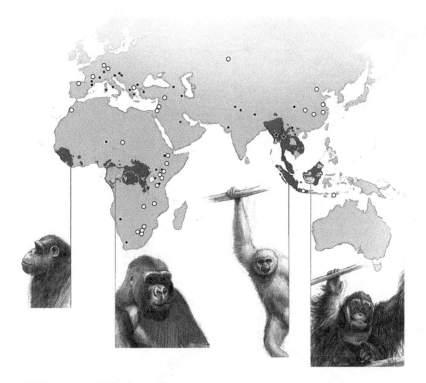

FIGURE 3. Geographic distribution of living apes.
Extant apes live in (or nearby) densely forested areas around the equator in Africa and Southeast Asia. Their distribution contrasts with the known range of regions where Miocene apes have been found (black stars; spanning about 23 to 5.3 Ma). Some regions may contain more than one site; contiguous regions are indicated with different stars if they extend over more than one political zone. The figure also indicates the location of these major hominin fossil sites (white circles). Two important "facts" to highlight: (1) thus far hominin fossil sites outside Africa have only yielded remains of the genus *Homo*; and (2) paleontologically speaking, most of Africa (west of the Rift Valley) remains highly unexplored.

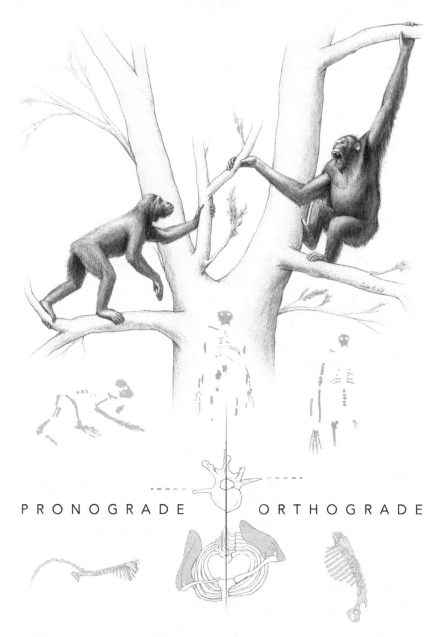

P R O N O G R A D E   O R T H O G R A D E

FIGURE 4. Pronograde and orthograde body plans.
On the left, the life reconstruction (top) and skeleton (middle) of the early ape *Ekembo* (about 18 million years ago [Ma]; Kenya), with the associated "pronograde" torso morphology as depicted from a macaque skeleton (bottom). On the right, the life reconstruction (top) and skeleton (middle) of the *Hispanopithecus* (about 9.6 Ma; Spain), with the associated orthograde torso anatomy as drawn from a living great ape (bottom). The middle center depicts the skeleton of *Pierolapithecus* (about 12 Ma; Spain), which shows some but not all the skeletal features present in living suspensory apes (e.g., orthograde torso, but shorter fingers). General differences between pronograde and orthograde body plan characteristics include the following: In opposition to most living monkeys, the modern hominoid orthograde body plan is characterized by the lack of an external tail (the coccyx being its vestigial remnant); the ribcage is mediolaterally broader and dorsoventrally shallower, with longer and more robust clavicles (collar bones) and dorsally placed scapulae (shoulder blades) that are more cranially elevated and oriented; and a shorter lower back and longer iliac blades. The vertebra in the center also shows differences in lumbar anatomy, including more dorsally situated and oriented transverse processes in orthograde hominoids. Fossil apes like *Pierolapithecus* show that hominoid evolution operated in a mosaic fashion, making it challenging to precisely determine where these fall in the hominoid evolutionary tree. The reconstructions, fossil skeletons, and skeletal elements are drawn to approximately the same size to facilitate anatomical comparisons.

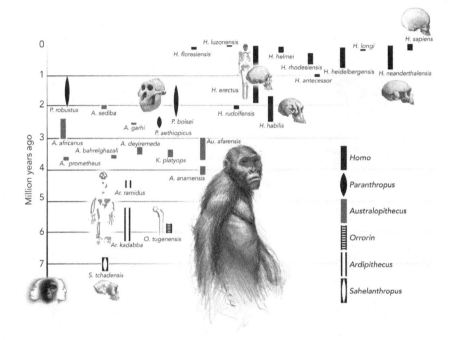

FIGURE 5. Hominin diversity across time.

The temporary range of each species is indicated without making any phylogenetic assumption other than the genus rank. The hominin reconstruction in the center represents "Lucy," the most iconic *Australopithecus afarensis* individual. Along with the answers in the book, some participants alternatively use terms such as "australopithecine" (involving the existence of the subfamily Australopithecinae, which is incompatible with considering the whole human lineage at the lower taxonomic rank of tribe Hominini) and "australopith" to refer to both *Australopithecus* and *Paranthropus*. In general, both genera are more archaic than members of the genus *Homo*. Authors refer to *H. neanderthalensis* as "Neanderthals" or "Neandertals" (both spellings are accepted common names). The figure depicts a speciose perspective on the human fossil record. However, many authors lump some of the species depicted (e.g., *Paranthropus* as part of *Australopithecus*, *Homo antecessor* with *H. erectus* or *H. heidelbergensis*, and the latter with early *H. neanderthalensis*). Others separate them (e.g., *H. erectus* from Africa as *H. ergaster*). Finally, "Denisovans" is the informal name for an extinct human lineage known from scarce fossil remains. Thus far it is differentiated based on ancient DNA.

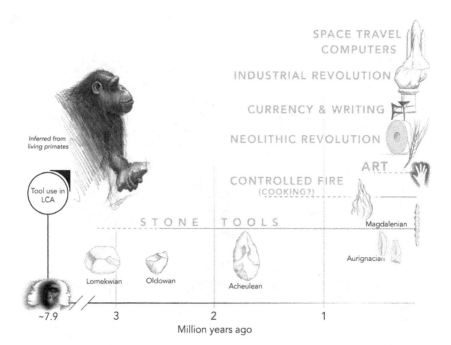

FIGURE 6. Cultural evolution.

Modern primates (and other animals) commonly use tools made of leaves, twigs, stones, or other naturally occurring objects. However, only in the human lineage did tool traditions become established cultural innovations that exponentially built up over time. It took more than 4 million years after the origin of the human lineage for the first potential stone artifacts to appear, and it was not until almost a million years later that purposive stone tools became more widespread in the archeological record. With the Neolithic revolution (about 12,000 years ago), *Homo sapiens* completely broke a balanced relationship with the environment. Horizontal dashed bars indicate the earliest evidence for selected evolutionary events (stone tools, controlled fire, art). The bars become solid when the archeological record indicates that the given innovation has become established. When and how the human language evolved is still a mystery. Some think it is a *H. sapiens* invention, the key to its evolutionary success. Others argue that the roots of human language can be found much earlier, perhaps when the Oldowan stone tool complex became increasingly widespread in the archeological record of Africa (around 2.5 million years ago). Even though these are not sophisticated tools, it requires a long time to master the skills necessary to make them. We learn by imitation, but we learn faster when following instructions. Perhaps some rudimentary (or just different) language was at play.

PART I

Prelude

ANSWERS FROM THE EXPERTS
LOOKING AT THE ANTECEDENTS
TO HUMAN EVOLUTION

1

DAVID ALBA

David Alba is the director of the Catalan Institute of Paleontology Miquel Crusafont (Barcelona, Spain) and a former coeditor-in-chief of the *Journal of Human Evolution*. David is an internationally recognized researcher with an extensive background in vertebrate paleontology, especially hominoid evolution (the ape and human clade) during the Miocene. One of his most relevant discoveries consists of the partial skeleton (nicknamed "Laia") of the tiny ape *Pliobates cataloniae* (about 5 kilos, or 11 pounds), which lived around 11.5 million years ago in northeastern Spain.[1]

YOUR BEGINNINGS

I was about twelve years old when I discovered Alfred Wegener's continental drift theory in my natural sciences textbook. I particularly wondered at his use of paleontological evidence (the same "mammal-like reptiles" being recorded in Africa and South America) to support his views. The truth is that I had always been fascinated by both living and extinct animals, but that very same day I naïvely decided that I wanted to become a paleontologist. For several years I had no idea what to do in order to become one. However, during high school I decided to pursue a bachelor's degree in biology because rocks were boring to me, and I was interested in the evolution of animals, after all. I entered the university determined to become a researcher after reading quite a number of popularizing books about fossils and evolution. I still remember arguing with my zoology professor during my first year at the university because he did not accept that birds were living dinosaurs. Then, during my second year at the university, I took a course in paleozoology that let me discover scientific literature in general and fossil primates in particular. This ultimately determined my will to become a researcher in the field of primate paleobiology. By then I had discovered the Paleontological Institute at Sabadell, and two years later I met my soon-to-be thesis advisor. After a short interview he agreed to become my supervisor (with a PhD dissertation focused on Miocene hominoids) once I finished my degree. That day signaled the culmination of a ten-year life journey (since my teenage decision to become a paleontologist), and I still remember it as one of the happiest days of my life.

GAME CHANGER

I have two answers for this question. From the viewpoint of the discipline of human evolution, I think the most important advance in the field has been the development of paleogenomics, which enables us to retrieve molecular information from relatively recent past organisms. This has shown, among other things, that different human species were able to successfully interbreed in the past. I don't find this surprising, as evidence from many other vertebrates has also shown that hybridization among species is the rule rather than the exception and that the tree of life is more of a network than we had previously thought. As a paleontologist working with Miocene fossils, I find it inconvenient that paleogenomics cannot go back further than about half a million years. In this regard, paleoproteomics offers the promise to go much further back in time (once the techniques improve in a few years), and I suspect that this approach will be the game changer in primate evolution during the next decade.

From a personal point of view, I read three books that changed the way I looked at human evolution when I was younger, and they played a significant role in my decision to become a paleoprimatologist. In chronological order, the first is Stephen Jay Gould's *Wonderful Life*,[2] about Burgess Shale fossils and the Cambrian explosion, which I read just before going to university. It made me realize that evolution is largely contingent and that humans are not an inevitable result of this process, but at the same time (and using Darwin's words) "there is grandeur in this view of life." The second book I would like to mention is Roger Lewin's *Bones of Contention*,[3] which I read during my first year at the university after having found it by chance at a department store with a huge discount. That book not only transmitted the excitement of fieldwork-based research but further instructed me about the controversies that surround the field of hominoid and human paleontology. And, more generally, it taught me that science is not a progressive way to truth based on discoveries, that interpretation of the evidence is central, and that disagreements among researchers are not only unavoidable but also necessary to guarantee the progress of the scientific enterprise. The third book I must highlight is the first edition of John Fleagle's *Primate Adaptation and Evolution*.[4] I discovered this book at the faculty library during my second year at the university, and for me it opened a gate to a brand-new world of accumulated knowledge on primate paleontology and evolution. This book more than any other showed me that it makes no sense to look at human adaptations outside the framework of primate evolution more generally, and it definitely determined my will to become a paleoprimatologist.

AMAZING FACT

For evolution in general: contingency and deep time. That living organisms, including humans, are the result of evolutionary changes that accumulated over millions and millions of years and that even though all the various steps involved in the evolution of a particular lineage (like ours) occurred for a reason, the factors involved are so numerous that predicting the particular outcomes a priori (such as the evolution of big-brained, bipedal apes like us) would have been completely impossible. The most fascinating thing about deep time is that it can make the unlikely become possible.

TIME TRAVEL

I would go to els Hostalets de Pierola in Catalonia (northeastern Iberian Peninsula), about 12–11.5 million years ago, to film a *Pierolapithecus* individual moving on the canopy (to show David Begun [chap. 3] and Carol Ward [chap. 44] that, unlike extant apes, it was not particularly suspensory) and to take some blood or hair samples of *Pliobates* to run DNA tests to confirm whether it is a hominoid and to see if it is more closely related to gibbons than to great apes. Although spotting a deinothere or a false sabertooth would be no less exciting.

DRIVING FACTORS

The human lineage is characterized by multiple characteristics, from adaptations to committed terrestrial bipedalism to huge encephalization, slow life history, enhanced dexterity, and reduced canine sexual dimorphism. It is difficult to pinpoint the most "essentially human" feature, and given the mosaic nature of evolution (not all human characteristics evolved at once and for the same reason), it also is difficult to isolate the "main" selection pressures that determined the evolution of these characteristics. It is generally recognized that, from the last common ancestor of chimps/bonobos and humans, bipedalism came first, followed by increased encephalization concomitant with adaptations to manipulation. By "freeing the hands," bipedalism opened new possibilities in regard to manipulation, and ultimately this probably had a synergistic effect on the subsequent relative brain size increase in the genus *Homo*.

So if I have to stress a single factor, I would highlight the environmental changes that occurred during the late Miocene in Africa that favored adoption of a semiterrestrial lifestyle, combining tree climbing with terrestrial bipedalism for horizontal travel. However, it is also important to recognize that bipedal adaptations were superimposed on an already orthograde body plan that had originally evolved because of selection pressures posed by arboreal locomotion and that was later co-opted for bipedalism. An orthograde body plan does not inevitably lead to bipedalism but is a necessary prerequisite, and this makes me stress that the contingent road leading to human evolution was paved much before we normally tend to believe, well within the Miocene.

FUTURE EVOLUTION

Homo sapiens is still evolving, that is for sure. Although studies in this regard tend to focus on ongoing selection pressures, I think it is more important to take into account the selection pressures that have been removed because of cultural factors (e.g., cesareans in medically advanced societies remove the selection pressures on birth canal size and newborn brain size). So culture and technology will surely shape the future of human evolution, especially by changing the environmentally driven selection pressures.

I have no doubt that humans will still exist in a hundred years, but I am not so sure about 100,000 years, much less after a million years. The frightening thing in this regard is that humans are the animal species that can alter the environment most dramatically, but thus far we have failed to modify it to our advantage (rather the contrary). So the same capacity that permits humans to inhabit almost all habitats on Earth, thanks to material culture, might ultimately determine our own extinction on a relatively short time span (at least on a geological scale). After all, paleontology clearly teaches us that all species are doomed to become extinct, either in absolute terms or by giving rise to new species (pseudoextinction). So in a million years modern humans will likely be extinct, at least as we currently conceive them. The question is whether we will have evolved into something different or become truly extinct.

Because of unavoidable demographic pressures, humans from the distant future might be forced to migrate to other planets (once technology allows for that, of course); this seems quite likely based on the dispersal capacities demonstrated by early *H. sapiens*. If humans colonize other planets, given the tendency of the members of our species to either annihilate or intermingle with individuals from other populations or even species (e.g., Neanderthals), I tend to think that diversification of *H. sapiens* into different descendant species is not very likely even in the long term, although I would not discount it altogether. Evolution into something that would be considered a new species is more likely, but predicting how it would look is a futile theoretical exercise given the contingent nature of evolution.

EVOLUTIONARY LESSONS

Our past should enlighten us about who we are and where we came from. As for the future of humankind, I think it is out of our control to decide where we would like to go at the species level. However, with regard to present individuals and societies, research in human evolution (as well as evolution in general) has a lot to contribute. The fact that these studies, like others devoted to basic science and unlike biomedical or technological research, do not have an immediate applicability doesn't mean that they are unimportant from a cultural viewpoint. Rather, they are essential.

Paleoanthropological research, for example, generates knowledge about the past evolutionary history of the human lineage and, in this way, helps satisfy one of the most fundamental and inherently human drives: curiosity. This is not a trivial matter, as humans we want (need) to know and our desire for knowledge has no limits. We tend to think of science as something different from the arts, but in fact they are different sides of the same coin: culture. Both the products of arts and the knowledge generated by science are fundamental constituents of human culture, not only accessible to scientists but at least in principle available for everyone to enjoy. We, as researchers, should not behave as esoteric priests who aim only to transmit our wisdom to the next generation of scientists; it is our moral duty to make current knowledge available to the society as a whole! So even if the aims pursued by the arts (such as beauty) are quite different from those of science (knowledge), in the end they both aim to satisfy humans' most intimate needs. With the difference, of course, that the scientific endeavor does not aim to evoke sentiments or feelings but to provide knowledge—in the case of human evolutionary studies, knowledge about our own origins from a biological viewpoint, what Huxley described as "Man's place in Nature"[5] in the book that inaugurated our discipline.

Jesus said "the truth shall set you free," and this has become a popular motto in academia. In fact, science cannot reach any absolute truth, although it aspires to progressively come as close to reality as possible. And, as is known, knowledge is power. Knowing about human origins might not provide comfort against our existential fears and uncertainties or teach us any moral lessons, and for sure it does not provide us with the meaning of life. However, at the very least it informs people about how and when we came to be what we are, as well as about our kinship relationships with our closest relatives, the apes, and the other living beings on Earth. In doing so, human evolutionary studies (similar to but more than any other scientific discipline) have the potential to empower people as freethinking individuals and thus prevent them from being manipulated on the basis of myths, superstitions, or plain ignorance.

SPECIAL?

We are certainly "peculiar" because of our great degree of self-awareness and the implications of knowing about our future death as individuals. However, I am not sure this makes us special. According to my dictionary, "special" means "better, greater, or otherwise different from what is usual." I think we fulfill the last condition because of what I mentioned previously. However, there is no absolute discontinuity as compared to other hominoids. There appears to be a considerable gap, for example, regarding self-awareness and cognitive abilities, but such a gap would be filled if we could inspect all the panoply of extinct species of our lineage. The ontogenetic development of a newborn child into adulthood is a good metaphor that helps us realize that human consciousness was probably acquired progressively over time and that the gap is only huge when we compare the starting point with the final end. There would be no big gap if we had access to everything that existed in between. In this sense, our closest relatives can help us realize this, but only to some extent; to actually fill in the gaps we must look at the fossil record.

RELIGION AND SPIRITUALITY

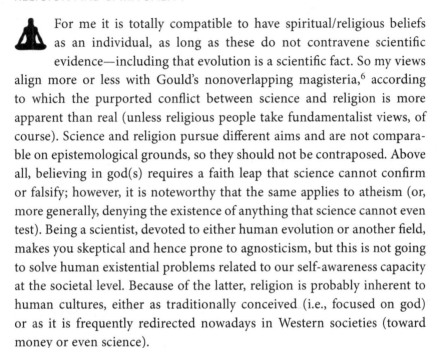 For me it is totally compatible to have spiritual/religious beliefs as an individual, as long as these do not contravene scientific evidence—including that evolution is a scientific fact. So my views align more or less with Gould's nonoverlapping magisteria,[6] according to which the purported conflict between science and religion is more apparent than real (unless religious people take fundamentalist views, of course). Science and religion pursue different aims and are not comparable on epistemological grounds, so they should not be contraposed. Above all, believing in god(s) requires a faith leap that science cannot confirm or falsify; however, it is noteworthy that the same applies to atheism (or, more generally, denying the existence of anything that science cannot even test). Being a scientist, devoted to either human evolution or another field, makes you skeptical and hence prone to agnosticism, but this is not going to solve human existential problems related to our self-awareness capacity at the societal level. Because of the latter, religion is probably inherent to human cultures, either as traditionally conceived (i.e., focused on god) or as it is frequently redirected nowadays in Western societies (toward money or even science).

Rather than trying to convince people not to believe in god because of science, we should try to convince people (e.g., antivaxxers) to trust in science for everything that is unrelated to their religion, spirituality, or moral beliefs. In other words, I do think that science should play a greater role in shaping the views of current societies, but if we conceive this as opposed to people's religious views, then attaining it will be more difficult than it should be, if it is possible at all.

INSPIRING PEOPLE?

Stephen Jay Gould, whom I mentioned in my responses a couple of times, has had a long-lasting influence on my training as a scientist. Unfortunately, he's no longer among us, but at least he left quite an extensive literature, including essays that treat several of the topics discussed here. As for the living, you might try reaching out to my friend and colleague Tomàs Marquès-Bonet [chap. 82].

2

PETER ANDREWS

Peter Andrews is an emeritus researcher in the Department of Earth Sciences at the Natural History Museum, London, where he was the head of Human Origins until his retirement in 2000. Using his unique background in African forestry, Peter studies how ecosystems change over time and their role in human origins. One of his primary interests is reconstructing the nature of the last common ancestor of apes and humans. Peter has published ten books and nearly two hundred articles in the scientific and popular press. His views and experiences are summarized in his book, *An Ape's View of Human Evolution.*[1]

YOUR BEGINNINGS

While I was working in the forestry department in western Kenya in 1966, I kept coming across sites with fossils protruding onto the surface, and I regularly took them to show to Louis Leakey in Nairobi. He was so enthusiastic and full of information that I decided there and then to try to work with him. If I got a degree in anthropology at Cambridge, Louis said he would support me to work on the fossil apes that had accumulated since his work began in 1951. I did this, with his help, and he supported my work for my doctoral thesis on Miocene apes.

GAME CHANGER

 Louis Leakey's book *Adam's Ancestors*[2] was a game changer for me because it integrated the study of hominin fossils with their environment and with stone tool technologies. I believe it is important to combine the different lines of evidence on human evolution and explain how each feeds off the other lines.

AMAZING FACT

 It is incredible how apelike humans are, particularly when you study fossil apes.

TIME TRAVEL

 I would travel to the Middle Miocene in Turkey, both because things were happening with fossil apes and because this was the beginning of the transition from fossil apes to modern apes and humans. The first "out of Africa" because it set the stage for later developments, both in apes and the first hominins.

DRIVING FACTORS

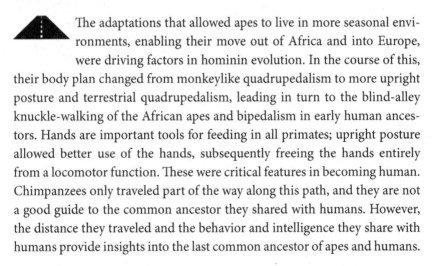 The adaptations that allowed apes to live in more seasonal environments, enabling their move out of Africa and into Europe, were driving factors in hominin evolution. In the course of this, their body plan changed from monkeylike quadrupedalism to more upright posture and terrestrial quadrupedalism, leading in turn to the blind-alley knuckle-walking of the African apes and bipedalism in early human ancestors. Hands are important tools for feeding in all primates; upright posture allowed better use of the hands, subsequently freeing the hands entirely from a locomotor function. These were critical features in becoming human. Chimpanzees only traveled part of the way along this path, and they are not a good guide to the common ancestor they shared with humans. However, the distance they traveled and the behavior and intelligence they share with humans provide insights into the last common ancestor of apes and humans.

FUTURE EVOLUTION

Control of the environment has already had an impact on human evolution, but control of human genetics and the genetics of aspects of the environment on which humans depend are going to be major influences over time. Each development made by future generations will be resisted in turn, but each will gradually become incorporated into populations in the manner of "two steps forward, one step back."

EVOLUTIONARY LESSONS

Evolutionary lessons can teach us how to interact with the environment and how to control genetic changes that will inevitably come about. Adaptations to the environment have been critical at all stages of human evolution, and they are not becoming less so as we learn to alter environments to suit our needs.

SPECIAL?

Biologically humans are not special, but we have taken advantage of bipedalism in a way that bipedal birds, dinosaurs, or a sprinkling of other mammals have not. This has allowed us to develop technologies to a greater degree than any other form of life. Fortunate mutations promoted development of speech and, through this, conceptual thought (the two are interdependent), and this makes humans special in the sense that we occupy another sphere in the world, what Teilhard de Chardin calls the "noosphere," the sphere of human thought.

RELIGION AND SPIRITUALITY

Evolution is neither compatible nor incompatible with religious belief. The study of human evolution is based on observation and interpretation of scientific evidence; religious belief is based on a concept in the human mind, in the noosphere.

ADVICE

Ask the questions that fascinate you, and do your best to answer them. Do not treat this as "just another job"; do not use it to build a career; do not seek to impose a personal view of life on others; and above all, do not use this to make yourself important in the eyes of the world.

INSPIRING PEOPLE?

Teilhard de Chardin tried to integrate evolution with religion. I do not think he was entirely successful, but at least he tried.

3

DAVID BEGUN

D avid Begun is a professor in the Department of Anthropology at the University of Toronto. Since the 1990s, David has been a leader in studying the evolution of the great ape and human family. Together with European and Turkish colleagues, his fieldwork efforts have recovered key fossil great apes (e.g., *Rudapithecus* from Hungary). Always looking at the big picture, David has made substantial contributions to our knowledge about the paleobiogeography and paleobiology of the apes that preceded the origins of the human lineage. His exciting and controversial ideas are summarized in his book, *The Real Planet of the Apes*.[1]

AMAZING FACT

It is not big brains, bipedalism, or dexterous hands that are the most amazing fact about humans, although these adaptations define humans. Many spectacular interconnections among organisms that have evolved within the diversity of life make these human innovations pale in comparison. What amazes me most is not a specific outcome but an aspect of the evolutionary process. The number of times that the same attribute has evolved independently is amazing. Wings and eyes have evolved many times (eyes in insects, marine invertebrates, vertebrates; wings in insects, bats, and birds), yet they are easily distinguishable in their details. They represent different genetic and anatomical pathways to the same function. In other cases, it is not obvious that independent evolution has occurred because even in the details the similarities among different groups with the same adaptation can be remarkable. Similar morphologies that have evolved independently in closely related taxa are called parallelisms. Often, parallelisms cannot be identified by subtle morphological differences but are revealed only with some independent information on evolutionary history. It amazes me that essentially identical morphologies could evolve independently. It is remarkable how many times the same adaptation has evolved in parallel in the great apes, and with so much morphological similarity that it is almost impossible to distinguish between parallel evolution and shared descent. Great apes are extremely similar in the morphology of their arms, chests, hips, and backbones, yet most of these attributes probably evolved independently in the lineages leading to orangutans and African apes. With careful and detailed analysis, it is possible to distinguish some of these morphologies, but in many cases the outcomes are so similar that only a reliable phylogenetic analysis allows us to conclude that they were achieved independently. Homoplasy, as this is called, is most commonly viewed as an impediment to our understanding of evolution because it misleads us into thinking that organisms are more closely related to one another. It makes it almost impossible to identify direct ancestors in the fossil record. Unless DNA can reveal relationships precisely, the ubiquity of parallel evolution makes it impossible to know who our direct ancestor is among the many candidates of Pleistocene *Homo*. Further back in time, it is challenging to decide which among many taxa are more closely related to the last common ancestor we share with chimpanzees. Frustrating, yes, but to me it is a fascinating and critical vindication of the theory of natural selection. If the environment is a selective agent—a major tenant of evolutionary theory—then one should expect similar solutions to the same problem, and this is exactly what we see in homoplasies. However, because homoplasy makes inferring ancestor-descendant relationships challenging, it is the principal source of disagreement among specialists about the precise details of the evolution of a particular group. It also fuels ill-informed efforts to discredit evolution by exaggerating the differences among researchers and ignoring the scientific legitimacy of debate and competing hypotheses. This may be one of the more important ideas to teach people who have questions about evolution. Evolution is real but incredibly complex, like math or physics. As a result, experts can have different opinions.

TIME TRAVEL

I would travel to the Late Miocene, of course. Whether it happened in Europe, Asia, or Africa, this is the time period during which humans diverged from the common ancestor we share with chimpanzees, the so-called last common ancestor (LCA) or missing link. I want to see what the

LCA looked like, how it behaved, and where it lived. I am sure I would be pleasantly surprised, observing some behavior or morphology that we could never have imagined from the fossil evidence. Perhaps they used tools like chimps do today, or maybe some of them were bipedal. Did they hunt? How much time did they spend in the trees? Did they hang below the branches, walk on top of them, or do some combination that we do not see today? Did they bed down at night in the trees as most great apes do today? Was there something in their behavior that presaged human origins? We know that their brains were as large as the brains in living apes. How clever were they? How did they use their hands? Who knows? I would love to be able to go there and see for myself.

FUTURE EVOLUTION

 Not to be too pessimistic, but we don't know if there will even be humans in a hundred years, and certainly we have no idea about anything a million years in the future. We can imagine how humans would adapt to various possible selection pressures to which we would be subject in the future. But without knowing the nature of those selection pressures, it is impossible to predict what will happen. We cannot do it based on what has happened in the past, such as the common idea that brains will continue to get larger because that is the trend in human evolution. There are good reasons to think that we have maxed out on brain size. Of course, we can imagine a situation in which random evolutionary forces such as drift or gene flow influence our future. An apocalyptic vision of humanity's future might include the survival of a small population somehow shielded from the effects of whatever wipes out the rest of us (war, famine, disease). This population might have certain attributes in higher frequencies than those in populations that did not survive. For example, a certain skin color, average height, or unique attributes such as few toes or small brains may help them survive under certain circumstances. Or a number of these groups could survive, each characterized by its own unique set of attributes. This could lead to speciation events and the evolution of several new species of humans. The point is that speculation about future human evolution lies in the realm of science fiction.

EVOLUTIONARY LESSONS

One idea that Steven J. Gould stated has stuck with me: "Human equality is a contingent fact of history." We could have ended up with several living human species with differing skills and capabilities, but that did not happen. There is only one human species—the one that survived. The one that won. In this light, it makes no sense to think of any one population of humans as having more, fewer, or different capabilities than any other group. The differences that are sometimes said to distinguish human groups (so-called natural aptitudes) result entirely from the circumstances of one's social context. People who grow up in an environment that values education and knowledge will seem smarter than those who do not, even though there is no native difference in intellectual capacity among human groups. People who grow up in a context that places a high value on athleticism, or whose circumstances favor the development of athletic skills as one of the few modes of social or economic advancement, will seem more naturally gifted athletically. There are no naturally "athletically gifted" human groups. Differences in body type (average height and weight, for example) or physiology (heart size, oxygen carrying capacity, metabolism, etc.) result from responses to climate, altitude, and other environmental factors and are not the consequences of selection for athletic abilities. This is one of the main lessons that has emerged from the study of human evolution.

SPECIAL?

On a fundamental biological and evolutionary level, humans are not special. The fossil evidence of human evolution, and the genetic evidence of continuity and similarity among all organisms, shows us that all organisms are interconnected and that nothing about the universe was created for humans. We are merely one of the trillions of trillions of consequences of evolution and other natural processes (gravity, quantum mechanics, chemistry, etc.). This is not to say that humans are exactly the same thing as other organisms. Sea cucumbers have specific adaptations that result from their evolutionary history and so do humans (and all other organisms). Those histories are unique, just as all organisms are unique. One of the consequences of human evolution is culture, the unique human capacity for complex behavior that has resulted in our technological prowess and our huge population. We are by far the most numerous large mammal on the planet. (Rats probably outnumber us, but they are small and who wants to count them?) There are many orders of magnitude more humans (~eight billion in 2022) than there would be if we had evolved to our current body size but without our current level of technology. Remarkably, we outnumber all of our large domesticated mammals and pets (cattle, sheep, goats, camels, horses, dogs, cats, etc.) combined. [David clarified that "we are outnumbered by rats and chickens, but not by domestic mammals cat-sized or larger."] These facts, our huge population and our technology, have placed enormous pressure on the environment. We are not special, but our behavior has resulted in an urgent ecological crisis. We are responsible for this situation, so we are responsible for fixing it. If worms were in a position to fix the imbalance in our ecosystem, they would probably do so—unless, like us, they prefer to act against their own self-interest.

RELIGION AND SPIRITUALITY

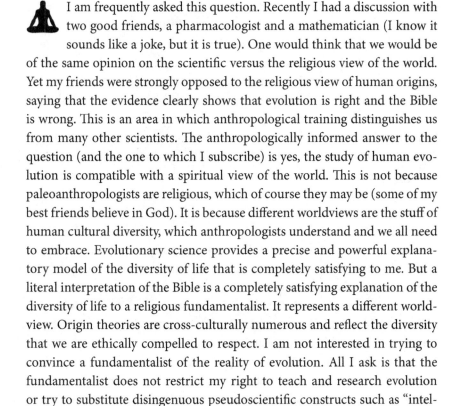 I am frequently asked this question. Recently I had a discussion with two good friends, a pharmacologist and a mathematician (I know it sounds like a joke, but it is true). One would think that we would be of the same opinion on the scientific versus the religious view of the world. Yet my friends were strongly opposed to the religious view of human origins, saying that the evidence clearly shows that evolution is right and the Bible is wrong. This is an area in which anthropological training distinguishes us from many other scientists. The anthropologically informed answer to the question (and the one to which I subscribe) is yes, the study of human evolution is compatible with a spiritual view of the world. This is not because paleoanthropologists are religious, which of course they may be (some of my best friends believe in God). It is because different worldviews are the stuff of human cultural diversity, which anthropologists understand and we all need to embrace. Evolutionary science provides a precise and powerful explanatory model of the diversity of life that is completely satisfying to me. But a literal interpretation of the Bible is a completely satisfying explanation of the diversity of life to a religious fundamentalist. It represents a different worldview. Origin theories are cross-culturally numerous and reflect the diversity that we are ethically compelled to respect. I am not interested in trying to convince a fundamentalist of the reality of evolution. All I ask is that the fundamentalist does not restrict my right to teach and research evolution or try to substitute disingenuous pseudoscientific constructs such as "intelligent design" for evolutionary theory. The fundamentalist (of any religion) and the evolutionary biologist both should know that they live in the same world and need to respect each other.

INSPIRING PEOPLE?

If most people do not say Darwin, it will only be because most of us think that everyone else would say Darwin is the most inspiring person. In addition to Darwin, I would like to have a conversation with Lamarck (Jean-Baptiste Pierre Antoine de Monet de Lamarck). Lamarck invented the concept of biology and came up with the first well-reasoned and logical theory of evolution. He understood the relationships among organisms—that is, the science known today as ecology—and he recognized that organisms respond physically and behaviorally to their environments. I think Lamarck might initially be shocked by the situation into which we have gotten ourselves, so out of balance with nature, but upon reflection he would probably see that it was inevitable. Lamarck endured, or so it is said, the disapproval if not the ridicule of his illustrious contemporaries such as Cuvier and even his mentor, Buffon. I would like to know more about this and how he felt. He died impoverished and forgotten in his time. He is remembered today—often for the wrong reasons (supposedly getting evolution wrong)—but at least he is remembered, and I would like to tell him that.

Lamarck is cited as having proposed that adaptations acquired during one's lifetime are heritable (the inheritance of acquired characteristics). I have twice read his masterpiece *Philosophie Zoologique*,[2] and he never says this, at least not quite so bluntly. He got most of evolution right: adaptation, ecology, change over time, geologic time, and deep time. All of these ideas were radical in his lifetime. Deeply held religious convictions about the literal truth of the Bible and the perfection of creation among influential scientists such as Cuvier and the emperor (Napoléon) worked against Lamarck. His daughter, Aménaïde Cornélie Lamarck, famously vowed (immortalized at the base of his statue in the Jardin des Plantes in Paris), "La postérité vous admirera, elle vous vengera, mon père" (You will be admired in posterity, it will avenge you, my father). Lamarck did have his disciples and was admired, even by Cuvier, for the depth of his knowledge of biodiversity and ecology.[3] I would argue that Lyell and Darwin are both indebted to Lamarck for his ideas on deep time, geological processes, and ecology. It is mainly Lamarck's mechanism of transformation that was incorrect (neither Lamarck nor Darwin in the first edition of *Origin* use the word "evolution"), and Darwin's pregenetic concept of blending inheritance was also wrong. Darwin's great advance was in the concepts of variation and natural selection, which proved to be correct and changed biology forever. Lamarck's vision did not. He recognized that if you cut off the tail of a mouse its progeny would still have tails. He never said that the giraffe's long neck is the direct result of the physical or emotional effects of stretching for food in the trees and was passed on to offspring. His actual mechanism is difficult to decipher, probably because he was not so sure. It involved "essences" or impulses that may be fluid or electrical (he knew about the chemoelectrical nature of the nervous system) that were recruited as a consequence of ecological contingency. If an animal is repeatedly required to carry out a particular task in order to survive, these impulses would have a direct effect on its gametes, which would be transformed to replicate the achievements of the parents. It is still basically wrong and lacks the brilliant insight of variation and natural selection, but Lamarck did understand that organisms respond to their ecological context, a central theme in natural selection. Today we know that external factors can in fact affect gametes (radiation, methylation, and other epigenetic processes). So Lamarck may not have been completely wrong after all!

PS: I have always been intrigued by the metaphorical connection between evolution and sports and would also like to meet "Joltin' Joe" DiMaggio, the Yankee Clipper, and ask how he feels knowing that seventy-eight years later his record of fifty-six consecutive games with a hit has yet to be topped. I grew up in New Jersey and am still a Yankees fan. I am sure Joe would be surprised that his record persists. Here is the evolutionary angle. An exceptionally good batter gets a hit once every three times at bat, and usually has four at bats in a game. For those unfamiliar with baseball, successfully reaching base on a hit more than 30 percent of the time means that you are among the most accomplished players in the game. In baseball, a 70 percent failure rate is a mark of great pride. But when you think about it, 30 percent survival is unusual for organisms, many of which experience well under 1 percent survival to adulthood of all offspring.

I sometimes wonder why selection (artificial in this case) has not worked to break records like those of DiMaggio. People capable of hitting a ball out of Yankee stadium come around only once in a generation (Josh Gibson, Mickey Mantle, Frank Howard). Why not more frequently? These athletes remind me of Richard Goldschmidt's "hopeful monsters." Variants that come along once in a generation and that represent game changers. Goldschmidt was ridiculed for this idea, although some have seen merit in it given recent advances in genetics.

It should be common for exceptional batters (and there have been many since DiMaggio) to have a hit at least once every game. Yet DiMaggio's record stands as perhaps the most unbeatable in baseball (maybe seven no-hitters as well). Some things in sports and evolution may simply not be possible, and these thresholds will never be crossed. I like that as a metaphor for evolution. I hope I do not overextend the baseball metaphor with this last thought, which is as true about evolution as it is about whatever it was Yogi Berra was trying to say: "You've got to be very careful if you don't know where you are going, because you might not get there."

4

BRENDA BENEFIT

B renda Benefit is a professor emerita of biological anthropology at New Mexico State University and specializes in the evolutionary history of Old World monkeys and apes. Since 1987, she has studied the 15-million-year-old primates from Maboko Island in a small gulf of Lake Victoria (Kenya). Her many discoveries include the most complete cranium of *Victoriapithecus macinnesi*, which bears insight into the last ancestor of all Old World monkeys.[1] The work of her team on *Kenyapithecus* (or *Equatorius*) *africanus* (depending on who you ask) provides essential information about why great apes were first able to adapt to open environments, thus being able to spread out of Africa.

YOUR BEGINNINGS

My ninth-grade biology class used the BSCS textbook in which evolution was the unifying theme. My teacher, Mrs. Zilberstein, had taken courses from Stephen J. Gould and played recordings of his lectures in our class. His lecture on the evolution of various types of symmetry made me cry because I thought it was so exciting. I knew at that moment that I wanted to study evolutionary biology. Two years later I was chosen to participate in an NSF-funded biology program for high school students at the University of Texas, Austin. We all worked on research projects and had to complete a minithesis by the end of summer. Many of us won the National Junior Science and Humanities competitions in our own states (Massachusetts for me) and presented our summer research at the national symposium. I was hooked.

GAME CHANGER

Early in my career as a college student my professor, Alan Walker, invited Richard Leakey [chap. 54] to speak about his discovery of KNM-ER 1470.[2] His lecture confirmed my strong interest in the evolution of apes and human ancestors and reinforced my desire to one day collect fossils and make discoveries in Africa.

AMAZING FACT

I am amazed that we cannot predict what the next human ancestor fossil to be discovered will look like. There is no predictable or neat pattern to human evolution. Because of this, we still do not know whether *Homo* evolved from *Australopithecus* or which species were ancestral to us. The more we learn, the more complicated our evolution appears to have been.

TIME TRAVEL

I would travel to Maboko Island 15 million years ago to see how correct our (my and Monte McCrossin's) reconstructions of *Victoriapithecus*, *Kenyapithecus*, *Mabokopithecus*, and the small apes are. I would take my video camera and lots of notebooks to learn as much as I could about their behavior and about the other fauna in the ecosystem. I think every paleontologist would like to learn firsthand if their work is correct.

DRIVING FACTORS

Humans are the canary in the climate change coal mine. Apes and human ancestors were highly sensitive to climate change, but contemporaneous monkeys had no problem adapting. During the Plio-Pleistocene monkey species survived with little change for three to four million years whereas human ancestor species became extinct and new species emerged. This pattern indicates that monkeys were more generalized and were able to withstand climate and ecosystem change much better than our closer relatives. What made the apes and human ancestors more specialized and sensitive to change?

FUTURE EVOLUTION

As I tell all of my introductory evolutionary anthropology students, thinking that there is a correct answer to a question like this proves that you know nothing about evolution, which is at its heart unpredictable.

EVOLUTIONARY LESSONS

One glaring evolutionary lesson is how vulnerable we are to the environment. At the end of the last ice age, we thought we could control the environment with agriculture, which eventually led us to think we could control it with technology. The lesson of our long historical perspective is that such ideas are nonsense and that climate and ecosystem change ultimately will lead to our extinction. I predict that *Homo sapiens* will be perhaps the shortest lived species on our human evolutionary tree. Malthus warned us that population growth could have horrible consequences for our species. If we can overcome our hubris, perhaps we can change our destructive practices. But I fear that it is too late.

SPECIAL?

There is nothing very special about us. I think we appreciate that more if we have a pet or study animal behavior. The adaptations that have allowed us to destroy the world around us are our high intelligence, language, and collective written knowledge. I suggest that these strengths will be maladaptive in the long run.

RELIGION AND SPIRITUALITY

Science and religion are different approaches to understanding the world and cannot be combined. Science has no room for mystical magical forces, and religion requires unquestioning belief and is by definition unscientific. Individuals can be scientists and have spiritual beliefs, but they simply cannot mix the two together. In the modern Trumpian fake-news world, I wish that religion could pass on ethical and moral values to a broader portion of the human population. Religion is one of the few things that may be keeping human greed and vice in check. Religion has a different purpose than science—it helps us to develop a conscience and gives us a sense of belonging. Social media does the opposite.

ADVICE

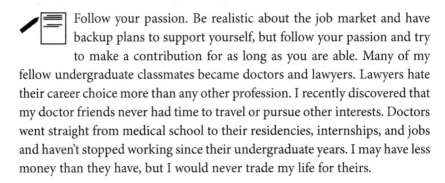 Follow your passion. Be realistic about the job market and have backup plans to support yourself, but follow your passion and try to make a contribution for as long as you are able. Many of my fellow undergraduate classmates became doctors and lawyers. Lawyers hate their career choice more than any other profession. I recently discovered that my doctor friends never had time to travel or pursue other interests. Doctors went straight from medical school to their residencies, internships, and jobs and haven't stopped working since their undergraduate years. I may have less money than they have, but I would never trade my life for theirs.

INSPIRING PEOPLE?

Two people inspired me: Alfred Wallace, because he did amazing things in Malaysia, and Von Koenigsberg, because he recognized *Mabokopithecus* based on a single tooth and was correct. His work on *Victoriapithecus* was great, and we could have a great discussion.

5

MICHAEL "MIKE" BENTON

Michael "Mike" Benton is a professor at the School of Earth Sciences at the University of Bristol (UK). He is a paleontologist with broad interests, including using fossils to study the diversification of life through time, the origins of key groups (such as the dinosaurs), as well as big extinctions (e.g., the end-Permian mass extinction). Mike's research led him to conduct fieldwork in Russia and China, and his research has been published in numerous high-impact journals. Still, for many of us, one of his most influential works is his seminal textbook *Vertebrate Paleontology*.[1]

YOUR BEGINNINGS

I think my interest began when I was a teenager. I had gotten interested in fossils and dinosaurs when I was seven (like all small children), but I did not grow out of it. The more I read, the more I realized that there were lots of gaps in knowledge and lots of chances for a new person like me to find a small niche somewhere and make some contributions.

GAME CHANGER

My game changer was the collective writings of Steve Gould. He showed me that paleontologists don't just collect the fossils (we do that too, and it is a very important task and duty). We also are in the prime position to apply new technology (e.g., CT scanning, SEM, mass spectrometers) and computation.

AMAZING FACT

 It is amazing that as many as five or six human species were living in eastern and southern Africa at one time, and they might have met. What would a tiny africanus say to a beefy boisei?

TIME TRAVEL

 I'd like to go back to the Cretaceous. This is long before the evolution of humans, but it was the time when terrestrial ecosystems were changing rapidly. Flowering plants had come on the scene, and they were stepping up the pace of life on land. This triggered the explosion of insect faunas, and then the insect-eating vertebrates including mammals. The first primates might just have seen a dinosaur as they peeped from behind a tree branch . . . and they survived the great mass extinction that killed the dinosaurs. And the rest is history, as they say.

EVOLUTIONARY LESSONS

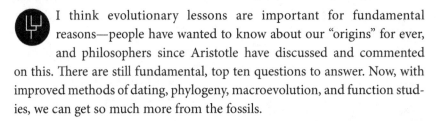 I think evolutionary lessons are important for fundamental reasons—people have wanted to know about our "origins" for ever, and philosophers since Aristotle have discussed and commented on this. There are still fundamental, top ten questions to answer. Now, with improved methods of dating, phylogeny, macroevolution, and function studies, we can get so much more from the fossils.

6

MATT CARTMILL

Matt Cartmill is a multifaceted anthropologist and a professor at Boston University. He is well known for his "visual predation" theory of primate origins and his extensive publication record on the evolution of primates, humans, and other animals. Matt also writes about the history and philosophy of evolutionary biology and science. He has written several books, including *The Human Lineage*,[1] which he coauthored with Fred H. Smith. Matt's body of work has earned him multiple honors and awards, including the AABA's Darwin Award for Lifetime Achievement in 2019.

AMAZING FACT

None! I believe that the emergence of humanlike intelligence was very probable given certain prerequisites. For me these include the initial appearance of life, multicellularity, bilateral symmetry, an internal skeleton, living on land and breathing air, endothermy, maternal care, very high neuron densities, and a way of life that depends heavily on making and using tools. Except for the first and last items, all these things have evolved more than once on this planet, so they aren't intrinsically unlikely. See Simon Conway Morris's [chap. 9] book *Life's Solution*[2] to learn more about this. If it hadn't been for an asteroid impact about 65 million years ago, some of the big flightless stem-group birds (aka "dinosaurs") might have become sapient before the mammals had an opportunity to produce us. Of course, we cannot test these conjectures unless or until we find other planets with life on them. Unfortunately, that may never happen. Here, too, I am inclined to agree with Conway Morris: life is probably exceedingly rare in the universe.

Regarding your follow-up question about accumulating contingencies as the main drivers of biological evolution: I think it is a mistake to think about evolutionary events that way, after the manner of Steve Gould, rather than as probabilities. The winner of a horse race is not always the horse with the longest and skinniest legs and the biggest heart—but if you bet that way, you will come out ahead, and every now and then you will bet on a Secretariat and be awestruck.

In assessing the likelihood of a so-called contingency, the thing to look for is the distribution of convergences. As for accumulation, remember that there are ten billion lines of evolution, all accumulating novelties. This greatly increases the probability that a particular string of novelties will accumulate in one of them, especially if similar novelties are appearing convergently in multiple lines under the driving forces of natural selection. Evolutionary novelties originate randomly, and there's no guarantee of survival for any accumulator; but over the long run you're more than likely to get fish and bugs, or at least things that look like fish and bugs. And once you get fish (those internal skeletons are important), you have a decent long-shot chance of getting talking animals if you wait long enough. Or so I like to think. Bring me five or six other planets with biospheres, and we'll check it out.

TIME TRAVEL

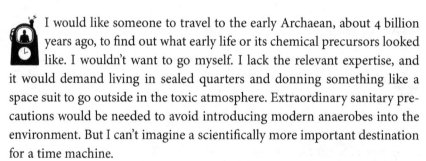 I would like someone to travel to the early Archaean, about 4 billion years ago, to find out what early life or its chemical precursors looked like. I wouldn't want to go myself. I lack the relevant expertise, and it would demand living in sealed quarters and donning something like a space suit to go outside in the toxic atmosphere. Extraordinary sanitary precautions would be needed to avoid introducing modern anaerobes into the environment. But I can't imagine a scientifically more important destination for a time machine.

DRIVING FACTORS

The main driving factor was a way of life dependent on making tools to a pattern. In *Darwin's Unfinished Symphony*,[3] Kevin Laland argues that such a way of life makes active teaching of young relatives adaptively profitable for the teacher and provides a context in which a useful ten-word protolanguage can emerge. I like this story. Once patterned tool making and protolanguage are in place, most of the other human peculiarities can be expected to grow out of them.

The fossil hyoid bones of ancient hominins reveal that more than 400,000 years ago some factor had already begun to reshape the human vocal organs from the primitive chimpanzee-like form seen in early hominins toward the morphology seen in modern humans and Neanderthals. What factor? Speech is the obvious candidate.

FUTURE EVOLUTION

If we're talking about natural selection, it depends mainly on which alleles promote having more babies. Humans won't evolve to become (say) taller unless tall people leave more descendants over the long run than short people do. The same holds true for our becoming shorter, smarter, stupider, darker, lighter, and so on. I don't know of any evidence that natural selection is pushing *Homo sapiens* in any of these directions today. And as soon as we start modifying our genomes to produce certain desired sorts of humans, all bets are off.

EVOLUTIONARY LESSONS

The only things we can learn from the past that will be helpful in the future are cause-effect relationships that can be expected to recur. Most of the key situations in human evolution aren't likely to recur in the future. Learning what our ancestors were like can give us some extra information on evolutionary processes and mechanisms, but it doesn't tell us anything about ourselves that we can't find out by studying ourselves. As I've said before, if it was proved that we are descended from rabbits, and not from monkeys, we would still prefer bananas to clover as an ingredient in pie.

So why study human evolution? For the same reason that we study cosmology—not to guide future actions but to understand how things got to be the way they are. Its value is its huge contribution to the scientific world picture.

SPECIAL?

Yes, we are. You don't hear crows or orangutans posing such questions. And yes, the only way we can tell what's unique about humans is by studying nonhumans. But when we do that, we find that we are less singular than many people like to think.

The nonhumans with the most forebrain neurons—apes and some birds—can make and use tools and learn simple, human-devised, me-Tarzan-you-Jane protolanguages involving arbitrary symbols (but little or no syntax). Our closest relatives, the chimpanzees, have gotten quite close to the rudiments of humanity. If some awful plague exterminated humans tomorrow, *Pan troglodytes* would be well positioned to explore the possibilities of becoming a tool-using ape living at the forest's edge.

And I don't think modern humans are not significantly different from some of the early *Homo* populations who overlapped in time with us. If that hypothetical plague had wiped out modern humans 60,000 years ago and spared the archaic peoples living farther north, I feel mortally certain that descendants of the Neanderthals and Denisovans would have gone on to produce pots, pyramids, and the Pythagorean theorem. In fact, I *know* they would have, because they did so; both groups are represented in the modern human gene pool. The Neanderthals buried their dead, did symbolic things with eagle feathers, and made crude jewelry out of eagle talons, painted shells, and fossils. It's conceivable that they were slower-witted on average than we are today, but the notion that they lacked the inexplicable divine spark of symbolic cognition that distinguishes humans from the beasts seems to me to be bogus.

In writing about the supposed uniqueness of human consciousness, I have tried to stick to the most basic question: Do other animals have subjective experiences? Even something as basic as that is difficult to determine. One reason is that the word "consciousness" can mean a lot of things. We don't have direct access to each other's minds, so I'm not sure exactly what aspects of consciousness all humans share. Maybe none. As I vaguely recall, being a three-year-old is a very different experience from being an eighty-year-old: simpler but less predictable, with brighter colors and stronger smells and flavors.

RELIGION AND SPIRITUALITY

 Like any other science, the study of evolution is compatible with any beliefs that can't be refuted by logic, mathematics, or observations. A lot of important beliefs fit that description.

MATT CARTMILL

ADVICE

For questions about our past, study genetics. That's where the fundamental discoveries about the histories, mechanisms, and origins of the traits of humans and other living things remain to be made. But the big questions about our future are not biological but geopolitical: Are we going to manage to have one, and if so, how?

7

YAOWALAK CHAIMANEE

Yaowalak Chaimanee is a research engineer at the French National Centre for Scientific Research (CNRS). She directs the French-Thai Paleontological Mission, a multidisciplinary research project (paleontology, geology, and paleoenvironments) that combines efforts from the PALEVOPRIM lab at University of Poitiers (a CNRS group) and the Department of Mineral Resources and Chugalongkorn University (both in Thailand). Over three decades, Yaowalak's research has informed our view about mammal evolution (especially primates), with continuous fossil discoveries published in prestigious scientific journals. Chief among her findings is that of *Khoratpithecus*, an orangutan relative from Myanmar and Thailand that live in Thailand 9 to 7 million years ago.[1]

YOUR BEGINNINGS

I began my career in the paleontology section of the Department of Mineral Resources in Thailand in 1985. At that time, everybody in the office worked on invertebrate fossils to support the Geological Mapping Unit. Therefore, we paid no attention to fossil mammals nor to their accuracy in dating continental deposits. I did not know that fossil mammals were found in Thailand until my boss received a mammalian tooth to identify one day. We decided to survey for fossil mammals, but it was several years before we discovered the first fossil mammal from the Krabi coal mine, which enabled us to establish the age of the basin in the Late Eocene. Later we found many mammalian fossils in Eocene, Oligocene, Miocene, and Pleistocene deposits around the country. I observed their changes through time from evolution, extinction, and radiation, and this interest led to my scientific study of fossil mammal evolution in Thailand.

AMAZING FACT

It amazes me that only one human species, *Homo sapiens*, exists today, because in Southeast Asia during the Late Pleistocene four species— *H. erectus*, *H. sapiens*, *H. floresiensis*, and *H. luzonensis*—have been documented, and the list is still open. This indicates the complexity of human evolution, and the situation is the same for earlier periods. The diversity of Miocene hominids is very high and was probably much higher than what we now know. Many interesting geographic areas have not yet been explored for fossil hominids. Advances in our knowledge is sluggish, yet humans plan to settle on other planets before we have unraveled our own history on Earth.

TIME TRAVEL

When we search for Pleistocene fossils in caves and fissure fillings in Thailand, we find a high diversity of fossil mammals, but the fossils are rather fragmented. I would like to travel back in time to the Pleistocene in Southeast Asia. It would be fascinated to see animals we have found as fossils—hyenas, giant pandas, and orangutans—roaming the country together with *Gigantopithecus* and *Homo erectus*. The occurrence of *Homo erectus* in Thailand is still in question, and so far we have found only one premolar of *Homo* sp. in a middle Pleistocene cave. I would be pleased to learn how these hominids shared the habitats and what kind of foods they collected. In addition, the savanna corridor enabled *Homo erectus* and other mammals to reach Java and other Indonesian islands. It would be great to follow this dispersal route along Sundaland and learn how *Homo erectus* settled Java so successfully.

DRIVING FACTORS

The two factors that usually drive mammal evolution are community structure and environmental changes. With so few fossils available, it is difficult to determine the role of either factor. The birth of the hominines lineage is an especially complicated subject. Our gaps in knowledge are also dramatic concerning the origins of chimps and gorillas, with almost no fossil record available to date or to reconstruct evolutionary changes. We have a better understanding of orangutan origins in the fossil record, but many problems remain such as the evolution of its suspensory adaptations.

RELIGION AND SPIRITUALITY

Every year, when starting our field trip in Myanmar, the chairman of the village prepares a sacrifice and brings us to pray for *nat*. Nats are spirits worshipped in Myanmar in conjunction with Buddhism. Every Myanmar village has a nat, which serves as a shrine to the village guardian. There are thirty-seven nats in total in Myanmar, and each area has its own nat. *Nat kadaw* is an old woman who can exclusively communicate with nat. During a brief ceremony, she will call the nat and ask us what we wish for; we always tell her "primate fossils, especially anthropoids." This traditional belief makes the villagers happy and encourages our team to work harder to find fossils. Once we found important new primate remains just a few hours after having requested the help of the nat. When we find primate fossils, the old chairman of the village is very happy and proud of his job. In Asia, we always mix faith with truth, but the spirit and science are never in conflict. This is an example of a positive effect of faith on research activity.

ADVICE

Always try to have a logical approach, and do not believe that fossils related to human evolution are different from other groups of fossils. Human evolution follows the same rules as those of other mammalian groups. Remember that fossils do not change over time, only interpretations may change, sometimes after new methodologies, new ideas, or new fossils emerge. Keep in mind that the fossil record is very incomplete and invest some effort to improve the fossil record of our human ancestors.

INSPIRING PEOPLE?

Gustav Heinrich Ralph (G. H. R.) von Koenigswald, a pioneer of human paleontology in Southeast Asia and China. He later discovered *Gigantopithecus* teeth in Chinese drugstores, making a significant contribution to Asian paleoanthropology.

8

GLENN CONROY

G lenn Conroy is professor emeritus at Washington University in St. Louis. Some of his best-known work relates to the paleontological evidence of African apes and humans during their divergence between 15 and 5 million years ago. In Namibia, together with Martin Pickford [chap. 18] and Brigitte Senut [chap. 39], they discovered the first fossil ape ever found in subequatorial Africa (*Otavipithecus namibiensis*, approximately 13 million years ago). This surprising finding was the "lucky accident!" of an expedition initially formed to find hominins at the Berg Aukas Mine.[1] Glenn explains his research in *Reconstructing Human Origins: A Modern Synthesis*,[2] coauthored with Herman Pontzer [chap. 86].

YOUR BEGINNINGS

As an undergraduate at the University of California, Berkeley in the 1960s, I was required to take a course that fulfilled the "natural sciences" requirement. At the time I had little interest in science and had really never even heard of anthropology, so I was looking for an easy course to fulfill the requirement. I was advised to take a human origins course offered by Sherwood Washburn, not knowing then that he was one of the most prominent physical anthropologists of his era. I was awestruck by the course. That course changed my life and my career trajectory.

GAME CHANGER

 Robert Ardrey's *African Genesis*[3] and Raymond Dart's *Adventures with the Missing Link*[4] were two of the most influential books of their day in changing the way (for good or bad) we looked at our own evolution.

AMAZING FACT

 That a small-brained, imperfectly bipedally adapted, virtually defense-less (in terms of weapons or canines), small-bodied hominin could ever survive on the African savanna is amazing.

TIME TRAVEL

 Achilles vs. Hector at Troy; Cortes entering Tenochtitlan for the first time; first mountain man rendezvous along the Green River; any time between about 1 million and 100,000 years ago because we really have no idea what was going on in human evolution during that period!

FUTURE EVOLUTION

 Extinct!

EVOLUTIONARY LESSONS

 The novelist Edward Abbey, whose book *The Monkey Wrench Gang*[5] became the mantra for the ecoterrorist movement in the 1970s, once wrote: "We are all ONE, say the gurus. Aye, I might agree—but one WHAT?" What, indeed, are we? Where did we come from? Did we spring full blown and whole during the sixth day of creation, like Minerva from the head of Jupiter, and were given dominion over all the other creatures of the Earth—or are we entangled with them, creatures great and small, through endless strings of time? The study of human evolution helps us address Edward Abbey's question.

ADVICE

 Learn a marketable skill—such as anatomy, genetics, statistics, biomechanics—and be cautious of "just so" stories about humanity's past or future.

INSPIRING PEOPLE?

Aristotle, Isaac Newton, and Alexander the Great.

9

SIMON CONWAY MORRIS

S imon Conway Morris is emeritus professor of evolutionary palaeobiol-
ogy in the Department of Earth Sciences at Cambridge University and
is a fellow of the Royal Society. Simon began his career studying the
origins and diversification of early animals during the "Cambrian explosion."
His study of the Burgess Shale fossils was popularized in Stephen Jay Gould's
Wonderful Life.[1] However, subsequently Simon disputed Gould's views in his
The Crucible of Creation.[2] Over the years, his research on evolutionary convergence, combined with a Christian perspective, has evolved into a unique
macroevolutionary overview on the cosmos. First presented in *Life's Solution*,[3] follow-ups include *The Runes of Evolution*[4] and, most recently, *From
Extraterrestrials to Animal Minds*.[5]

YOUR BEGINNINGS

If there was ever a "moment," it was when, at about the age of seven,
I was given an album by my mother, and the challenge was to stick
large stamps showing prehistoric animals into their correct spaces.
Something clicked. I dimly recall having a similar one with the flags of
various countries, but somehow this didn't turn me into a vexillologist.
Thereafter it was as a schoolboy collecting fossils on the Dorset coast, in
the Cotswolds, and the Atlantic coast of France (with intervals for *vin ordi-
naire*) that a hobby turned into something more dangerous. Not least when
we trespassed into quarries, sometimes being chased away, but in other
cases being welcomed by crafty old gaffers who knew all about fossil shark's
teeth, even if they thought they were some sort of bird bill. Mind you, I was
hardly a successful collector, routinely walking over a patch of ground that

moments later yielded to my companion reptile bones by the ton. Off to university and a somewhat irascible lecturer telling me, "There's this chap Whittington at Cambridge, doing work on the Burgess Shale. Why not drop him a line?" So, in a sense, a succession of "moments" have led me to the vastness of Xinjiang, the endless landscapes of Mongolia, and the tundra of Greenland.

TIME TRAVEL

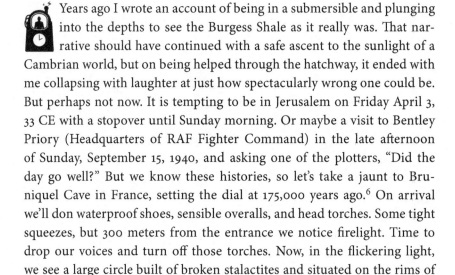 Years ago I wrote an account of being in a submersible and plunging into the depths to see the Burgess Shale as it really was. That narrative should have continued with a safe ascent to the sunlight of a Cambrian world, but on being helped through the hatchway, it ended with me collapsing with laughter at just how spectacularly wrong one could be. But perhaps not now. It is tempting to be in Jerusalem on Friday April 3, 33 CE with a stopover until Sunday morning. Or maybe a visit to Bentley Priory (Headquarters of RAF Fighter Command) in the late afternoon of Sunday, September 15, 1940, and asking one of the plotters, "Did the day go well?" But we know these histories, so let's take a jaunt to Bruniquel Cave in France, setting the dial at 175,000 years ago.[6] On arrival we'll don waterproof shoes, sensible overalls, and head torches. Some tight squeezes, but 300 meters from the entrance we notice firelight. Time to drop our voices and turn off those torches. Now, in the flickering light, we see a large circle built of broken stalactites and situated on the rims of the low walls a ring of fire. Our expert nods, undoubtedly Neanderthals, she announces. We strain our ears. Surely they are not singing the closing section of *Die Walkurie*?

I think Neanderthals were also straying into metaphysical territory. Such seems also to be the case with the evidence of flensing of wing bones of bird raptors for their primary feathers, and presumably subsequent employment for decoration. Here, too, we seem to see symbolic thinking. And the list goes on to suggest that Neanderthals are cognitively equivalent to us. But are they? One snippet I find particularly interesting is the paucity of small mammal bones among Neanderthal sites, hinting that they did not use the wind/frost resistant pelts. With us the story is the other way round. So was Neanderthal cognition in one way or another "deficient," or did they inhabit a parallel universe of consciousness? In any event, the report from Bruniquel is fascinating.

FUTURE EVOLUTION

Exactly the same as has been molding our evolution since at least 100,000 years ago and maybe a bit more. Farewell to that dreary Darwinian slog, and let's throw open the doors to the Lamarckian world. Now we stand on the threshold of a glorious country permeated with meaning, where intentions are valid and reasoning is not only joining the dots in those eureka moments but glimpsing the inner order of the universe. We speak and just as quickly tell jokes. Everything becomes a metaphor, an analogy, and symbols carry enormous power. We are no longer animals and are on a new trajectory that most likely was initiated in the Oldowan. We have always loved technology because tools are no longer inanimate but direct extensions of our cognition. So, too, we soak up gossip, but when we really want to say something useful, we write poetry and compose music. So what's next? We can reply with total confidence: "Exactly the same, but unimaginably different." Should we be worried? Hardly. Let's borrow that time machine and fast forward our ancient hominin and allow him to celebrate 100,000 birthdays in a few minutes. On arrival, and I know just the spot in Dorset to welcome him, he'll need a bit of time to get his bearings, but probably not as long as you might expect. After all he'll share our insatiable curiosity, and before long we'll learn as much from him as he from us. Now let us leave our newly arrived companion to enjoy a crackingly strong gin and tonic while we step into the time machine and whizz through another hundred thousand years into the future. We open the door and sure enough not a Morlock in sight, not a sniff of any Eloi. Still us, but you might find the end result not quite what you had been led to expect by those who pronounce "We'll turn into cyborgs" or "That's the end of history." No indeed, still us, but unutterably different. Never mind, now let's take a jump another million years into the future. Now you'll have to look pretty hard, but we'll be there. Just visit the rose garden and listen to the laughter of the children.

SPECIAL?

Only humans could possibly understand these questions, so isn't the answer pretty obvious? Not that we need to be diverted by weasel words like "special," let alone "pinnacle of creation." Given there is an evolutionary continuum between ourselves and the great apes, it would be passing strange if they did not throw light on our cognitive basement.

But it's pretty dim and dingy down there, and stumbling around in the half-darkness isn't going to tell us very much. Wittgenstein famously said that if a lion could speak we wouldn't have the foggiest idea what it was saying. Wrong; if a lion spoke, it would no longer be a lion. Certainly animals vocalize, and often very noisily, but thinking that these are the rudiments of language is simply special pleading. No more are these sounds proto words, nor does any animal show the least aptitude to string them into sentences let alone export these noises into new contexts. Animals will play but never tell a joke. Animals use tools, but why has no animal ever used one tool in order to make another? To them a cumulative culture is inconceivable. Indeed as various people have pointed out, animals may have local traditions, but they neither know nor care that they are traditions, let alone cultures. Animals learn, and they'd be in a sorry state if they couldn't. But none teach—well, with the exception of tandem-running in ants. Only if an animal could see into the mind of the pupil would teaching work. Ask any juvenile chimpanzee who spends several years observing his mother crack open nuts but never receives a murmur of advice, let alone encouragement. Incipiently prosocial and showing nascent empathy? Tell that to an orphaned chimpanzee; at best the orphan will receive grudging assistance from its kin but never from another member of the troop. But surely they have emotions? Most likely, but when you see retaliation, don't think it is anything to do with spite any more than any animal has the notion of revenge. No, no, our closest relatives are sturdy individualists, or to be less charitable (a totally alien concept to any animal), in reality they are much closer to sociopaths. The gap is real: Ask them to reason, to join the dots, to understand cause and effect, and they have not the vaguest intuition that such worlds exist. Animals know that their chums have intentions, but never can they step into their metaphorical shoes and see the world through the eyes of another. If they could, they would have teams rather than mobs and would tell us how all the world's a stage.

But there is a vital exception. Let the animal be reared in human company, ideally with nonstop encouragement and training. Oddly they become conspicuously smarter, and all those features that we see as closing the gap—mirror self-recognition, primitive "language," and mathematics—emerge, at least after a fashion. But in the wild their counterparts drift through their meaningless and atemporal lives in blissful ignorance. These enculturated animals tell us something much more important. What other species would bother training rhesus monkeys to entrain with a metronome for a year and still accept their failure as somehow interesting? No wonder our stories are full of talking animals, but never do they show the least inclination to tell stories.

I was particularly struck by the paper on mirror self-recognition (MSR) by Eric Saidel[7] who notes that among nonhumans MSR simply isn't a specieswide phenomenon and that, by and large, enculturated animals perform much better. In humans MSR is part and parcel of our cognitive realm and so is analogous to language, whereas in nonhumans it is a local skill that finds no wider context (ever seen an animal using a heliograph?). Even avoiding the pitfalls of anthropomorphism/centrism, these days I am more struck by how fundamentally animals "just don't get it." If they did, they would speak, teach, hypercooperate, and use cumulative culture. They don't, they can't, and they are not in the least bit interested. Why should they? To them our world is invisible.

To date I have no satisfactory explanation for our uniqueness. Flimsy evidence suggests the separation began with the australopithecines, but it might have been more recent (but surely by the time Acheulian tools were being made). So, too, arguments such as increasing population size and brain size increase are regularly wheeled out, but I am less persuaded.

RELIGION AND SPIRITUALITY

 A circular question. If we lacked "a spiritual or religious view of the world," we wouldn't be human. Not necessarily any worse, but certainly not us. Like music, which summons us to infinite depths of emotion, and language, in which we find ourselves in a story not of our making, so too it is far from coincidental that the spiritual is a human universal. In its full luminosity, we stroll across a glass floor above the existential abyss. It is not surprising that this is an area where scientists tend to stumble. I remember reviewing an excellent book by a distinguished biologist that captured all the excitement of his career. But toward the end he lost the plot and was suddenly swept away by the sheer emotion of seeing the vehicle that miraculously had ferried men back from the Moon. Instead of seizing the moment and recognizing how such a feeling was just as genuine as any scientific truth, the moment slipped through his fingers. As I wrote then, if I look into the eyes of my lover and insist that she has crystallins like nobody else, the only result is the sound of the bedroom door slamming shut.

ADVICE

 Read voraciously, and far outside your area; listen to experts but not too closely; ignore social media; and try to write with a fountain pen unless you are a journalist. Remember that everything you publish will either be wrong (no disgrace there) or with luck a tiny contribution to a much bigger question that only the next generation will intuit. When you realize that you are bored to death at conferences, stop going, and don't add too much tonic to the gin.

INSPIRING PEOPLE?

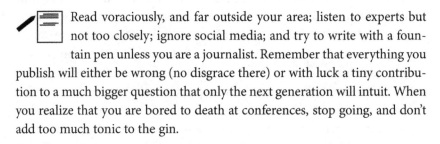 G. K. Chesterton; he knew a thing or two about caves.

10

ERIC DELSON

E ric Delson is a professor of anthropology at Lehman College and at the Graduate Center of City University of New York (CUNY). Since its founding in 1991, Eric has been the director of the New York Consortium in Evolutionary Primatology (NYCEP), a multi-institutional graduate training program in biological anthropology. NYCEP reflects Eric's wide range of interests in learning about human (and primate) evolution from a 360-degree perspective. Eric's extensive research includes more than two hundred scientific articles, several books and monographs covering everything from Pliocene monkey evolution to human phylogenies, as well as novel three-dimensional morphometric tools to model ancestral morphologies. As this book goes to press, Eric is retiring from teaching and administrative work to concentrate on research, mainly based at the American Museum of Natural History.

YOUR BEGINNINGS

I enjoyed science in high school and in college planned to study physics, which turned out to be very difficult. As I finished the last required course in the physics major, I took a class on human evolution "for fun." William Howells, who taught the class, allowed me to take it without the required prerequisite, and I was hooked. For my last three semesters, I took classes in physical anthropology, Paleolithic archaeology, Pleistocene geology, and biology. I wrote term papers on *Ramapithecus*, Pleistocene climate, Oldowan tools and (in a class on primate behavior) the fossil record of African and Eurasian cercopithecid monkeys. Howells had me work as a lab assistant, curate the teaching collection, assist with his study of the

Kanapoi distal humerus (the first specimen of *Australopithecus anamensis* ever found) with Bryan Patterson, and serve as teaching assistant when he taught the same course on human evolution again; he even allowed me to teach a class on fossil primates. Because there were so many researchers interested in human fossils (almost more than the number of such fossils), I decided to focus my career on the evolution of cercopithecids, and I continued to study human fossils when possible.

GAME CHANGER

 Early Man, a book in the Time-Life science series, was written by F. Clark Howell and appeared in 1965.[1] In it, Howell explained his approach to paleoanthropology, the integrated study of human paleontology, prehistoric archaeology, and the geological background (including dating, stratigraphy, and other aspects) of the study of a site. I read it just as I was taking Paleolithic archaeology courses, and it helped convince me to become a paleoanthropologist and recognize the right way to study that field.

TIME TRAVEL

As a science-fiction buff, this question really grabbed me. I can't choose just one destination, so here are several:

(a) The Sima de los Huesos (Pit of the Bones) is a section of the Atapuerca cave system in northern Spain where hundreds of bones representing at least twenty-five individuals have been found. The excavators think the inhabitants (early ancestors of Neanderthals who lived some 400,000 years ago) threw corpses down a "chute" deep inside the cave in some kind of ceremony. Others think the bones may have floated in on flood waters or otherwise arrived naturally. I'd like to travel there to see what really happened and to observe these early humans interacting with each other.

(b) A similar situation existed in the Dinaledi Chamber of the Rising Star cave system in South Africa. The humans who lived in the area were members of the species *Homo naledi*, which looks like early *Homo* of 2 million years ago but has been dated to between 335,000 and 225,000 years ago. Hundreds of bones representing numerous individuals were again found at the end of a long drop-off that can only be reached today by an arduous climb through the cave, but it was probably accessed another way at the time. I would travel there to see if they actually threw corpses into the chamber and also to see what these strange humans looked like.

  (c) At several other sites in South Africa much earlier humans, termed australopiths, lived, and some scientists have suggested that two different species were found together. At Sterkfontein (perhaps 2.4–2 million years ago), most fossils are identified as *Australopithecus africanus*, but some have been suggested to represent *Australopithecus prometheus*; it is not clear whether they actually lived at the same time and place or replaced each other across the landscape over time. Later fossils from Sterkfontein (about 2–1.8 million years ago) are identified as early representatives of *Homo*, who made stone tools. Across the Blaauwbank Valley at Swartkrans, within the 2–1.6 million years ago time span, members of a bulkier species named *Paranthropus robustus* are thought to have lived alongside or near other early *Homo* individuals, and it is not clear if one or both species made stone tools. If I could travel there and move forward in time from 2.4 to 1.6 million years ago, I might be able to see if these species were different, if they interacted, and who made tools.

  (d) The first site that has yielded well-preserved fossils of *Homo* outside Africa is at Dmanisi, in Georgia, dated to about 1.8 million years ago. There is a great deal of morphological variation in the five known skulls, and some authors think that more than one species lived there. It would be interesting to look at a larger population living there to see whether they were members of a single species or if more than one species lived there over time and space.

  (e) More recent populations are also of great interest. For example, the *Homo floresiensis* inhabitants of the Liang Bua cave on the Indonesian island of Flores are colloquially known as "hobbits" because of their small body and brain size. Their morphology indicates that they may have had an ancient divergence from other humans (about 2 million years ago?). They may have become reduced in size over time as do many large mammalian species isolated on islands where food is at a premium. Some authors have suggested that they are merely modern humans with some degenerative disease, but recent dates show that they lived between 100,000 and 50,000 years ago, well before modern humans seem to have reached this area. I would be interested in observing these humans and their way of life and to see if they persisted long enough to interact with *Homo sapiens* once they reached Flores.

  (f) Of similar age and size are the less well-known *Homo luzonensis* from the Philippines, and it would be interesting to observe them as well.

  (g) Another little-known species of *Homo* are the Denisovans, mainly known from a few bone fragments found at the Denisova Cave in Siberian Russia and one lower jaw from a cave on the high-altitude Tibetan Plateau in China. Study of the DNA of these fossils shows that the Denisovans are most closely related to the Neanderthals (and farther from *Homo sapiens*), but some modern populations also have small amounts of similar DNA. A visit to Denisova would reveal what they actually looked like and confirm that they interacted with the Neanderthals who also lived locally, between about 100,000 and 50,000 years ago.

(h) The Neanderthals (*Homo neanderthalensis*) are far better known and lived across western Eurasia from about 400,000 (see [a] above) to 40,000 years ago, when modern humans appear to have replaced them in a short period of time. Visits to several Neanderthal sites during this interval would reveal how they behaved, how "modern" they were in cognitive abilities, and eventually how they interacted with the incoming *Homo sapiens*. At the French site of St. Césaire, one Neanderthal fossil is known alongside stone tools called Chatelperronian, which look like those made by later modern humans, and it would be interesting to learn about the relationship between these species.

(i) In addition to these questions about fossil human species, I am also interested in looking into questions about the monkey fossils I study. For example, I spent parts of seven summers excavating the site of Senèze, in central France, where mammals came to drink from a lake inside a volcanic crater about 2 million years ago. I would love to visit there and see what it actually looked like. One of the animals known from that site is a large monkey called *Paradolichopithecus*, related to the macaques of Asia and North Africa (and Gibraltar). Similar monkeys are known from fossil sites in Spain, Romania, and Greece, and others (called *Procynocephalus*) from India and China, all dated between 3.5 and 1.5 million years ago. Some of my colleagues think they are really the same animal, but I don't, and I'd like to watch them all to see if they are different. I also would like to visit a site in India called Kurnool, which yielded a tooth I identified as *Theropithecus*, similar to a species that lives today only in highland Ethiopia. I am not sure how that animal (a young juvenile based on its little-worn tooth) got to India, and that would be fun to track down.

EVOLUTIONARY LESSONS

 I am not at all sure that human evolutionary studies can be helpful for people today, but I find it fun.

ADVICE

I advise students thinking about studying paleoanthropology to learn all they can about the fields that fit together to make that integrated discipline: biology, especially evolutionary aspects, ecology, and comparative morphology; geology, especially topics like dating methods, stratigraphic correlation, climate reconstruction, and taphonomy; paleontology (the intersection of geology and biology), especially of mammals; statistics and other mathematical approaches, and computer science; and the major components of biological anthropology (human paleontology, skeletal biology, genetics, and behavior) and prehistoric archaeology (Paleolithic studies, stone tool manufacture, lifeway reconstruction and excavation techniques). After learning about these concepts, a student can put them together and decide which to emphasize in a career in paleoanthropology.

INSPIRING PEOPLE?

I'd like to discuss these ideas with Thomas Henry Huxley, "Darwin's Bulldog." He was more interested in human evolution and its paleontological and comparative morphological basis than was Darwin, and he seems to have been far more outgoing in his interpersonal communication. It would take a while to explain a lot of the background, but I imagine he would catch on quickly and have some fascinating reactions.

11

MARC FURIÓ

M arc Furió is a paleontologist based at the Department of Geology at the Autonomous University of Barcelona (Spain). He is an expert in mammal evolution and extinction, especially shrews and other small, short-lived animals whose adaptations can accurately track environmental changes through time. Marc's research combines fieldwork at different international localities with lab-based imaging techniques. He is committed to science outreach at all age levels, focusing on the future of humanity and the dangers of current unsustainable economic practices. His ideas are summarized in his book *La Especie Humana*[1] coauthored with Pere Figuerola.

YOUR BEGINNINGS

As a child, I was fascinated by science (outer space, present-day fauna, dinosaurs, human evolution, and archeology). One of my oldest memories is from when I was four or five years old and my parents were trying to explain to me what a "galaxy" was. I had heard about it on TV, and it was simply amazing in my mind. When I was nine years old, we had to make a collection of minerals at school, and I fell in love with geology. Later on, minerals drove my interest toward chemistry and fossils. After that, and already studying for my college degree, my passion for fossils broadened my curiosity in sedimentology and ecology. And ecology, nowadays, makes me feel curious about many other scientific and social disciplines (apparently unconnected but all linked somehow), including medicine, psychology, and economics. I cannot pinpoint exactly the moment I decided to do research; I simply arrived here. I have always wanted to know. . .

TIME TRAVEL

 I am interested in the ecological role played by humans, so I have three responses to this question, depending on the purpose of the "visit."

As a scientist, I would use this scientific "field trip" to see with my own eyes how mass extinction works (in a fast-forward mode). The Permian (about 252 million years ago) or the Cretaceous (about 66 million years ago) are probably the best choices to get an idea about the next steps we will experience in the current ecological collapse driven by humans. The impact of a giant asteroid in a world ruled by dinosaurs seems more exciting than staring in front of some volcanoes in Siberia for a million years, but both of them would provide essential clues to the main breakpoints when ecosystems collapse on a planetary scale.

As a curious person, a second option would be a jump one thousand years into the future. I would love to discover that all my guesses about the destiny of humankind were absolutely wrong and that we still exist and live in a wonderful and healthy planet Earth full of plants and animals, with no environmental or social conflict.

As a personal experience, a third option would be visiting the "virgin" Europe around 100,000 years ago and have a chance to live in a Neanderthal clan (previous accepted application, of course). If I could adapt my body and biology to "neanderthalize myself," I would not hesitate to make a rather long visit to the transition from the Middle Pleistocene to the Late Pleistocene. For many reasons, there is no other moment in human evolution in which I think I could find a better fit.

FUTURE EVOLUTION

Unfortunately I am not optimistic about our future. As a species, we humans are closer to becoming extinct than to taking a next step in evolution. Many scientists from all fields of knowledge are warning about a "perfect storm" coming in this twenty-first century. Global warming is close to reaching the point of no return, most continental and marine ecosystems are collapsing (or on their way to collapse), and the most significant energetic and material resources (fresh water, oil, coal, metals, uranium, fisheries, biomass) have peaked and are dropping down. We humans are depleting all kinds of resources at the very same time that our global population grows at a rate of about one million people every four or five days. [Marc invited me to check worldometers.info/world-population and do the numbers.] Two more decades of these trends will be sufficient to force the starvation of billions of humans, and the consequences will be massive migrations and extremely complex international conflicts. The magnitude and range of these dramatic consequences are yet to be seen.

If a rapid collapse of all the complex societies does not blow away all humans, some surviving isolated groups could develop different evolutionary strategies in just a few centuries. In such case, a new cladogenesis is a likely option, with an overall simplification of societies and several different human species adapted to their corresponding regional conditions. It is difficult to describe the specific adaptations humans may display in the future because evolution has been shown to be unpredictable. However, we can easily imagine remarkable changes in metabolic rates, dentition, and digestive tracts to survive on the few edible plants and insects available in each region, together with an overall shortening in life expectancy. It seems quite unlikely to me that the humans of the future will be the cyborgs represented in many sci-fi movies, whose perfectly combined biological parts and technological devices extend their lives roughly forever.

EVOLUTIONARY LESSONS

Tracking the fossil record of the last 3 million years, we found that human biological evolution was linked to technological evolution.

From early stone chopping tools more than 2.5 million years ago to the Neolithic tools of a few thousand years ago, change is evident. And from the Neolithic to present-day technology, the improvement is simply astounding. However, it was not technology that enabled humans to survive and take over the world—it was easier access to energy. For instance, scavenging first and hunting later provided access to different and more energetic feeding resources. Controlling fire provided protection against colder climates and made the conservation and digestion of many feeding resources easier. The more advanced were the tools used by humans, the less dependent we were on the unpredictability of the environment. Consequently, technology was key to ensuring a year-round higher energy supply, which enhanced larger brains and increased the chances for offspring survival when compared to the rest of the animals.

Humankind's overwhelming domination of energetic resources has a related significant drawback. The huge global demand for energy and materials to maintain our complex societies (or "technosphere") is now threatening the sustainability of global ecosystems, and we absolutely depend on this equilibrium. The study of mass extinction events in the past (Ordovician, Devonian, Permian, Triassic, and Cretaceous, many millions of years before the rise of any bipedal primate) indicates that when ecosystems collapse on a global scale there is little hope for the survival of large-bodied, complex biological entities such as us. Technology will not be enough to save our future. The lesson is quite clear: human success will depend on technology only when the problems derived from its use do not overcome its benefits.

SPECIAL?

Yes. There is "something special" in some humans: a "magical mix" of critical reasoning, logical thinking, and abstract imagination. That mix provides the capacity to create new things deliberately, not by chance. But according to some researchers, this something special belongs to only about 10 percent of the world population. Many *Homo sapiens* simply behave like trained bipedal apes accomplishing their daily routines (which is not bad in itself). I am aware that this is a harsh statement, but the more we learn about animal behavior, the less special most humans look. Every time somebody says that "(we) humans have stepped on the moon," I like pointing out that such a milestone was not carried out by "all humans." It was reached by a selected group of extremely smart people who planned, designed, and executed the trips to our satellite. The other 99.9999 percent of the human population (me among them) have not been involved in those missions and probably do not have the skills necessary to participate. We all have something that makes us unique (an ability or a physical trait), but we do not all have the something special that makes a real difference between us and other primates.

With respect to the second question, absolutely yes: the study of present-day apes provides extremely valuable data. I enjoy analyzing most of the news in terms of animal ethology. Look around you: apelike behavior explains almost 90 percent of all the social and politic human movements in the world. The social reaction of many people is usually driven by instincts (fear, revenge, compassion, group belonging, survival, etc.) rather than by logical thinking, which is supposed to be one of our most distinctive characteristics.

RELIGION AND SPIRITUALITY

Literal interpretations of ancient sacred texts are difficult to accommodate within the paleoanthropological finds of the last century. Today science can provide a logical explanation for many transcendental questions about our evolution as a species. However, others questions will remain unsolved for a long time. Who knows if some secrets of human evolution will ever be unraveled. A spiritual or religious view of the world can help cover these gaps. After all, science and religion operate in different dimensions: the former works with facts and logical thinking whereas the latter is based on beliefs and faith. As a final remark, the more we know about the evolution of our species and nature, the closer we get to the deity or the divinities behind this wonderful creation. Our own existence is truly an unexpected and unlikely miracle in the universe.

12

DAN GEBO

Dan Gebo is a biological anthropologist based at Northern Illinois University, where he spent a thirty-two-year career studying and teaching the evolution of locomotor adaptations in early primates, apes, and humans. Although his research focused on the anatomy of the foot, Dan has worked on a wide range of topics and anatomical areas, including the functional morphology of *Morotopithecus* (about 21 million years ago), the first ape showing anatomical similarities with living apes in the vertebral column, shoulder, and femur. His research has been highlighted in the *New York Times*, and his teachings are summarized in *Primate Comparative Anatomy*.[1]

YOUR BEGINNINGS

There was not one eureka moment that led me to choose research, but in my sophomore year in a California high school I took a biology class that changed my life. I fell in love with the idea of animal adaptations and, of course, with biological evolution. From then on I wanted to understand animals and ultimately to study primate evolution. My second teenage decision point occurred a little more than a year later. As a high school senior in the state of South Carolina, I was able to take a second class in biology that was pretty much an independent study. I appreciated the free time to read; I read about human evolution and, in particular, about Louis Leakey. I worked hard in that class, and I was sold on studying human evolution as a graduating high school senior. I wanted to do fieldwork and find fossils that would explore our evolutionary history. I am sure that the adventure as well as the science motivated my teenage thoughts. Later my interests shifted more toward primate versus human evolution, but fieldwork and fossils were always a central theme. The one take-home lesson from my youth that I remembered later as a professor was to always make an extra effort to make a good impression when teaching first-time students in introductory science classes such as Human Origins, Becoming Human, or the Introduction to Physical Anthropology. Who knows what might happen?

GAME CHANGER

The obvious answer has to be the work of Raymond Dart and Robert Broom. The fossil specimens from Taung were clearly a game changer, but it was decades later before paleoanthropologists appreciated the australopithecine fossils. The idea that humans started out with big brains was upside down. We thought then, and often do today, that human success was wrapped up with our large brains and our intelligence. This is true later in human evolution but not at the beginning, as Dart tried to explain in 1925.[2] Although Dart made the discovery and described the fossils from Taung—thereby receiving the bulk of the credit—I think Robert Broom is the key scientist to put the first step of human evolution on the right path for subsequent generations of paleoanthropologists. Broom found the first adult specimens of australopithecines, and he wrote a long monograph that convinced Sir Arthur Keith as well as other top paleoanthropologists that human evolution started out with smaller ape-sized brains.[3] Broom's work cemented australopithecines as key participants in our views about early human evolution.

The Dart/Broom work on australopithecines and its subsequent history has influenced my career. The key message is to consider all of the known evidence for a fossil, consider different evolutionary viewpoints, and select the best supportable answer even if it goes against the mainstream. My work on eosimiids would certainly fit this scenario.

AMAZING FACT

The most amazing fact or incident concerning human evolution in my lifetime were the fossil discoveries from Flores Island. Who would have dreamed of discovering a dwarfed species of a fossil human living in the same time period as late hominids such as Neanderthals and modern humans? We can conceive of many intermediate fossils between the major groups, but the tiny hominids from Flores (*Homo floresiensis*) are quite remarkable and certainly a discovery no one could have predicted. A fossil discovery like the one at Flores is one reason I like to collect fossils: no matter how much you think you know, you always find something that does not fit the known pattern. With another similar find at Luzon Island in the Philippines, you would think paleoanthropologists would be mentally prepared for dramatic surprises in the human fossil record, but we are an old-school traditionalist group of scientists.

TIME TRAVEL

 I have worked on all of the geological epochs during the Tertiary, so it is a difficult to pick only one stop. The questions posed in this project deal generally with human and ape evolution, but I would like to travel farther back in time to the beginning of primate evolution: the Early Eocene, some 55 million years ago. I am interested in the initial origin of all primate lineages, and the Early Eocene is the warmest period planet Earth has ever seen, with crocodiles and palm trees populating the Arctic Circle. Primates were moving biogeographically across the northern continents (North America, Europe, and Asia), and none live in Africa, Madagascar, or South America, as they do today. In the Early Eocene, we have several known fossil species for the two primate suborders (Strepsirhini and Haplorhini). I would particularly like to observe the most primitive fossil primates in each suborder: *Teilhardina, Archicebus, Donrussellia, Cantius, Marcgodinotius,* and perhaps *Altanius.* How different were these early fossil primates in terms of diversity? They set the pattern for subsequent primate lineages, and I would like to watch them move and feed, and learn whether they lived in groups. The behavioral ecology of the early fossil primates could provide the driving force for primate adaptations we have come to study.

DRIVING FACTORS

The key evolutionary factors that begin the transition to the human lineage occur twice in ape evolution. The first major change is when primitive arboreal "apes" (proconsuloids) change their monkeylike quadrupedally adapted horizontal bodies into upright-oriented arm-swinging (brachiating) apes. This clade of brachiating apes includes the living apes: gibbons, orangutans, gorillas, chimpanzees, bonobos, and humans. To be a brachiating ape, you need a wholesale remodeling of the proconsuloid body plan, especially above the waist. Arms get longer; the thorax is compressed, being wide side to side but narrow front to back; the shoulders are positioned on the sides of the body and the clavicle is elongated to meet the shoulder joint; for the hands to face forward, the upper arm bone, the humerus, must twist (we call it medial torsion); we must fully extend the two bones in our forearm; and fingers are elongated (later shortened in the human lineage). This short list of skeletal changes does not include necessary soft tissue changes: like tacking the heart and lungs down into the thorax cavity or the large intestine in the abdomen to keep them from falling downward in an upright thorax, nor the many other muscular features that had to change for this new orthograde body plan. These upper body changes allow our shoulders to rotate above our head. The motion is called circumduction, and it is unique to the living apes, including humans. Brachiating apes use their long arms and this unique shoulder motion to swing through the trees and to hang below branches to feed. Humans have the same upper body plan as the living apes, which is why we can hang on a jungle gym or a tree branch for quite some time.

The second major change occurs among the lineages of African apes (gorillas, chimps, bonobos, and humans). All of these lineages moved to the ground ancestrally. Terrestriality is the second major adaptive event prior to human bipedalism. Moving to the ground has always been a risky maneuver among primate lineages, and few species in any primate group have opted for this ecological solution to long-time survival. Only Old World monkeys, as a group, make this substrate switch work, and many of these monkey species went back into the trees. An important question today is why did any African ape opt for terrestriality? It is not a common niche for arboreally adapted primates. What is it about the rain forests in Africa in the Middle to Late Miocene that lured these arboreally adapted African ape species to choose a new survival strategy? It obviously worked because gorillas, chimpanzees, and bonobos are alive today, and all predominately move using terrestrial quadrupedalism. We often describe this movement as knuckle-walking because these African apes must fold their evolutionarily long brachiating-adapted fingers to walk on the ground. Long arms and high shoulders relative to short legs and lower hips reflects their brachiating ancestry. Their long arm history gives them a back oriented at forty-five degrees, a less efficient orientation relative to that of monkeys or fossil proconsuloids. One connection between the knuckle-walking quadrupedal African apes and bipedal humans is that we all strike the ground first with our heels when we stride forward. We call this heel-strike plantigrady, and this foot feature is linked with African ape terrestriality. The African ape–human lineage needs only to stand up and walk on two legs with all of the subsequent changes to the human lower limb relative to the pelvis and legs of African apes. Human bipedalism is certainly a uniquely human phenomenon.

The human lineage needs one other major change from that of the African apes. We needed to modify our teeth so that our back teeth (molars and premolars) are large relative to our small front teeth (incisors and canines), a reverse condition relative to the African apes. These dental modifications document an evolutionary shift toward a brand new diet for our ancestral lineage and a move away from the softer fruits consumed by chimpanzees and bonobos. Large molars with thick enamel point to a diet of harder edible items that require a larger grinding surface to process. I assume that a wholesale dental switch like this is made to avoid competition with other species of sympatric African apes. No matter what the ultimate driving force may be, australopithecines clearly demonstrate humanlike dental changes and characteristics for bipedalism. The "answers" to this critical shift between African apes and humanity lies in the fossil record of pre-australopithecines.

INSPIRING PEOPLE?

If I had a chance to talk to one individual from the past, I would enjoy meeting and talking to John Napier. Our discussions would not center on the human evolution questions discussed here, although bipedalism or human hand function might enter into a far-ranging discussion. I would like to talk to John Napier about his views on primate locomotion and how primate locomotor systems changed over time. I have talked to his students, but it would be fun to ask "the mentor" about living and fossil primates. I would especially like to talk to him about the thinking behind his classic article with Alan Walker on vertical clinging and leaping. John Napier left a huge impact on all later students who studied locomotor evolution in primates, and it would be an insightful experience for me to have a moment with him.

13

JAY KELLEY

Jay Kelley is a researcher at Arizona State University's Institute of Human Origins who studies the evolution of hominoids (the ape and human clade). Jay approaches his research from a big-picture perspective, but his expertise lies in dental biology. Dental anatomy can inform the life history in living and extinct species and is key to elucidating species-level taxonomy. Jay is also an active field paleontologist. He was recently involved in excavating the terminal Miocene site of Shuitangba (about 6 million years ago) in Yunnan Province, China, which has yielded an intriguing juvenile ape cranium assigned to the species *Lufengpithecus lufengensis*.

YOUR BEGINNINGS

It was at the beginning of my senior year in college. Well, I wasn't actually a senior; I had fallen behind my class due to a couple of incompletes and a dropped class here and there. I wasn't aware of this technicality, but my draft board certainly was and reclassified me 1-A. Within a couple of weeks I was ordered to report for my physical—Vietnam, here I come. (At that time, one could get a deferment from the draft by attending college, but eventually this was rightfully ruled to be unfair. I was in the last class of graduating high school seniors who could do so.) It was the same 1-A threat that kept me from leaving college to hit the road after my freshman year to get some seasoning. In one of the great ironies of those times, a conscientious objector to the war was attached to the chaplain's office at my college for his alternative service, and he made it his role to advise students on how to avoid the draft. I was able to throw up enough roadblocks to sufficiently delay being drafted until Henry Kissinger ended our involvement in the war (sort of) and hence the draft.

Wait, how did I get onto this? What was the question? Research, right. All of my alternative activities and lack of focus during my first three undergraduate years turned into an intense focus in my senior year. I had certainly enjoyed my biology courses and done reasonably well (at least in these), but I became completely absorbed my senior year, particularly in courses oriented around pursuing independent study rather than coursework, which I have never liked. That final year wasn't quite enough to pull my overall GPA above 3.00, but it was enough to earn a couple of good letters of recommendation, and combined with solid GREs I was accepted into a master's program in evolutionary ecology. This, plus switching to paleobiology, ultimately led me (after a few years and another detour or two) to the anthropology doctoral program at Yale.

GAME CHANGER

For a book, it has to be *On the Origin of Species*.[1] The implications for our own evolution were obvious, immediate, and impactful, as events following its publication demonstrated. During the first quarter of my turnaround senior year in college I took a course in evolution. It was a small class structured around reading the *Origin* cover to cover, two or three chapters a week (and it had to be a facsimile of the first edition). The class was held in a small, nineteenth-century-like lab with floor to ceiling glass doors along one side—always open in southern California—set in the side of a wooded hill. The instructor, John Stephens, would sit on a lab stool and begin musing on the week's reading to open the discussion. We each had to pick a topic from the book to revisit in light of current knowledge and write a paper that was the sole basis for our grade in the course. Although I had no special interest in human evolution at that time, it was without a doubt the key experience that eventually translated into my pursuit of a study of human evolution.

AMAZING FACT

Few things in the study of human evolution can meaningfully be called "facts." One example of a fact is that the evolution of our lineage is demonstrably like that of any other mammalian lineage. Although a superficially mundane fact, this is an eye opener and something to be pondered by those who might be receptive to human evolution but know little or nothing about it, or have lots of misapprehensions about it, and think that human evolution must somehow be special. It is crucial to make a clear distinction between "what happened" and "why it happened" because these are very different sorts of questions.

TIME TRAVEL

 I would travel either to the nineteenth century, the great age of naturalist explorers, or about two hundred years into the future to piss on the graves of climate change deniers—at least those who are still above water.

DRIVING FACTORS

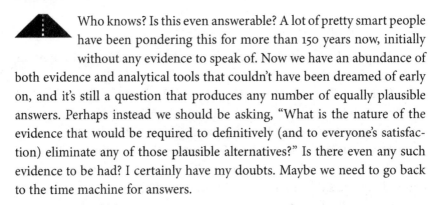 Who knows? Is this even answerable? A lot of pretty smart people have been pondering this for more than 150 years now, initially without any evidence to speak of. Now we have an abundance of both evidence and analytical tools that couldn't have been dreamed of early on, and it's still a question that produces any number of equally plausible answers. Perhaps instead we should be asking, "What is the nature of the evidence that would be required to definitively (and to everyone's satisfaction) eliminate any of those plausible alternatives?" Is there even any such evidence to be had? I certainly have my doubts. Maybe we need to go back to the time machine for answers.

FUTURE EVOLUTION

Who knows, but there's an episode from the 1960s television show *The Outer Limits*, guest starring David McCallum, that does as good a job as any I suppose in marching us sequentially through the second of these questions. [In that 1963 episode, titled "The Sixth Finger," the character played by David McCallum enjoys the benefits of having an extra finger on each hand and an enlarged brain.]

RELIGION AND SPIRITUALITY

 I would frame this more broadly and ask how is being a scientist compatible with having a spiritual or religious view of the world? The two realms rely on completely different and nonintersecting epistemologies. That being the case, they are in principle completely compatible because they don't bear on each other. In my view, however, the problem for a scientist is that engaging in two different epistemologies means being inconsistent, and inconsistency opens the possibility of arbitrariness in deciding which questions belong to which realm. Inconsistency and arbitrariness should be anathema to every scientist. To me, being a scientist is a frame of mind, not a matter of doing science things. That frame of mind can't coexist with some other frame between which one jumps back and forth depending on what one is pondering. I'm sure my position on this is partly a matter of my brain chemistry; I seem to entirely lack the capacity to generate a psychological state that feels the need for some immaterial reality.

ADVICE

 If you feel compelled to provide an answer to every question, or an explanation for every phenomenon, and if you can't deal with perhaps unyielding uncertainty, go into another line of work.

INSPIRING PEOPLE?

 Graham "Brian" Chapman [actor with the lead role in two Python films]. The answers might not be particularly insightful, but they would be hilarious, I'm sure, and that's worth something!

14

YUTAKA KUNIMATSU

Yutaka Kunimatsu is a professor at the Department of Business Administration at Ryukoku University (previously based at Kyoto University and Kyoto's Primate Research Institute). During his thirty-plus-year career, Yutaka has contributed extensively to our knowledge about hominoid evolution. Especially relevant are the discoveries of *Nakalipithecus* and other primates (about 10 million years ago) from Kenya,[1] which inform us about ape and monkey evolution in the Late Miocene of Africa, a key but poorly understood time period preceding the emergence of the human lineage.

YOUR BEGINNINGS

I do not remember the exact moment when I decided to do research. It was not a single moment of conversion but a lot of small pieces of experience in my early life that gradually influenced me and eventually led me to be a researcher. For example, when I was a young boy, I was fascinated with dinosaurs and other prehistoric animals, as many young boys were and are in many countries all over the world. One day in my boyhood, I found a Japanese translation of *The Dinosaurs* by W. E. Swinton[2] in my uncle's house (he lent me the book). I also remember finding a Japanese translation of the *History of the Primates* by W. E. Le Gros Clark[3] in the small public library in my hometown. I have to confess that both of these books were quite difficult for a boy to understand. In addition, I loved to read science fiction books, which taught me "the sense of wonder." My favorite science fiction book in my high school days was *Inherit the Stars*, written by James P. Hogan,[4] a story about human evolution. But at that time I did not expect that I would become a physical anthropologist in the future.

GAME CHANGER

 Among recent discoveries, I like to choose the discovery of Ardi in Ethiopia. [Yutaka refers to the nickname of the 4.4-million-year-old *Ardipithecus ramidus* skeleton ARA-VP-6/500 uncovered by Tim White (chap. 45) and his team.][5]

AMAZING FACT

 What amazes me is the fact that multiple species of hominins coexisted in the past but now we are alone.

TIME TRAVEL

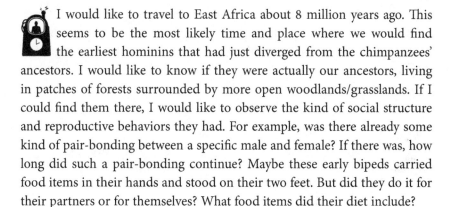 I would like to travel to East Africa about 8 million years ago. This seems to be the most likely time and place where we would find the earliest hominins that had just diverged from the chimpanzees' ancestors. I would like to know if they were actually our ancestors, living in patches of forests surrounded by more open woodlands/grasslands. If I could find them there, I would like to observe the kind of social structure and reproductive behaviors they had. For example, was there already some kind of pair-bonding between a specific male and female? If there was, how long did such a pair-bonding continue? Maybe these early bipeds carried food items in their hands and stood on their two feet. But did they do it for their partners or for themselves? What food items did their diet include?

DRIVING FACTORS

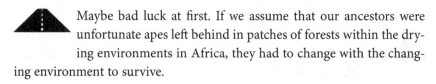 Maybe bad luck at first. If we assume that our ancestors were unfortunate apes left behind in patches of forests within the drying environments in Africa, they had to change with the changing environment to survive.

EVOLUTIONARY LESSONS

Human evolutionary studies teach us the factors and backgrounds in the past that have shaped our present nature. The more we understand why and how we have become what we are today, the better,

I hope, we can control ourselves, taming the dark side of our nature. In addition, human evolutionary studies show us different aspects of the stream of time beyond our daily life. In recent years I begin my lecture on human evolution by telling my students that I want them to acquire a sense of time counted in units of a million years through learning the history of human and primate evolution. In the class, I mention the tragedy of the big earthquake and tsunami that attacked the northeastern part of Japan on March 11, 2011, and the consequent accident at the Fukushima Daiichi Nuclear Power Plant, which forced many thousands of people to leave their homes. It is far beyond human abilities to stop earthquakes and tsunamis, but we could have reduced the damage caused by the nuclear power plant accident. If the decision-makers in Tokyo Electric Power Company and the Japanese government had been familiar with thinking in units of a million years, they might have listened more seriously to cautions from scientists and could have made a better decision on the safety planning for the nuclear power plant. The nuclear power plant accident became more terrible because the backup electrical facilities of the plant had not been moved to higher ground despite warnings from scientists about the occurrence of gigantic tsunamis in the past.

SPECIAL?

Any animal species is special in its own way. Yes, comparison is the basis for understanding. To understand in what aspects we are special, we need to understand the nature of the apes and other primates.

RELIGION AND SPIRITUALITY

Japanese researchers probably have less difficulty with this than European/American colleagues who have a Christian cultural background. The majority of Japanese people are Shintoists-Buddhists, and neither religion minds if humans have evolved from other animals. Here is an answer adjusted to the Christian cultural context: If we define "God" as the one who created this world, the universe is consequently the direct document written by "God" itself, and science is a better tool than the Bible to understand the true message (euangelion) from "God," isn't it?

ADVICE

 Be patient and optimistic.

INSPIRING PEOPLE?

 The late Junichiro Itani, one of the founders of Japanese primatology.

15

LAURA MACLATCHY

L aura MacLatchy is a professor and associate chair in the Anthropology Department at the University of Michigan. She is a paleoanthropologist interested in the evolution of primate locomotion with an emphasis on hominoid (apes and humans) origins. One of her main research interests is the origin of the habitual upright posture and terrestrial bipedality in the human lineage. Laura's work on one of the first apes, *Morotopithecus* (Uganda; about 21 million years ago), is exploring essential information about the earliest steps in the hominoid lineage. She complements paleontological work with studies of chimpanzees in Uganda and with lab-based research.

YOUR BEGINNINGS

The periodical *National Geographic* had a substantial impact on my childhood. In its pages, I read about the research of Richard Leakey [chap. 54], who unearthed hominin fossils in Kenya, and Jane Goodall, who studied chimpanzees in Tanzania. From reading about their discoveries, I knew that I either wanted to ride around on a camel in Turkana, Kenya, looking for fossils, like Richard Leakey, or traipse around the African bush with binoculars observing chimpanzees, like Jane Goodall. It was not until much later that I realized the study of our fossil relatives and the study of our living nonhuman relatives are interconnected disciplines and that you can engage in research on both as an anthropologist.

GAME CHANGER

 The comparative approach is central to the study of evolution, including human evolution. This approach argues that one of the best ways to determine whether natural selection is responsible for a given trait, or set of traits, is to investigate whether different organisms living under similar circumstances evolve along similar lines. It predicts that a four-legged mammal that spends increasingly more time in the water may, after millions of years, come to look like a fish, which is just what happened during whale evolution. Matt Cartmill [chap. 6] used the comparative approach to great effect in the early 1970s to overturn the so-called arboreal hypothesis of primate origins.[1] The arboreal hypothesis posited that the grasping fingers and excellent depth perception of humans and other primates evolved so our ancestors could successfully navigate in the trees. However, Matt Cartmill pointed out that many other kinds of arboreal creatures—squirrels, opossums, etc.—don't have good depth perception or grasping hands or feet. Using the comparative approach, he suggested that trees per se did not select for primate traits. Instead, he argued that grasping extremities allow an animal to move about trees in a particular way, namely, by holding onto *small* branches (as chameleons do) rather than walking on top (or up) large branches and trunks using claws (like squirrels). He also noted that forward-facing eyes with good depth perception are found in nocturnal predators such as cats that hunt using vision. This led him to propose that forward-facing eyes and grasping extremities enabled the earliest primates to be visual predators on small branches, a niche still occupied by many living primates including mouse lemurs and, critically, by other vertebrates such as chameleons and some marsupials. Matt Cartmill's "visual predation hypothesis" continues to be discussed and debated to this day, and his elegant and adamant use of the comparative approach is in large part why anthropologists and biologists look across a range of animals today to understand the origin and function of many key human traits. For example, elephants have been studied to explore empathy, crows for tool use, and kangaroos for bipedalism. In my own research, I have been inspired to study a range of fossil and living apes to piece together why the upright trunk of humans (and all other apes) evolved.

AMAZING FACT

I am amazed that evolution has led to a creature with an incredibly rich life of the mind as well as a deeply felt emotional life. We are lucky to have the kind of consciousness and cognition that enable us to appreciate our own existence, to both marvel at and seek to understand the world, to love and feel joy, and to share our emotions and thoughts (thanks to language) with each other.

TIME TRAVEL

 We know rather a lot about fossil hominins like the australopiths, who were bipedal, smallish brained, and lived in a range of environments about 4 million years ago. But we don't know what the ancestral primates that gave rise to these early terrestrial bipeds were like. Thus, I'd like to travel to East Africa about 7 million years ago in some kind of flying vehicle that would allow me to zoom over the forests and woodlands and track down the so-called last common ancestor (LCA) of humans and apes. Were the LCAs living in large social groups, and what were their relationships with one another like? Did they use bipedality, either in the trees or on the ground? Did they eat mostly fruit and leaves in the trees, or did they also forage terrestrially? Did they share food and help one another raise their young?

DRIVING FACTORS

 Based on the *existing* fossil record, the first apes to become "set apart" exhibited diagnostic locomotor and dental features. The locomotor distinction involved moving bipedally on the ground, and the dental shifts included a reduction in canine height and a concomitant, or slightly later, enlargement of the cheek teeth. But what *factors* drove these changes? A general consensus holds that these shifts most likely reflect natural selection for organisms that were better able to thrive in a more open, seasonal environment than had previously occurred in Africa. Climate data from sources such as deep sea cores indicate a cooling trend starting around 14 million years ago that resulted in both increased seasonality in rainfall and forest fragmentation. Although many hypotheses have been proposed to explain how these environmental trends are linked to hominization, two remain particularly influential today. The first, proposed by Peter Rodman and Henry McHenry,[2] suggested that forest fragmentation made increasing demands on apes, who would have to travel farther on the ground each day, crossing gaps amid the trees, to acquire food. It follows that any small tweak to the locomotor anatomy (e.g., the shape of the knee joint) that increased energetic efficiency might be strongly selected for. Measurements of the oxygen consumption of humans and chimpanzees as they walk on treadmills have demonstrated that human walking is more efficient than an ape's, regardless of whether the ape walks quadrupedally or bipedally. Thus, this energetic hypothesis has some intriguing support. A second influential hypothesis has taken many forms and focuses on the benefits that bipedalism reaps by freeing the hands for carrying. In one version, advocated by Owen Lovejoy,[3] a shift in reproductive strategy is tied to increasingly patchy resources. In this scenario, ancestral males bonded to particular females, mating with and provisioning them by carrying food to them, thereby increasing the survivorship of their babies, who would, incidentally, inherit parental genes predisposing them toward pair-bonding. Lovejoy has also tied the decrease in canine size to this increase in pair-bonding; comparisons across primates show that the canines of males in monogamous species are shorter than those of males that mate with many females. However, other researchers, such as Bill Hylander, have shown that canines are also reduced in primates that have flat faces and short jaws, and such jaws are more powerful when chewing.[4] Thus a purely dietary explanation for reduced canines and enlarged molars is also possible. To test these hypotheses, we must obtain more data from the fossil and geological record. High-resolution environmental data from fossil sites where hominins are found are needed to reconstruct parameters such as rainfall patterns and vegetation to document in more detail the effects of seasonality on food resources. Equally important will be the discovery of many more, well-dated hominin fossils so we can determine more precisely the nature and timing of bipedalization and dental changes, and whether they occur in tandem with environmental shifts or are inferred shifts in foraging and social structure.

SPECIAL?

Unquestionably, the answer to this is yes. However, many of our "special" abilities, including our capacity to use language, make tools, learn through observation, and share with one another, have antecedents in our primate relatives. Studying the capacities of apes helps us home in on what is truly different about humans.

RELIGION AND SPIRITUALITY

The capacity for religiosity and spirituality are unique human universals and products of evolution, perhaps not unlike our unique capacity to create and appreciate music. As a biological anthropologist, I applaud attempts to understand how and why these unique capacities evolved. However, engaging in religious pursuits, like composing music, is not scientific.

16

SALVADOR MOYÀ-SOLÀ

Salvador Moyà-Solà is an ICREA research professor at the Catalan Institute of Paleontology Miquel Crusafont (Spain), which he founded and directed from 2006 to 2017. He studies the origin and evolution of hominoids (apes and humans) from a paleontological perspective. In particular, Salvador is trying to reconstruct the ancestral ape from which modern hominoids evolved. His fieldwork focuses on the Vallès-Penedès Neogene basin (near Barcelona) where, together with his team, he discovered and described the partial skeletons of *Pierolapithecus catalaunicus* (about 12 million years ago)[1] and *Hispanopithecus laietanus* (about 9.5 million years ago).[2] These are the best-preserved fossil great ape specimens in all Eurasia.

YOUR BEGINNINGS

I can't remember a specific moment when I decided to do research. In fact, with the long perspective that I have now (I am in my sixties), I have the feeling that it was my curiosity for the natural word that, in an insensible way, led me to be a researcher. Somehow, I have the intimate impression that, in one way or another, my brain was already structured to try to understand how the natural world works. In fact, I never decided to be a researcher, it just happened. I never imagined any other work to which I wanted to dedicate my life.

DRIVING FACTORS

If by driving factors we understand the selective pressures that have led evolution onto the road leading to the human lineage, the question is not easy to answer. Two main adaptations define the human lineage—bipedalism and a large brain—but these characteristics did not appear at the same time in the evolutionary history of our lineage. Furthermore, bipedalism is not only a question of feet. In my opinion, bipedalism and humanlike skilled hands appeared together as a complex adaptation to exploit the shrub stratum of the forest, an ecological niche not exploited by other primates. It was a new adaptive primate type, but the brain did not play a relevant role in this. The ultimate reason this new type of adaptation was so successful is probably related to the dramatic ecological changes during the last millions of years that started at the end of the Middle Miocene (14 millions of years ago): a process of cooling that had, as a direct consequence, reduction of the tropical rain forest (vastly extended in Eurasia and Africa in the Miocene) for more open areas. For primates adapted to and in fact depending on forests, this time exerted strong selection pressure due to a reduction of their ecological niche and the increase of interspecific competition between primate forms adapted to similar ecological niches. These selective pressures determined the need for a new way to exploit the environment in primates: bipedal walking with dexterous hands to harvest the shrub stratum of vegetation. But the brain does not enter into the scene until much later. It was between 3 and 2 million years ago that the first evidence of a clearly larger brain appears, together with the first evidence of purposive and systematic lithic industry. This phenomenon takes place, again, in full intensification of the change in climate. Open spaces expanded more and more, and humid forests were increasingly smaller in size. The same selective forces (climate change toward cooling and interspecific competition) favoring the apparition of open countries in previously forested areas were probably behind the selection of a new bipedal primate with a very peculiar characteristic: a large brain with high cognitive abilities. In other words, a primate with a different and effective way to exploit the new open areas. Under this perspective, the main driving factors that set an ape apart from the road leading to humans was global climate change. Interestingly (not to say dramatically), it is the product of a past climate change (us) that is causing another climate change, which may be, at the same time, our end.

EVOLUTIONARY LESSONS

 This is a very, very important issue. If the high cognitive abilities of humans have some advantage, it is the ability to anticipate, predict, or in other words, foresee the future. This is not magic or witchcraft; it is knowledge. Knowing how and why the phenomena that surround us happen, and knowing the dynamics of the natural world around us, enables us to make forecasts about the future. Why is this important? Let me provide an example, although it is not directly related to human evolution but rather to the possible end of the human odyssey!

Concerns about climate change and its consequences are part of the daily news in the media today. The influence of human activities on climate change is already a fact and not just an opinion, no matter how much some politicians deny it. Many scientists have proposed changes in human activities to reverse climate change and its consequences. The use of clean energy, changes in diet and ways of harvesting, and many others. Politicians, pressed by public opinion, propose measures to reverse the dramatic situation of the planet Earth, but there is no general agreement between countries (e.g., the Kyoto protocol failure) on how to proceed. However, this is not the most dramatic situation. Paleontologists and other biologists know very well the mechanism of evolution in island ecosystems, isolated from the continent. In islands, where there are no terrestrial predators, populations of herbivorous mammals grow out of control, which results in overpopulation and the dramatic reduction of available trophic resources. When resources run out, the final result of this process is the mass mortality of herbivores! Most of the population dies, and only those who, for one reason or another, have some advantage over others survive. This ecological dynamic is the main selective force in insular evolution, but now it explains something more dramatic.

Earth is, in fact, a large island in space where the *Homo sapiens* population is growing exponentially out of control. The final outcome of this phenomenon is written: overpopulation and mass mortality. With the exception of a few scientists, nobody today mentions that our population must be controlled, that it is the only way to avoid the catastrophe! In fact, it is the only mechanism that can save the planet and the human species, but no politician has put this option on the table. Why? I believe it is because population control goes in the opposite direction to current economic philosophy: economic liberalism and permanent economic growth. If the economy does not grow, it will sink. If the population does not stop growing, the planet will sink. What is the solution to this dilemma?

SPECIAL?

This is an interesting question, and the answer has important consequences, although perhaps not for the reasons we might think at first glance. If we observe chimpanzees and humans, we can easily observe that, apart from superficial aspects, we are very similar (grosso modo). Bones, muscles, organs, brain, behavior, and feelings are part of both, and of any animal with a nervous system. In fact, humans and chimpanzees share 99 percent of their DNA! Any test of statistical similarity with the usual 95 percent confidence interval would accept that we are equal! But it is not true. A few changes in DNA can have important consequences in all aspects of a species: morphology, behavior, and culture. However, humans and chimpanzees have many similarities, including culture. The differences are important, but the similarities are too. This can be extended to the entire animal world. "Animal beings" have the same basic structure, and many of them also have a culture. Human beings are "special" (different) in that we have a larger brain in relation to our body size (mass), a fact that gives us greater cognitive abilities than other animals. Yes, human beings are special animals, probably in the same way than chimpanzees are special, but for different reasons. Under the concept of special we have, in my opinion, developed too much of a high opinion of ourselves. And this special place in which we have put ourselves is leading us to consider the rest of our evolutionary partners as mere merchandise created for our use and enjoyment. If we do not change our mentality, it will soon be too late for the megafauna and the biodiversity of planet Earth. My children may be the last generation to see elephants and chimpanzees alive.

RELIGION AND SPIRITUALITY

Devoting one's life to research, in the way I have done it, implies asking why things are happening. The mind of a researcher works with data, hypothesis, and probabilities. The brain is structured to work with this information, and what is left out of this framework is part of a different sphere that is less accessible and difficult for researchers to define. There is no evidence that humans have a soul (it's a pity!), and religion is no more than a way humans had (and some still have) to explain the surrounding world and our origin. There are many aspects of human beings, and of the vast universe, that we humans are not yet able to imagine and, therefore, understand. The universe has plenty of surprises, and we cannot discard any possibility. But to accept it, we need evidence. There is still a lot of work for future researchers!

INSPIRING PEOPLE?

Robert Broom (1866–1951), a Scottish South African doctor and paleontologist and a professor of zoology and geology at Victoria College, Stellenbosch, South Africa. His work in the field of paleoanthropology significantly changed our minds about human origins.

17

MASATO NAKATSUKASA

M asato Nakatsukasa is the head of the Physical Anthropology Laboratory at Kyoto University. He studies primate evolution and is especially interested in the divergence of each living hominoid lineage. He leads paleontological excavations in different localities in Kenya and other countries. His work at Nakali (northern Kenya; about 10 million years ago) is particularly relevant because it has revealed a large diversity of primates at a poorly known time. Chief among the discoveries at the site (together with Yutaka Kunimatsu [chap. 14]) are the remains of the large-bodied ape *Nakalipithecus*, which could be closely related to the African ape and human clade.

AMAZING FACT

Our ancestors were an aberrant ape species that deviated from the ancestral arboreal habitat to the ground, relying on upright bipedality. I predict that such a drastic change (or shift) of survival strategy may bring short-term evolutionary success, eventually becoming a dead end. However, the human lineage has survived for several million years, which is amazing and interesting.

Bipedality caused an irreversible departure from the basic mammalian body design in human anatomy. I'd compare it with the hoofed-toe in ungulates or chiropteran's patagium. Such changes constrain the direction of adaptations and potentially increase the risk of extinction in a new environment. Although the evolutionary success (radiation) of artiodactyla/perissodactyla or chiroptera are really awesome, successful cases are rare when you consider the number of mammalian orders. Terrestrial bipedalism certainly increased the risks of predation and injury. It must have been a big bet for K-strategists (species whose populations fluctuate at or near the carrying capacity [K] of the environment), and the nonspecialized gastrointestinal tract restricted a range of exploitable food resources even on the ground (unlike the case for grazing mammals). The evolution of humans appears as a chain of unlikeliness (including the obstetrics solution for encephalization).

TIME TRAVEL

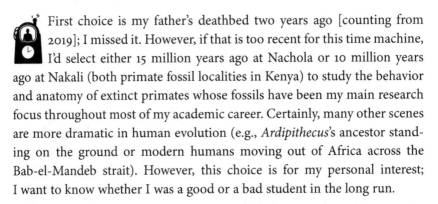 First choice is my father's deathbed two years ago [counting from 2019]; I missed it. However, if that is too recent for this time machine, I'd select either 15 million years ago at Nachola or 10 million years ago at Nakali (both primate fossil localities in Kenya) to study the behavior and anatomy of extinct primates whose fossils have been my main research focus throughout most of my academic career. Certainly, many other scenes are more dramatic in human evolution (e.g., *Ardipithecus*'s ancestor standing on the ground or modern humans moving out of Africa across the Bab-el-Mandeb strait). However, this choice is for my personal interest; I want to know whether I was a good or a bad student in the long run.

FUTURE EVOLUTION

Future evolution will not be shaped by UV radiation, thermoregulation, endurance running, or information and communications technology (ICT) adaptability. Cities are overpopulated and connected by a rapid, large-scale, worldwide transport network. Large cities face growing challenges to supply clean drinking water and to dispose of garbage and human wastes properly. In some areas, the expanded interface between humans and animals (both domestic and wild) is rapidly increasing the risk of outbreaks of zoonotic disease. In this environment, an immune system resistant to various pathogens will be a major target of selection in the near future.

EVOLUTIONARY LESSONS

 Human evolutionary studies do not help us in the same way as engineering (e.g., electric vehicles, AI, nanomachines). Curiosity is human nature, and every ethnic group has developed legends and myths to explain its origins and history. Human evolutionary studies are cooperative activities to produce a common story of humans' deep past by objective and verifiable methods. In busy modern societies, we may survive for most of time without recalling our legends (or Casals's cello or Picasso's paintings). However, could we stay calm if we were to miss impressive music and arts (and our history too)?

SPECIAL?

 No doubt, humans are special. The scale of our niche construction in only two centuries has been tremendous, but this has resulted in catastrophic effects on the Earth's ecosystem and environment. Putting technology aside, we are a primate species with a large body size, extreme altriciality, and a slow life history, which we more or less share with our close relatives (of course, we should not forget our unique feature—upright bipedality). To understand characteristics of our anatomy, physiology, and psychology, comparative study of living primates is a sound methodology unless we forget that there are no "living fossils" in modern primates.

INSPIRING PEOPLE?

The late Dr. Junichiro Itani, one of the forerunners in primatology. He had deep insights into human nature and the evolution of human society.

18

MARTIN PICKFORD

M artin Pickford is a paleontologist born in England, raised in Kenya, and currently based in France. He was lecturer-researcher and chair of paleoanthropology and prehistory at the Collège de France for many years. Martin continues to be an active field paleontologist and surveys diverse sites in Africa. His expertise ranges from geology to Cainozoic mammal evolution, including paleoanthropology. Among other surprising discoveries is *Orrorin tugenensis*,[1] one of the earliest known members of the human lineage (Kenya; about 6 million years ago), which he found together with Brigitte Senut [chap. 39].

YOUR BEGINNINGS

When visiting the Muséum National d'Histoire Naturelle, Paris, in 1964, I was amazed to see a fossilized toad from the Palaeogene of Quercy, complete with petrified skin and eyes. This "discovery" intrigued me. How could soft tissues turn to stone? I started reading about fossils, and this began my lifelong interest in geology and paleontology. The "taphonomy" fascinated me, although I did not come across this word until much later in 1972.

GAME CHANGER

The discovery of *Orrorin* in 2000 was a fundamental game changer in paleoanthropology. Prior to this finding, it was generally accepted that hominids split from the apes about 4.5 million years ago and evolved in the savanna, and that prehominids were quadrupedal terrestrial creatures. *Orrorin*, at 6 million years ago, expanded the calendar of human origins: it lived in the dry evergreen forest (not savanna), was fully bipedal, but still climbed trees. We now consider that the African ape and human lineages most likely split between 12 and 10 million years ago. This discovery was welcomed by many colleagues, but it also produced a goodly share of unfortunate comments of a sociopolitical nature from certain colleagues in Kenya and the United States. No need to go into details, but these paleo politics led to difficulties that interrupted our research program, thereby stalling further progress.

AMAZING FACT

Many animals bluff other species. For example, the coluber plays dead and smelly to avoid being eaten by a fox, thereby bluffing a potential predator—but it never bluffs other colubers. Humans are the only known species that lies to members of its own species—not just at the level of individuals but also at the level of small groups, communities, and societies—using not only vocal but also visual and gestural behaviors. This plus the fact that we can potentially detect that someone is lying are amazing (although this trait is not perfectly developed; if it were, lying wouldn't work, nor would political campaigns or most publicity or propaganda).

Humans lie to obtain an advantage, and by definition this means the advantage is gained over other humans. But if a lie is detected, is it still advantageous? Yes and no. The person (or group) detecting the lie can denounce it, thereby nullifying the advantage, or can keep the knowledge quiet for later denunciation, thereby turning the initial advantage into a subsequent disadvantage (politics). In the long spans of evolutionary time, this kind of mental manipulative behavior proved successful for some groups (although not in every case because of the possibility of detection), and it eventually led to selection for greater intelligence. Larger brains enhanced our ability to invent new lies, which was required as old lies were easily detected. Better (but imperfect) abilities to detect lies and the ability to mentally store information for later use increased this advantage. Without the preliminary stages of lying via gestures and sounds, human language as we know it may well not have evolved.

This "virtual" behavior probably began some 8 million years or so ago as a survival adaptation, but over the long term it has led to the evolution of human-type intelligence, calculation, language, and politics and an infinite capacity to delude ourselves (e.g., acceptance by billions of people of the existence of an everlasting, immortal, virtual power in the sky who determines almost all aspects of our existence).

TIME TRAVEL

 We need to carry out more research on the period 12 to 6 million years ago in Africa at large (not just the Rift, the SA caves, and the Maghreb). Current research efforts are focused on less than 1 percent of the expanse of the continent, meaning that we are ignoring or neglecting a vast amount of potential evidence.

DRIVING FACTORS

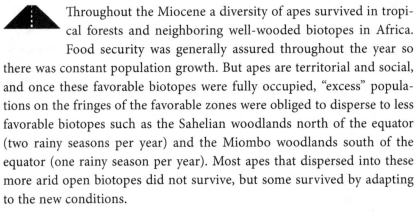 Throughout the Miocene a diversity of apes survived in tropical forests and neighboring well-wooded biotopes in Africa.

Food security was generally assured throughout the year so there was constant population growth. But apes are territorial and social, and once these favorable biotopes were fully occupied, "excess" populations on the fringes of the favorable zones were obliged to disperse to less favorable biotopes such as the Sahelian woodlands north of the equator (two rainy seasons per year) and the Miombo woodlands south of the equator (one rainy season per year). Most apes that dispersed into these more arid open biotopes did not survive, but some survived by adapting to the new conditions.

Survival involved both genetic and nongenetic changes. Among the nongenetic changes were the exploitation of new kinds of food resources (underground plants and insects, scavenging, hunting small animals, etc.) and the acquisition of new intestinal biota, new skin biota, new nervous system biota, new blood system biota, and changes to the immune system (different diseases were found in the newly inhabited zones). But above all adaptation to the more open environments involved genetic changes. Long-term occupation of these areas, especially Miombo Woodland, led to severe selection pressures in diverse domains: life cycle parameters (prolonged childhood, delayed maturation, and prolonged life span ensuring enhanced group memory); selecting for accumulation of subcutaneous fat reserves during wet seasons (seasonality of food resources—Miombo woodlands have a single annual dry season seven months long); gastrointestinal changes related to broadening of the dietary base to include foods not generally consumed by apes; skeletal changes related to different arboreal substrates (trees further apart, necessitating frequent traverses on the ground) and the paucity of horizontal branches leading to increased orthogrady (vertebral column oriented vertically when in the trees, bipedal locomotion while on the ground); sleep and repose on the ground led to a reduction of body hair, among other things, to minimize parasite infestation and avoid distractions caused by "velcro-grass"); selection for an increase in body dimensions, especially stature when standing upright (increase in femoral and tibial lengths, locomotor efficiency, rearrangement of the foot bones to reduce divergence of the big toe and to produce a shock-absorbing arch [instep] of the foot); and enlarged home ranges necessitated having enhanced mental maps of territory and comprehension of seasonal fluctuations in resources. Furthermore, large home ranges would select for enhanced hearing to detect distant resources (e.g., birdsong in distant fruit trees), enhance aural direction-finding (inner ear complexes placed further apart, which required a broadening of the skull base), and accurate detection of predators while on the ground. Vision would also be enhanced (long-distance visual prospecting for resources, predators), but olfaction would remain rudimentary (nose not close to the ground where odor trails are concentrated but essential for close-range encounters such as mate selection and bonding within each group). In brief, natural selection of the human bauplan occurred in a habitat that would have been hostile for most quadrupedal apelike primates (baboons are an obvious exception, but they are not apes). Sexual selection probably played an important role almost from the beginning (hidden estrus, enlarged easily visible external penis, prominent breasts, facial hair, for example). There are many other aspects to this evolutionary story, but only a few leave any signs in the fossil record: skeletal changes reflecting bipedality, increased brain size, orthognathy, and dentition (small canines, microdonty).

Most ape groups that dispersed out of the tropical forest and neighboring well-wooded zones into the more open seasonal biotopes did not survive. But the "experiment" was undertaken millions of times. In the immense span of geological time, one or more groups did eventually adapt to the new conditions, and they were able to spread out over vast areas (comprising at least 15 percent of the surface of the continent).

Once adapted to this new biotope, groups could disperse to other neighboring biotopes devoid of hominoids (Mopani Woodland, wooded savanna, and bushland), and eventually they did so. But that is another story.

FUTURE EVOLUTION

 Humankind will probably become extinct well before 1 million years have passed if we continue to carry on as we have since agriculture and pastoralism were "invented" and then "money" became the virtual power!

Our current medico-scientific and agro-alimentary practices have resulted in changes in bacteria, viruses, and other micro organisms (plasmodia, for example), fungi, plants, and insects, resulting in the evolution of "resistant" forms. At the moment we're keeping slightly ahead in the game, but in the long run we're doomed to failure. Evolutionary processes are far more potent than humankind's tinkering efforts with antibiotics, bactericides, acaricides, antimalarials, fungicides, fertilizers, and biotechnology.

EVOLUTIONARY LESSONS

 For millions of years, early hominids were an integral part of the ecosystem, interacting with other species and the ecosystem at large in a way that Darwin called natural selection (allied with sexual selection). However, since the development of agriculture and pastoralism (and medicines among other cultural "improvements") in the past few thousand years, we have profoundly upset nature's balance. Food security is a constant limiting factor for all animal species. During periods of food security populations increase, but during periods of food insecurity populations decrease, a consequence of which is differential survival, which over the medium and long term results in evolution. Humans developed agriculture and pastoralism, and thereby entered a long period of food security and an ever-expanding population, occupying greater and greater areas of the Earth's surface. This demographic factor has led to deterioration of the ecosystem and degradation of soils, waterways, and oceans, which has had negative outcomes for a vast number of plant and animal species. The invention of virtual power structures (money and its copassengers—organized religion and political parties) has aggravated the situation (mining, deforestation, energy sector industries, water scarcity, poaching of elephants for their tusks or rhino for their horns for financial gain not for nutrition, and national and international politics). Understanding this might get a few people to reconsider the place of humans in the world, but I don't have much hope that human society will change its ways until it is forced to do so by nature.

SPECIAL?

No, we're not special, unless you are lying to yourself. We've built a society that postulates that humans are special (i.e., we are outside or above nature in some way or other: special selection by a divine power, things like that), but each of us is born, may reproduce, can fall sick, sweats, defecates, dies, and will be recycled by micro-organisms. In the long run, nature will prevail when the human lineage becomes extinct, which is virtually a certainty. More than 99 percent of animal species that existed during the last 4 billion years have become extinct. Studies of other primates may help some people to relativize the place of humankind in the world, but I don't have much hope that people who consider money and power to be the sole reason for existing will change their minds. Whether the human lineage will become extinct without issue or whether it will morph into another lineage is a moot point. I think the former is more likely than the latter.

RELIGION AND SPIRITUALITY

No conflict. Everyone can believe what they want to believe (twenty-four angels can dance on the head of a pin; little green men live on Mars; existence of an eternal, divine power figure), but in science we need to get away from the all-too-human tendency to "lie," not only to other members of the species but also to ourselves. Scientists should make observations and then make deductions on the basis of those observations in an honest way, leaving aside issues of money, sociopolitics, careerism, and religious belief. I see evidence of evolution every day, everywhere, so for me it is not belief. Religion, in contrast, is based on unobservables (beliefs) at one level or another and therefore falls outside the domain of science. Incompatibilities arise when people try to reconcile the two domains.

ADVICE

Don't let religion or money or sociopolitics affect the way you carry out your scientific studies, which is almost impossible given that humans are sociopolitical animals. At the very least, scientists need to minimize the effects of nonscientific and antiscientific factors. There is plenty of evidence in recent literature that a lot of currently active paleoanthropologists are compromising their science for financial and sociopolitical reasons. Many of them tend to "REACHE" for funds and for notoriety rather than for accurate research. The result is substandard scientific output—fake news, in fact, often accompanied by intense press coverage purveying exaggerated, inaccurate, poorly supported views of human origins and related topics. Such people are unwittingly providing fuel for religious-minded people and are weakening science, not only by disseminating dubious and unsustainable results but also by diverting funds from genuine scientists, many of whom find themselves marginalized or underfunded.

Unfortunately, the current fashion by many colleagues to publish their papers in so-called high-ranking journals in order to boost their science citation index has led to the installation of powerful editorial and refereeing lobbies that favor like-minded colleagues and are hostile to genuine scientists who do not belong to the same sociopolitical networks. This development, which is politically inspired (and has serious commercial undertones), is eroding scientific endeavors worldwide. My advice would be to avoid submitting articles to such journals. A good article is good, no matter where it is published, whereas a poor paper is poor even if it is published in the journal with the highest impact and is accompanied by intense press coverage and internet exposure.

19

DAVID PILBEAM

D avid Pilbeam is the Henry Ford II Research Professor of Human Evolution at Harvard University, a curator emeritus of paleoanthropology at the Peabody Museum of Archaeology and Ethnology, and a U.S. National Academy of Sciences member. David is interested in a wide range of topics involving human and primate evolution, especially the relationships of the controversial Miocene apes. Since the mid-1960s, David has produced numerous essential publications related to hominoid evolution. Especially relevant was the discovery of a nearly complete face and jaws attributed to *Sivapithecus sivalensis* (Pakistan; about 9 million years ago) that is still complicating our understanding of great ape evolution.[1]

YOUR BEGINNINGS

In the spring of 1962, when I was an undergraduate at Cambridge, I was in the library of the Cambridge Philosophical Society (an excellent place to work, never anyone else there) to read Elwyn Simons's *Postilla* article on *Ramapithecus*.[2] At the time, I wasn't thinking of an academic career, but I had stubbed my toe on a question that's been with me ever since—the origins of hominins (as they now are)—and I became fascinated with it. After graduating from Cambridge I went to Yale as a graduate student of Elwyn's, still without serious thoughts of an academic career. Things turned out differently.

GAME CHANGER

Robert Ardrey's *African Genesis*[3] had a big effect on me, as it did on many. But being immersed in fossils from the time I went to Yale (1963), then back to Cambridge (1965–1968), before settling again at Yale kept me focused on that question.

I had been at Yale when J. T. Robinson visited in early 1970, and I heard him say emphatically that *Ramapithecus* was not a hominin. I began fieldwork in the Potwar Plateau of Pakistan in search of more hominoids, and by the second half of the 1970s I had strong reservations about *Ramapithecus*'s status.

Paleontology reveals its secrets slowly; even apparent mega finds often turn out to be not so mega, so I see it as an accretionary process, with few saltations.

AMAZING FACT

 That a species can continue to be "part of" nature yet at the same time appear to be "outside nature." We are members of communities but also detached from them. How did that happen?

TIME TRAVEL

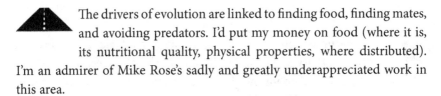 I'd pick the Late Miocene (7 to 6 million years ago) of tropical Africa, maybe where forest grades into woodland, to observe the ancestors of chimps and humans immediately after they became separate lineages. Would they resemble each other closely? Or was one or both already different from their common ancestor?

DRIVING FACTORS

 The drivers of evolution are linked to finding food, finding mates, and avoiding predators. I'd put my money on food (where it is, its nutritional quality, physical properties, where distributed). I'm an admirer of Mike Rose's sadly and greatly underappreciated work in this area.

FUTURE EVOLUTION

 Given how natural selection works, this is impossible to answer.

EVOLUTIONARY LESSONS

 Believing that evolution has no direction (hence no interesting predictions are possible); it is contingent. Example: *Australopithecus*

was not predetermined to evolve into *Homo*: such an ancestral condition was arguably "necessary," but it would be difficult to argue that it was both "necessary and sufficient," i.e., that the relationship between the two was causal. It is important that people understand that "history"—including evolutionary history and therefore paleoanthropology—is not predictable. Of course, it is always possibly to explain events when viewed retrospectively, but such explanations are rarely unanimous!

So, there is no "direction" to history or evolution, and that's one important reason for studying them.

SPECIAL?

 Humans are different. We have cumulative culture and cultural behaviors that can change rapidly, and in that sense we are special. But as far as the basics of genetics, physiology, other aspects of phenotype, no we aren't.

I believe, especially, that the African apes, notably *Pan*, can provide particular insights on the behavioral and cognitive parameters of the *Pan-Homo* ancestor.

RELIGION AND SPIRITUALITY

 I am a nonbeliever, but I understand and respect the beliefs of those who find all, some, or no compatibility.

ADVICE

 Start by learning math and probability theory, then move on to the numerous other necessary fields. But never forget that new insights are thought up—not thought out.

INSPIRING PEOPLE?

Probably Charles Darwin, but I'd like to have Alfred Wallace involved in the conversation as well.

PART II

Beginnings

ANSWERS FROM EXPERTS
FOCUSING ON THE EARLY STAGES
OF THE HUMAN CAREER

20

LESLIE AIELLO

eslie Aiello is president emerita of the Wenner-Gren Foundation for Anthropological Research and professor emerita at University College London. Born and educated in southern California, she spent most of the thirty-plus years of her academic career at University College London. Her highly influential work on the evolutionary history of human anatomy and bioenergetics has earned multiple highly prestigious awards, including the AABA's Charles R. Darwin Lifetime Achievement Award. Leslie continues contributing to the public outreach on science and human evolution.

YOUR BEGINNINGS

For me it was a multistep process: the moment I decided on anthropology, the moment I decided on paleoanthropology, and the moment I decided on research into the evolution of human adaptation. The first two decisions happened as a direct result of participation in excavations. A field school on a Basket Maker site in Cedar City, Utah, in 1963 hooked me on anthropology and particularly on the magic and excitement of discovering the past. A summer-long excavation at the site of Solvieux in France with Jim Sackett (with appearances by Sally and Lou Binford and François Bordes) in 1968 sold me on paleoanthropology and convinced me that my interests were in our ancestors rather than in the tools they left behind. Research did not come until later. I loved teaching anthropology and realized that I could only continue to do so with a doctorate. There was no better place for this in the 1970s than London. I found myself in a heady environment surrounded by Michael Day, John and Prue Napier, Peter Andrews [chap. 2], Chris Stringer, Theya Molleson, Bernard Wood [chap. 46], and many others. Research couldn't help but became a fascination.

GAME CHANGER

Without a doubt it was the discovery of Lucy in 1974. This was a game changer for the field; this one fossil destroyed the long-held assumption that there was a feedback relationship between the evolution of bipedalism, tool use, and brain size. We now knew that bipedalism preceded the evolution of brain size and the Oldowan (the earliest known tool culture at the time) by more than 1 million years.

This sent my career in two separate directions. It was the reason Chris Dean and I wrote *An Introduction to Human Evolutionary Anatomy*.[1] There was no other resource to help anthropology students understand the detailed comparative anatomy necessary to interpret Lucy and the wealth of fossils coming out of East Africa at the time. It also sent me on a search for other explanations for hominin brain evolution, resulting ultimately in "The Expensive-Tissue Hypothesis."[2]

EVOLUTIONARY LESSONS

The main lesson of human evolution is that the way we live today is very strange when compared to the lives of our evolutionary ancestors. A look at a population curve shows how our species has taken over the world since the industrial revolution. The agricultural revolution about 10,000 years before that had its own impact on the human body and on our relationship with the natural world. We were hunters and gatherers for eons before that, and large-brained hominins living in small groups may have been the key to long-term survival. There was enough room to move when confronted with environmental change, and the large brains provided the necessary flexibility of adaptation. We still have our brains and flexibility of adaptation, but both may be obstacles to continued survival. There are no more unoccupied places to move to, and we have the cultural ability today for major destruction and catastrophic warfare. Recent events have also shown us that large population sizes and our interaction with the natural environment are recipes for pandemics that can be equally devastating.

Essentially, population increases and environmental control since the agricultural revolution have resulted in a very different world from the one we inhabited for the great majority of our evolutionary life span. We are facing challenges to existence now that we did not encountered in the past.

RELIGION AND SPIRITUALITY

I once had a doctoral student who was a Catholic nun. She had no problem with the compatibility of human evolution and her spiritual beliefs. She believed that evolution by natural processes resulted in the physical body, and God was responsible for the soul. This was an eye-opener for me; it clearly demonstrated that there need not be an insurmountable conflict between the fossil evidence for human evolution and a spiritual view of the world.

ADVICE

Don't spend all of your time studying what is already known in anthropology. Major contributions come from reaching outside the discipline. Paleoanthropology is a discovery discipline, and new fossils are certainly important. However, the major advances have come through analysis and interpretation, and many of these insights have come from other fields (e.g., genetics, isotope studies, morphometrics, imagining, energetics, dating, and many more). Learn techniques that will give you an edge in the interpretation of the evidence we currently have and that will continue to appear, sometimes from unexpected areas.

Also, at the graduate/postgraduate level, seek out the leading people working in the specific analytical area you want to pursue. Try to be at the cutting-edge as early in your career as possible.

INSPIRING PEOPLE?

Obvious people would be early pioneers such as Darwin, Huxley, Wallace, and Haeckel. But I would like to be able to talk with Kenneth Oakley again. Oakley worked at the Natural History Museum in London and is best known for exposure of the Piltdown Man hoax in the 1950s, using fluorine dating. I knew him at the end of his career when he lamented that his work had had no real relevance in the modern world, and if given the chance, he would have pursued other career directions. I would like to show him what has happened in the field of human evolution over the past forty years since his death and, in particular, how generations of young colleagues are advancing the field while at the same time emphasizing outreach and the broader relevance of our fascinating discipline. It would have made him both happy and proud.

21

BERHANE ASFAW

Berhane Asfaw is the codirector of the Middle Awash Research Project (Ethiopia), the Rift Valley research service manager, a founding member of the Ethiopian Academy of Sciences, and a member of the U.S. National Academy of Sciences and the World Academy of Science. Since the 1980s, Berhane has played a crucial role in developing paleoanthropological research at the National Museum of Ethiopia and protecting the incredible heritage of his country (even putting his life at risk). Berhane's research teams' contributions include, among others, identifying some of the oldest Acheulean stone tools and *Homo erectus* remains from Ethiopia.[1] He has also described new early species of the human lineage, such as *Australopithecus garhi*[2] and *Australopithecus ramidus*.[3]

TIME TRAVEL

 My wish is to see how the different hominids that lived between 3 and 2 million years ago interacted. Was their encounter peaceful? How did they behave toward each other? Were they copying innovations from each other?

DRIVING FACTORS

The ability of hominids to carry food may have been one major driving force that led hominids on toward the road to the human lineage. This consistent behavior of hominids is one pivotal factor that might have paved the way toward hominization. This behavior may have helped the offspring reach maturity with a better chance than the other apes that do not have mothers who are supported by the male. This process also might have nurtured social behavior to develop to a higher level of socialization.

FUTURE EVOLUTION

Evolution is not predictable; we can't forecast the path of evolution. There is always an opportunistic element. We cannot know what kind of gene will emerge accidentally through mutation that may be beneficial and accumulate in a new age for the hominids to better exploit or conform to the environment and reproduce better or that may be damaging and threaten the species to extinction. The modern human lifestyle is more susceptible to genetic mutation. We are exposed to all kinds of security apparatus that scan our body using doses of X-rays. They may be considered low-dose rays, but we don't know their cumulative affect on us. We use all kinds of electrical ray-dependent machines in our households, such as microwave cookers, and artificial chemical components abound in our surroundings. All of these may have some cumulative affect on us. These exposures will definitely create some new genes, and we have no idea what they will bring for us. In addition, we do not know what environmental challenges our species will face in the future. As a result, some human variations will survive and proliferate through the process of natural selection, and eventually humans may be transformed into what we may recognize as a new species.

SPECIAL?

Humans are special because we can heavily manipulate our environment to fulfill our needs for food, shelter, etc. The study of our closest living relatives will definitely help us understand the roots of our own behavior. They are among the many clues we can use to reconstruct our behavioral and biological evolutionary path.

ADVICE

Paleoanthropology is a fascinating field of research. Anybody engaged in this research will uncover new information over time. You will be pleased with your work, and it is never boring. This is a field that you will enjoy your whole life long.

22

ANNA "KAY" BEHRENSMEYER

nna "Kay" Behrensmeyer is a curator of vertebrate paleontology in the Department of Paleobiology at the Smithsonian Institution's National Museum of Natural History and a codirector of the Evolution of Terrestrial Ecosystems Program at NMNH. She's also a member of the U.S. National Academy of Sciences. Kay studies the geological and taphonomic context of human evolution. She discovered a famous artifact site in a volcanic tuff that is a signpost for the age of fossils in the Koobi Fora deposits in Kenya, and that tuff is named after her (KBS Tuff, about 1.9 million years ago).[1] For this and other research on rocks and fossils, she was named one of the "50 Most Important Women Scientists" by *Discover* magazine.

YOUR BEGINNINGS

From as early as I can remember, I was curious about the natural world and especially about rocks and fossils, which were common on the Illinois farm where my brothers and I roamed on weekends. I read books about fossil brachiopods and corals, asked questions, figured things out for myself—but I did not have a name for this then. It was just the fun of discovering and learning. I had lots of pets. I wanted my mice and guinea pigs to have special coat colors, so I picked out certain individuals to mate and kept records of the colors of the offspring—my first foray into experimental research. My mother put a stop to it when she discovered a population of twenty-seven guinea pigs in our basement. I had to agree that things had gotten a bit out of hand! Looking back, I realize now that I was encouraged by my parents, aunts, and teachers to follow my curiosity and to learn by doing. When I settled on a PhD dissertation topic on the taphonomy and paleoecology of East Turkana in northern Kenya (then East Lake Rudolf), it was the first time I remember saying to myself, "This is scientific research." I had to figure out how to do it because there really weren't any guidelines for this kind of study. I wanted to test whether the vertebrate fossils preserved in different fluvial and lake margin environments accurately recorded the ecology of the Pleistocene land-scapes inhabited by early hominins. Working out the methods and learning how to formulate and test hypotheses laid the foundation for my career in taphonomy and paleoecology. Mentors and role models—Vince Maglio, Glynn Isaac, Bryan Patterson, Mary Leakey, and Richard Leakey [chap. 54]—asked tough questions along the way, forcing me to think and to defend my ideas while also helping me build confidence in my ability to do scientific research.

GAME CHANGER

As a beginning graduate student in 1969, I discovered stone artifacts at East Turkana that we originally thought were 2.6 million years old. This was exciting at the time because it pushed back the date of human use of technology by more than 500,000 years. The dating did not hold up to further analysis, however, and was revised (after much controversy) to about 1.9 million years old, only a little older than Olduvai Bed I. For its major impact on understanding human evolution, the discovery of the Lae-toli hominin trackways was the biggest game changer for me.[2] There was no disputing that the tracks were made by two-footed hominin individuals (although the number of them is debated), or their geological age of about 3.7 million years ago. The hominins were crossing terrain freshly covered by volcanic ash and found with a mammal fauna different from what usually is associated with hominin fossils in East Africa. To me this was particularly interesting because it showed more habitat diversity and adaptability of *Australopithecus* than anyone expected. Laetoli preserves a wealth of other evidence for the paleoecology of the Pliocene in East Africa, from fossil termite mounds and other trace fossils (tracks of a three-toed horse!) to gastropods, birds, beetles, and raindrop impressions. This reinforced my conviction that the paleontological record has much more to offer about where and how hominins lived that goes beyond the traditional focus on who they were and their evolutionary connections.

AMAZING FACT

All life forms have creative potential in their genetic code that can generate new ways to manipulate other organisms and environments, so in this sense humans are part of a very long evolutionary tradition. But few species have changed the whole planet to expand their reproductive success, and none has done this as rapidly as *Homo sapiens*. To me that is the most amazing aspect of human evolution. From my viewpoint as a natural scientist and a paleontologist, it was our unique brain-hand-language adaptive complex that set us on course for this planet-scale evolutionary success story. Combined with our primate social heritage, these attributes enabled us to utilize all kinds of resources—plants, animals, rocks, caves, water, wind—whatever came to hand, leading over millions of years to the best and the worst that characterizes humans today. Our creative potential was honed by environmental challenges in Africa, which probably included bottlenecks that some of our ancestral populations barely survived. Natural selection turned those survivors into the master manipulators of life and the environment. In many past cycles of boom and bust, we were able to exploit our ecosystems until populations grew too large.

Now we are seeing the success of our species manifested in a booming population on the scale of the entire biosphere. It appears that the only limits to our ability to control our environment are powerful forces of nature that match that global scale: microbes, climate change, volcanic eruptions, and meteorite impacts. Scientific understanding of the processes that shaped us, the Earth, and the biosphere is a unique and wonderful achievement of our species. Using this knowledge wisely can guide us toward decisions that will help sustain the networks of life that we depend on. But our primate social and cultural legacies operate in much smaller arenas and are shaped by imperatives to belong to a group—family, community, tribe—and (especially when resources are scarce) to win out over other groups. Finding ways to navigate the tension between this heritage and the collective wisdom of our species is likely to determine the next phase of human (and biosphere) evolution.

TIME TRAVEL

I would like to hover in my time machine over the lower Omo River Valley in southern Ethiopia and observe the interactions of two hominin species during a climate-driven transition from resource abundance to resource scarcity around 1.5 million years ago. I can imagine a

beautiful gallery forest, grading into lush thickets and more open vegetation away from the big meandering river. Although it is the dry season, the
continuous corridor of forest is full of life. Trails pass through it from the
grasslands farther away, and I can see antelope, elephants, large suids, and
baboons making their way through the dense vegetation to drink at the river.
I look for hominins and glimpse small groups in shady places under the
trees. Has *Paranthropus* joined other C4 (grass) feeders on the trail to water
from the grasslands? Is the *Homo* group that is lounging in the shade nearby
perhaps thinking of a meat meal? Do the two hominin species acknowledge,
ignore, or avoid each other? Only a time machine could answer such questions. Mine has a special dial that allows me to fast forward 10,000 years to
another part of a precession-driven climate cycle. The riverine forest is much
reduced, with broken trees and trampled thickets, and the river is a series
of muddy pools with hippos and crocodiles vying for water and space. The
grasslands are dried out and partially burned. There are sounds of baboons
in distress, also elephants. Species are competing for the few resources that
remain, and inter- and intraspecies conflicts are frequent and sometimes
deadly. Partial carcasses lie scattered near the remaining waterholes. How do
the hominin species interact during such times, and how do they compete
with nonhuman primates and other members of their vertebrate community,
all in need of food and water? What about hungry carnivores?

We cannot know the answers to these questions unless a time machine
could take us into the days, hours, and minutes when we could observe what
was going on between species. But a different type of time machine—one
that can look at time scales accessible in the paleontological and geological
record—could provide clues about long-term outcomes of short-term behaviors and environmental stressors. Modeling is a kind of time machine that
allows paleoanthropologists to scale up the consequences of different selection pressures over evolutionary time, including the impact of environmental bottlenecks and intensified interspecies competition. These times of more
intense pressure from natural selection were—and are—critical factors driving our evolutionary trajectory as well as that of many other species.

SPECIAL?

Decades of learning have reinforced my earlier belief that, in many
ways, humans (*Homo sapiens* and its immediate ancestors) are not
as different from other life forms as many (most?) people think.
Almost all the special things that used to "make us human" have weakened

or fallen away with the growth of knowledge about other species, past and present. Among these are bipedality, tool use, large brains, consciousness, empathy, complex sociality, intelligence, and an ability to project into the future and remember the past. We now know that there were many bipedal species of hominins, that other species utilize tools, and that brain size is not well-correlated with behavioral complexity. New understanding of mammals, birds, fish, and insects based on objective scientific research and the revelations of modern media show that evolution has created all kinds of complex forms and behaviors that rival or surpass our own. What does this leave that makes us special?

The adaptive advantage of female menopause in our species, which is a unique attribute compared with our close primate relatives, was a discovery that made a big impression on me. This finding underscored the importance of grandmothers and social groups (especially related individuals) in successful child-rearing, and it also highlighted the need for science to focus on female as well as male contributions to human evolution. Given this example, I strongly suspect that understanding what makes us special will continue to change as more diverse minds and approaches tackle this question.

Most of our superpowers come from brains wired for creative thinking, especially under stress, allowing flexibility, rapid adaptation to change, and group strength through social bonding. These traits characterize our species across the entire spectrum of gender and ethnicity in the global population. From my perspective as a paleobiologist, looking across billions of years of evolution, our behavioral flexibility and ability to imagine past and future are what make humans a special form of life. Complex language, art, abstract thinking, higher mathematics, technological prowess, scientific knowledge, and cultural complexity all follow from neural capabilities shaped over millions of years by spatially complex and ever-changing environments. Whether our special powers are up to dealing with the rapid, global-scale changes we are causing remains to be seen.

23

RENÉ BOBE

R ené Bobe is a biological anthropologist and paleobiologist at Oxford University's School of Anthropology and Museum Ethnography. He has studied the impact of climate and environmental change on human evolution for over thirty years. Research projects include the paleo-ecology of early members of the human lineage in Ethiopia and Kenya and the origins of South American primates. Recently he has been codirecting paleontological research in the Tertiary deposits of Gorongosa National Park in Mozambique and the conservation of its modern landscapes. René uses information from the past to better understand the present (and the other way around).

YOUR BEGINNINGS

I went to graduate school with the idea of doing research, and my ideas were broadly within the fields of biological anthropology and paleontology. I think it was reading Rob Foley's book, *Another Unique Species*,[1] that guided me toward doing the kind of research I do today. In that book, Foley provided concepts, ideas, and approaches to thinking about human origins in an ecological and evolutionary framework. At the same time, I was discovering Jonathan Kingdon's volumes on *East African Mammals*.[2] We are another unique African mammal, and ecology and evolution are intertwined and fundamental to understanding how we came to be the species we are today. My excitement in thinking about human origins within an ecological context has never diminished.

TIME TRAVEL

 I would have many misgivings about going into the future, even though the temptation of curiosity is strong. So allow me to go back to the past. Many time periods would provide a compelling journey, including the time of hominin origins. But I think I would settle for a journey to the Turkana Basin about 1.8 million years ago. I would expect to find four hominin species in the landscapes of the Turkana region: *Homo habilis*, *Homo rudolfensis*, *Homo erectus*, and *Paranthropus boisei*. How did these species interact with each other? How did they interact with the fearsome predators living at the time? Some of these hominins may have behaved in ways very familiar to us, but others did not. Could some of them talk to each other? I would expect to find a lush landscape teeming with life, with diverse megaherbivores, large predators, many species of primates, and different antelopes in specific vegetation zones. How did the hominins really fit into these landscapes? How did they move about, and how did they acquire their basic resources? If possible, I would travel from the Lake Turkana region to the west into the tropical rain forests of the Congo Basin to find the ancestors of gorillas and chimpanzees. I would expect to find apes similar to the modern ones, but I could be surprised. In the excursion from the Turkana Basin to the Congo forests 1.8 million years ago, would I find ghost lineages of African apes, including hominins, for which we have no fossil record?

DRIVING FACTORS

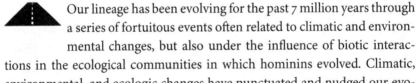 Our lineage has been evolving for the past 7 million years through a series of fortuitous events often related to climatic and environmental changes, but also under the influence of biotic interactions in the ecological communities in which hominins evolved. Climatic, environmental, and ecologic changes have punctuated and nudged our evolutionary trajectories. At the time of hominin divergence from other African apes in the late Miocene, the Earth was undergoing profound changes in climate and in the distribution of vegetation types with the expansion of tropical grasslands (C_4). Like most apes, these earliest hominins inhabited woodlands and forests but probably had access to the ecotones between the woodlands and more open habitats. Vegetation mosaics were a prominent feature of the East African Rift Valley, where many of the earliest hominins have been found. Early hominins such as *Australopithecus* began to thrive in

the ecotones and vegetation mosaics of eastern Africa, where they had access to trees but were in proximity to grassy areas. I think *Australopithecus* evolved some of its key anatomical and behavioral characteristics by adapting to environments that would have been marginal for earlier apes. During the Pliocene, *Australopithecus* expanded its range of resources to include items that could be found in more open areas, but they also continued to exploit the woodlands. Evidence from Dikika and Lomekwi suggests that by 3.4 million years ago hominins may have been using stone tools with some regularity. This shift by *Australopithecus* in the use of resources and ways of acquiring food set the stage for the evolution of a very different kind of African ape.

FUTURE EVOLUTION

All living organisms evolve, and humans are evolving and will continue to evolve. The impact of modern humans on Earth's systems is on a scale approaching the largest disturbances that the planet has suffered in its long geological history. We are changing the composition of the atmosphere and driving global warming. We are destroying the ecosystems that in the past allowed our species, and many other species, to evolve and flourish. We are polluting our sources of clean fresh water, our sources of food, and the air that we breathe. Thus, we are destroying the systems that give us life. Many people live in unhealthy, crowded conditions. We are unleashing plagues that are taking a heavy toll on human populations. These pressures on the planet and on human populations may result in selection pressures that shape our evolution. These pressures are shifting, sometimes acting on some systems and other times on others; sometimes in one direction and at other times in the opposite direction. Although we will continue to evolve, the direction of our evolution is impossible to predict. Most important, however, we should realize that the environmental destruction underway is not inevitable, at least not yet. Our patterns of behavior and the political systems that guide our behavior can still be changed for the sake of a healthy planet. It is up to us.

EVOLUTIONARY LESSONS

To understand human evolution, we need to understand the climatic, geological, and biological context in which we evolved (i.e., the matrix of our evolutionary history). We need to understand how climate changed in the past, and what its consequences were. What were the impacts of severe climatic changes in the deep past? What levels of climatic variation can be considered natural? We also need to know how geological processes shaped the landscape of our evolution. For example, how did formation of the East African Rift system and volcanism shape the landscapes of our ancestors in eastern Africa? We also need to appreciate how our ancestors related to other species, sometimes with mutually beneficial ecological interactions, sometimes through competition or predation.

Human evolutionary studies help us comprehend the history of the ecosystems in which we evolved, from details of food/energy acquisition to the broader climatic context of the Earth system. An appreciation of how and why we have evolved is indispensable when we tackle current issues of climatic change, environmental degradation, and loss of biodiversity. Knowledge of our evolutionary history and its context provides a baseline for evaluating the current and future challenges we face as a species. It would be ideal if this knowledge translates into wisdom and enables us to make the right decisions to solve, or at least ameliorate, current environmental problems.

INSPIRING PEOPLE

 I would like to ask my son, now seven years old, these questions when he becomes an adult.

24

TIM BROMAGE

Tim Bromage is a professor at the New York University College of Dentistry and is the director of its Hard Tissue Research Unit. Tim gathers insights into development and evolution from living (and extinct) organisms by studying bone and tooth biology. In addition, as a visual artist, Tim implements his research imagery into his craft. His research has included paleontological fieldwork on the shores of Lake Malawi, where he discovered remains from several members of the human lineage (*Homo rudolfensis* and *Paranthropus boisei*). His contributions have earned him multiple prestigious awards, including the Max Planck Prize in the Life Sciences.

YOUR BEGINNINGS

Mrs. Vincent, my sixth grade teacher, gave her students freedom to design research projects in science education. Having no reason other than curiosity, I asked for and was given permission to plant some pumpkin seeds in a small patch of dirt outside the back door to the classroom, which faced the playground. My project was to determine whether sunlight filtered through different colored cellophanes would have an influence on the growth of the plants. Lifted on sticks, I placed different colored cellophane tents over small patches of seeds, and I left one patch open to the sun. I was in awe of the difference in the heights of the plants after only one week: plants under cool colors, especially green, were stunted; plants under warm colors were taller even than those in the sunlit patch. From that moment and going forward, I began thinking about research questions that were inspired by nature.

GAME CHANGER

I did not understand it at the time, but I responded emotionally to the model system described in *Earliest Man and Environments in the Lake Rudolf Basin: Stratigraphy, Paleoecology, and Evolution*,[1] which presented an interdisciplinary approach to the study of human evolution. This system of research was an important and necessary step in the evolution of research strategies, not just for paleoanthropology but for many disciplines, including organismal life history. I read the work of a variety of specialized researchers on the multifarious topics around the theme of early human evolution.

I regret that paleoanthropology has been unable to take the step that naturally follows on from interdisciplinary approaches, which is to adopt an *integrative* approach. An attempt was made to take this step at the University of California, Berkeley, by placing paleoanthropologists into a new Department of Integrative Biology, but the hoped for integration did not succeed as planned. A study of this failure has highlighted the semantic difference between interdisciplinary and integrative approaches. *Interdisciplinary approaches* bring two or more fields to bear on a subject. This brings collegiality and efficiency to scientific research, but it does not necessarily result in significant cooperation, sharing, or a material impact on one subject arising from the activities of another. *Integrative approaches* do not exhibit a discrete disciplinary interface; rather, the subjects are miscible and data collected by researchers is used by multiple investigators. Interactions between researchers are promoted that lead to investigative research at all scales and in all directions of science, with links that connect reductionist areas of interest otherwise contained within separate research domains. Many important discoveries in the human evolutionary sciences are forthcoming, but for these discoveries to be interesting and broadly applicable to the sciences orbiting the field requires paleoanthropology to get busy being more integrative.

AMAZING FACT

No facts about human evolution amaze me. Facts represent the foundation of knowledge and are important and necessary, but they are boring. No hyperbole over the facts that typically accompany a new fossil find has any meaning for me. However, if we were to ponder the meaning of these facts at all scales, from individual to ecosystem, new known unknowns and questions emerge.

These new *questions* in search of new facts is what I find amazing, not the facts themselves. The time we spend dwelling on facts stunts the field and breeds the presence of conservative self-images in paleoanthropology that refuse to let "facts" go. The perceptual biases of such wardens of facts cloud our concepts, prevent us from seeing other more useful perspectives, and greatly reduce the pace of progress in the field.

TIME TRAVEL

 All complex systems, whether physical, biological, or social, have certain characteristics in common at all scales. They have many parts, and these parts are diverse, connected, and interdependent within a hierarchical branching network that typically obeys a mathematical power law. Finally, complex systems adapt and evolve. If we had mapped this network properly, we could use it to understand how the system works; in our case, how human evolution works.

As a heuristic device, I would make a network map of human evolutionary history and make predictions about the future, to which I would take my time machine. Any differences between my predictions and the future reality—of which there are sure to be many—can be used to inform the map and improve our ability to understand our evolution. A jump time of at least 100,000 years should yield useful information on which to predict and retrodict our evolutionary history.

DRIVING FACTORS

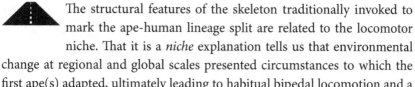 The structural features of the skeleton traditionally invoked to mark the ape-human lineage split are related to the locomotor niche. That it is a *niche* explanation tells us that environmental change at regional and global scales presented circumstances to which the first ape(s) adapted, ultimately leading to habitual bipedal locomotion and a lineage leading to humans living today.

However, this explanation falls short. All evolving organisms and environments are complex systems, and an oversimplified locomotor niche model can be built with knowledge we already have. For instance, if the locomotor niche changed, then so did the dietary niche, so teeth and their supporting structures were also driven to adapt by the same environmental factors. We see no claims for origins of the human lineage observed from characteristics of the teeth, but this is exactly what must have happened. So much knowledge, and so little understanding.

FUTURE EVOLUTION

It is debatable whether humans will exist 100,000 years from now. All signs at present indicate that we will not. But, for the sake of argument, let's say that somehow humans get on track and manage to maintain an adaptation to their surroundings. Certain trends are likely to continue. For instance, continued reductions in the mechanical properties of foods will carry on the already existing trend for diminished jaw size and tooth crowding. Culture will intercede to mitigate forces deleterious to our biology and, in the example given here, that means the emergence of craniodental professions to correct or compensate for the harm we do to ourselves by eating such processed food.

EVOLUTIONARY LESSONS

We have many questions about worldly phenomena that require information collected over a long span of time. Even when we have this data, we remain challenged by the complexity of those phenomena. For example, climate models are deep enough to provide an approximate five-day weather forecast with some accuracy, but this forecast is unlikely to be right on the mark because weather systems are complex. If we struggle to be reasonably good at forecasting the weather five days from now, how can we think and hope that anything we say about regional or global climate change over decades will be anything but fantasy? The problem we have in describing climate change over long periods of time is that we have little more than the here and now on which to model our predictions.

Enter the past. In the past, time is on our side. We have all the time in the world, as they say, plenty of it. If we want to ask questions, acquire knowledge, and gain an enhanced understanding of climate change from a multitude of perspectives, we have access to any amount of time in the past we need to examine these changes. That understanding gained, we can apply the lessons learned to forecast possible futures of climate change that are consistent with the depth and refinement of our knowledge of the past. In fact, phenomena such as climate change can only reliably be examined in the context of deep history.

Evolutionary medicine is another example. Think of our bodies as memory storage devices. Deep human evolutionary history is contained within our physical structure, physiology, and behavioral repertoire, which can be read out and used to understand medical problems and to guide treatment. For instance, the system we have evolved for digesting starch is highly adaptive, particularly in marginal environments. But starchy foods have a high glycemic index and are readily available and inexpensive, leading to obesity particularly among the poor. Unfortunately, in the United States we do not recognize the evolutionary history of our special starch digestion capacity, and progress on bringing evolutionary biology to the landscape of medical research supported by the National Institutes of Health (NIH) is impossible. What else can I conjecture but that the "E" word detracts from NIH's perceived relevance to the medical usefulness of evolutionary predispositions to conditions and disease. Some human evolutionary scientists have gone into medical research and have, for the most part, disappeared from paleoanthropology.

SPECIAL?

There is nothing about the biology or culture of humans that is in any way special compared to other animals. Particularly in comparison with other primates, we are scaled appropriately in our tissue/organ/body characteristics, including our brain size, and in our behavioral repertoire, which is a classic nonlinear outcome of our brain size. We do have unique attributes, as do *all* organisms, and all obey first principles of physics and biology that have been vetted by natural selection and are, for the most part, adaptive under particular environmental conditions.

RELIGION AND SPIRITUALITY

We can safely say that existence is complex, and here we confront ourselves with the very foundation of human complexity; that is, the interplay between art, science, and the mystic and, indeed, our ability to comprehend it. When a single artist, scientist, or mystic believer practices, we witness their craft as a *personal* endeavor. Their pursuit gives them all the sensations of life: pleasure, frustration, memories, sadness, sleepless nights, fulfillment as a human being, and more. Their behavior, or how they practice their craft, may change through time as they acquire new experiences, but their search for truth and meaning endures. All artists, scientists, and mystics, in the absence of threats to their existence, will keep their talent, their passion, and their faith targeted along a path they set for themselves. Individuals are the *linear* threads of humanity that test both themselves and us all with their novel ideas about the world, leading us ever forward into the present. This understood, we must ask ourselves how the linear behaviors of individuals could be transformed into the nonlinear phenomena of complex systems.

A complex system combining art, science, and the mystic is a shared pattern of behavior that gives meaning to life and the world. Shared patterns of behavior are more commonly appreciated as the glue that holds human societies together; that is, an idea we call *culture*. These three parts, when combined, stimulate society to undergo a phase transition that, in fact, gives consequence to what it means to be human. Our culture and, indeed, our humanity will degrade if one part is taken away or perverted.

For example, if art is removed from the system, the identifying links of color, form, and pattern that define everything from our personal adornment to our architecture, and that binds humans together, will disappear. *The human need to belong is lost in the absence of art.*

If science were removed from the system, our powers of observation would fail us miserably, and our responses to an ever-changing world would be illogical, worthless, and lead to our demise as a race. *The human need to survive loses hope in the absence of science.*

If mysticism were removed from the system, our comprehension of the force that binds us and the universe together would be chaotic, and human societies would descend into anarchy. *The human need for transcendental order is unraveled in the absence of mysticism.*

The relationships between art, science, and the mystic are rooted in the science of complexity. It is from this perspective, one of science, that our shared patterns of behavior make sense, providing substance to our understanding of existence and creating the multiplicity of cultures in the world that enrich all humanity. No society that rejected any one of these three requirements of humanity has ever endured. Our challenge is to be cognizant of the need to develop each of these areas of human existence equally, without bias toward one or another.

ADVICE

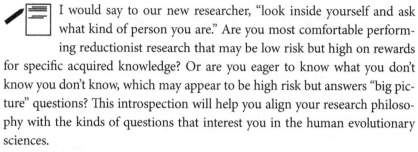 I would say to our new researcher, "look inside yourself and ask what kind of person you are." Are you most comfortable perform-ing reductionist research that may be low risk but high on rewards for specific acquired knowledge? Or are you eager to know what you don't know you don't know, which may appear to be high risk but answers "big pic-ture" questions? This introspection will help you align your research philoso-phy with the kinds of questions that interest you in the human evolutionary sciences.

If this research is to be performed with a mentor, it is also rather important that the research philosophies of you and your mentor agree. When looking for a mentor, you have a couple of options. On way is to undertake a research internship with your chosen mentor for a semester or two; it won't take long to know whether there is alignment or not. Another way is to look at the titles of all scientific papers published by the mentor. If the line of research continues uninterrupted from PhD to the present day, this is a reductionist researcher. If publications bounce around from topic to topic, now and then finding coher-ence between topics in some papers, this is a big picture researcher.

INSPIRING PEOPLE

 Max Karl Ernst Ludwig Planck (aka Max Planck).

25

JEREMY "JERRY" DeSILVA

eremy "Jerry" DeSilva is a paleoanthropologist in the Department of Anthropology at Dartmouth College. He studies the evolution of locomotor behavior in hominoids, primarily focusing on the foot and ankle anatomy in the early stages of human evolution. Jerry has also studied wild chimpanzees in western Uganda to better understand the tree climbing capabilities in fossil species. His interest in the evolution of human bipedalism has led him to study other related topics, such as the evolution of neonatal brain size in relation to obstetrics. Jerry's ideas are presented in his book *First Steps*.[1]

YOUR BEGINNINGS

I was the kid who turned over every rock and fallen log looking for bugs, salamanders, and snakes. I distinctly recall the rush of wonder I felt looking at the craters on the moon through an old telescope my parents purchased at a yard sale. I was fascinated by the natural world and how it worked; I was well on my way to becoming a scientist. In middle school, I was bored with my assignments and became a bit of a troublemaker, especially during trips to the school library. The librarian and I clashed: She didn't like me, and I wasn't a fan of hers. One day she cornered me in the hallway, and I assumed I was in trouble again. She said nothing and handed me a piece of paper. On it she had written a question: "How many moons circle the planet Jupiter?" I didn't know and told her so. She smiled and said, "Find out." There was no internet in 1987, so I went to the library and looked up the answer in an astronomy book. On the back of the paper I wrote: "14, but the space probe Voyager recently found 3 more, so 17." The next day I handed her the note, and before she even read it she handed me another: "What is the largest lake in the world?" I didn't know and told her so. "Find out," she replied once again. This back-and-forth went on for much of the year. It was addicting. Questions have answers, and those answers lead to more questions. The world is *knowable*. No one hands me little slips of paper with questions on them anymore, and answers to the questions I work on now cannot be found in a book in the library. Those are the questions I love the most now—ones no one has the answers to yet. But the answers to many of those questions are indeed out there, in the bones of our ancestors, many still buried in ancient soils, waiting to be discovered. My passion for research began with that exercise with my librarian and continues today. We should never stop asking questions about our world, and it doesn't hurt to revisit ones we think we've already answered—seventy-nine moons are now known to circle Jupiter.

GAME CHANGER

Australopithecus sediba was a game changer for me.[2] I had just completed my PhD thesis on the foot and ankle of *Australopithecus* and was asked by Lee Berger and Bernhard Zipfel [two paleoanthropologists working in South Africa] if I would be interested in applying my findings to the foot fossils of a new 2-million-year-old species from Malapa cave, South Africa, they were calling *Australopithecus sediba*. I enthusiastically agreed. I knew what the foot and ankle of *Australopithecus* looked like, and I assumed *sediba* would be no different. I planned to plug the data into graphs I had already made and have publication-ready figures in no time! But what I saw in those bones shocked me. The heel was way too small and apelike. Parts of the ankle were all wrong for an *Australopithecus*, and a bone in the middle of the foot didn't look like any *Australopithecus* I had ever seen. This foot was different and must have been from a species of *Australopithecus* that walked in a different manner than we do today and differently from other early hominins.[3] Because of *Australopithecus sediba*, I've spent the last decade reanalyzing foot bones from our earliest ancestors. They reveal that different species of early hominin were walking differently from one another—there were all sorts of bipedal experiments going on millions of years ago! But *Australopithecus sediba* is a game changer for me for two other reasons. First, the first fossils were found by a nine-year-old (Lee's son Matthew Berger) in a region of South Africa that has been searched for hominin fossils for nearly a century. In other words, bones of a brand new species of *Australopithecus* were under our noses this whole time. If we are still making discoveries like these in our proverbial backyard, can you even imagine how much more must still be out there waiting to be discovered? Second, instead of restricting access to these new fossils—as was standard in our field at the time—Lee Berger and his colleagues at the Evolutionary Studies Institute in Johannesburg, South Africa, made the fossils, casts of the fossils, and digital copies of them available to the entire world. Making these fossils open access has shifted the culture of our field to be more open, more collaborative, and more scientific.

AMAZING FACT

I'm amazed that we are here at all. As paleoanthropologists, we have the great good fortune to study the remains of long extinct human relatives and ancestors. In the Miocene, there were dozens upon dozens of ape species, first in Africa and then eventually in Asia and even Europe. We have fossils of extinct apes *Morotopithecus*, *Pierolapithecus*, *Ouranopithecus*, and *Oreopithecus*. Apes were once quite numerous and speciose. But today apes are extremely rare—found only in refugia forests in Africa and Asia—and their numbers are dwindling rapidly. The only ape that is not rare is the bipedal one, *Homo sapiens*. Paleoanthropologists have discovered more than twenty distinct species of upright walking hominins that lived in the Plio-Pleistocene, and not all of them were on the direct line to us. There were all kinds of evolutionary experiments going on. It is tempting to think that these now extinct apes and hominins were failed versions of us. But I disagree; they were beautifully adapted to the environment they lived in at the time they lived. Think about *Paranthropus boisei*, an extraordinary hominin that lived in eastern Africa from around 2.5–1.5 million years ago. It could walk on two legs like us and had enormous cheek teeth and chewing muscles to digest its fibrous plant diet. But this magnificent cousin of ours became extinct. More recently, there is evidence that our own species, *Homo sapiens*, shared the landscape with up to five other populations or species of upright walking hominins: Neanderthals, Denisovans, the small-brained *Homo naledi* and *Homo floresiensis*, and the newly discovered mysterious *Homo luzonensis* from the Philippines. Today it is just us, *Homo sapiens*. This fact is sometimes used as evidence of our superiority and dominance. Instead I suggest that the dozens of extinct hominins point to our serendipity and continued vulnerability as a species on this changing planet.

TIME TRAVEL

 I would set the dials of my DeLorean time machine to 8 million years ago and go on a safari, searching for the last common ancestor of humans and chimpanzees. We know, from genetic studies and from fossils of the earliest hominins, that our branch of the family tree split from the lineage leading to the genus *Pan* (chimpanzees and bonobos) around this time. However, we still don't know what this animal looked like. This may be surprising to many who might assume that humans evolved from a chimpanzee-like animal and that humans have slowly and gradually stood from a hunched over, knuckle-walking ape just like the classic image seen so frequently on bumper stickers and T-shirts would imply. But we have few fossils from this time period, and what little we do have indicates that this ancient ancestor may not have knuckle-walked. So how then did this animal move around its world? Did it walk through the trees using hand-assisted bipedalism like an orangutan does? Or perhaps it was smaller and moved more like a gibbon? Some researchers hypothesize that the common ancestor was more like a big, tailless monkey, without the specializations of a modern ape. Still others advise us not to overthink this and to stick with the model of a knuckle-walking, chimpanzee-like ancestor. The truth is, we don't know what this common ancestor looked like, what it ate, what kind of environment it preferred, and more interesting for me, how it moved. We therefore have no idea how upright walking got started because we aren't even sure what locomotion preceded it! I may not have access to a DeLorean, but I am lucky enough to be a paleoanthropologist. The fossils we discover are sort of like time machines, allowing us to travel back to long ago times and reconstruct what life was like for our ancestors. With luck, someday we will have fossils from this time period and will better understand the nature of the common ancestor we share with chimpanzees.

DRIVING FACTORS

 Upright walking. Mammals are wonderfully diverse in how they move from point A to point B. Some hop; others leap or fly. There are swimming mammals and swinging mammals. Most mammals move on all fours—walking at a slow pace and galloping to go faster. But of the nearly five thousand species of mammals, only humans habitually stride around on just two legs. Other mammals, such as bears or chimpanzees, *can* occasionally move on two legs, but only humans do it all the time. Hints of upright walking can be found in the earliest hominin fossils, dating back 7 million years ago. Throughout our evolutionary history, this unusual form of locomotion set the stage for everything that made us human. It liberated the hands from the duties of locomotion, freeing them to make and use tools. These tools allowed us to extract more energy-rich food from our environment, stimulating brain growth in early members of our own genus *Homo*. But walking on two legs also makes us vulnerable to predation and to injury. We are slow and unstable on just our two feet. On a landscape full of predators, I cannot imagine how our ancient ancestors could have survived unless they looked out for one another. I think that the traits of empathy, compassion, and conspecific care were selected for in the context of bipedal walking. More than any other animal, we display tremendous intragroup compassion—we take care of one another. It could be, therefore, that one of the most wonderful and mysterious aspects of the human condition arose out of our vulnerabilities, being a slow bipedal animal on a difficult to survive landscape.

26

STEVE FROST

S teve Frost is a paleontologist and morphometrician who studies human and primate evolution and broader aspects of evolutionary theory in the Department of Anthropology at the University of Oregon. Steve is an expert on Old World monkeys and the relationship between their evolution and global climate change. His research has had profound implications for contextualizing human origins. Steve combines multiple fieldwork projects in eastern Africa and Europe with analytical research, applying novel three-dimensional shape quantification techniques that link fossil anatomy and locomotor behavior.

GAME CHANGER

I don't think there have been many game changers. Even ideas that have changed people's thinking in major ways—biological evolution, relativity, quantum mechanics, plate tectonics—were more incremental in their day than they seem now. In my opinion, science is just more cumulative than that. We tend to teach the history of science in a relatable, human narrative format (person A, proposed hypothesis X, to explain phenomenon Y, through observation Z), so out of necessity we ignore all of the related and relevant research, breakthroughs, and dead-ends that occurred around the same time that give any particular discovery meaning. No major discovery or paradigm shift on its own is enough to "change everything." It can't. The body of knowledge needed to appreciate any aspect of science is simply too large.

What has been most inspiring to me are works that have synthesized relevant information from across diverse disciplines and brought them to bear on particular problems, which is harder and harder to do with the increasing amount of knowledge in different areas. In my case, one instance was Vrba's habitat theory, which appeared in the *Journal of Mammalogy* in 1992.[1] Although the same ideas were developed in many papers from the 1980s to the 1990s, Vrba's paper was the best because it spelled out each hypothesis clearly, and more important, included its predictions. Vrba's habitat theory integrated the available ideas of macroevolution, paleontology, and climatic data. Although current evidence has falsified the ideas of climatic forcing and turnover pulses as being important in mammalian evolution, this elegant, falsifiable, and explicit set of hypotheses got me interested in the Neogene, Africa, and cercopithecids.

Our most current data are not consistent with these hypotheses, or at least their predictions, but this is almost beside the point. I can't emphasize enough how important it is that Vrba laid out her hypotheses in such a coherent and testable fashion. This is to her credit and was far more important than "getting it right." This is how science is supposed to work. In my opinion, too many of our hypotheses about human origins are not testable or are so vague in their predictions that nearly any data point will be consistent. Making clear, testable hypotheses is difficult to do in the historical sciences because, in nearly all cases, any observable effect can have multiple possible causes.

AMAZING FACT

The sheer amount of information humans have been able to acquire about our own evolution is amazing, everything from fossils, trackways, comparative genomics, ancient DNA, immunology, proteomics, geochronology, sedimentology, oxygen isotopes from benthic foraminifera, continental dust in cores from the sea floor, sapropels, paleosols, carbon isotopic data from tooth enamel, growth increments, and paleopathology. It is incredible.

TIME TRAVEL

 Somewhere from which we have little or no fossil record, ideally one of the Gondwana fragments, such as India before contact with Asia; Madagascar any time between the Cretaceous and Holocene, or central Africa in the Paleogene.

DRIVING FACTORS

 I don't think we'll ever really know because so much about an organism's biology, ecology, and environment just doesn't get preserved (paleontologically, comparatively, or genetically). We have a very limited amount of information about these earliest hominins and their close relatives; the odds of finding just the right piece of the puzzle this long after the fact strike me as astronomically against us. We like to think canine reduction or other facts are important, but that is just bias toward what we can see. Is it information? Yes. Is it what we need to know to solve this puzzle? Probably not. I'm not saying we can't learn a lot about early hominins, and I'm not saying we shouldn't try. But I remain very skeptical of any answer to a question like this. I'm also skeptical of their being just one primary factor; I suspect there would be many.

EVOLUTIONARY LESSONS

I have a lot of caveats to this question. First, some things about human nature, health, and society probably can be at least partially informed by human evolutionary studies in the broadest sense (i.e., not just by paleoanthropology alone). Second, it depends on what you mean by "helpful" and to whom. People are already using a lot of what we have learned about how evolution has shaped human cognition and other mental processes. For example, some have learned that we don't respond as well to facts as to emotions. People who want us to behave in certain ways (e.g., charities, political parties, advertisers) have learned to provide a specific emotionally charged example rather than lots of facts. Tremendous resources are being spent by tech companies to keep us "engaged" with their content, often using what has been learned about how our minds have evolved. I'm not sure this counts as helpful, but it is certainly marketable.

RELIGION AND SPIRITUALITY

I don't think they are compatible. That said, I don't think it matters because we all have contradictory beliefs. Human minds are great at holding contradictory beliefs and functioning just fine.

27

YOHANNES HAILE-SELASSIE

Yohannes Haile-Selassie is the director of the Institute of Human Origins at Arizona State University. He studies the earlier stages of human evolution. Through his fossil discoveries, Yohannes has significantly contributed to what we know about human evolution. Since 2005, he has been directing the Woranso-Mille Paleontological Project in the Afar region in Ethiopia. Some of his most remarkable discoveries include a partial skeleton of *Australopithecus afarensis*,[1] the Burtele foot (with an opposable big toe),[2] and most recently a complete cranium of *Australopithecus anamensis*.[3] Yohannes has named two early human species: *Australopithecus deyiremeda*[4] from the Middle Pliocene and *Ardipithecus kadabba*[5] from the Late Miocene.

YOUR BEGINNINGS

I don't know exactly when I started this process, but it was definitely after I finished my undergraduate degree in history. My interest in human evolution began with a course on early man I took during my sophomore year. However, this interest almost died when I finished school and was hired by the Ethiopian Ministry of Culture as a historian. After three years, I transferred to the paleoanthropology laboratory at the National Museum in Ethiopia, and it was there that I had the opportunity to see actual fossils and to meet researchers who were studying human evolution. I accompanied them into the field as an antiquities officer, and I became really interested in human evolution when we started finding hominin fossils, which I quickly realized was a big deal! This was a few years before I went to grad school.

Clark Howell was not my primary advisor in graduate school (Tim White [chap. 45] was), but Howell was on my committee. And he and his office, and his library of books and reprints, were available to me for the entire time that I was in grad school. He was available if I wanted to talk to him, and his encyclopedic knowledge was a valuable resource for me that he had no reservations about sharing. He was a great role model in our field and remains so for those of us who knew him closely.

AMAZING FACT

 The intricacy of human evolution is amazing. Have you ever heard of another animal that increased its brain size exponentially through time (relative to body size, of course)?

TIME TRAVEL

 I would travel to the Mid-Pliocene (3.6–2.6 million years ago) to see how multiple hominin taxa were running around on the landscape, how they managed to do that, and to see what triggered the origin of our genus.

FUTURE EVOLUTION

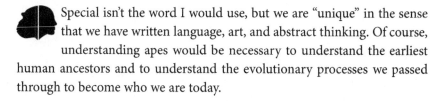 The environment will continue shaping human evolution, as always. But I doubt that our species will last another 100,000 years! Our brain power and invention will end up being our enemy.

SPECIAL?

 Special isn't the word I would use, but we are "unique" in the sense that we have written language, art, and abstract thinking. Of course, understanding apes would be necessary to understand the earliest human ancestors and to understand the evolutionary processes we passed through to become who we are today.

SPIRITUALITY AND RELIGION

 As a person who studies human evolution, I'd say that these two ideas are not compatible. If a person studies human evolution and at the same time has spiritual/religious views, it is up to that person to make them compatible. It would be like having two different faces on a coin. Which side of the coin you show would depend on who you are talking to.

ADVICE

 This is a really interesting area to study, and if it is your passion, go for it. But you won't make a lot of money in this field.

INSPIRING PEOPLE

 F. Clark Howell.

28

ASHLEY HAMMOND

A shley Hammond is the curator of biological anthropology at the American Museum of Natural History. She oversees the curation and stewardship of one of the most sensitive collections of human remains in the world. Ashley seeks to understand ape and human evolution, especially in eastern Africa, where she collaborates with colleagues from the National Museums of Kenya in field projects. Her laboratory work centers on the evolution of locomotor behaviors, especially bipedal locomotion, which is the hallmark of the human lineage. She is a pioneer in implementing novel analytical techniques to fragmentary fossils.

YOUR BEGINNINGS

I was a junior in college when I participated in a month-long field school run by the University of the Witwatersrand in South Africa. This experience was transformative and thrilling. I saw the Little Foot skeleton in situ in Sterkfontein, examined the Taung child and other original hominin fossils at the university, excavated at the legendary site of Swartkrans, and met famous paleoanthropologists. I returned to the United States to enter my senior undergraduate year with the conviction that I *must* pursue paleoanthropological field research in Africa. With a bit of bright-eyed naiveté, I would start out on my path by applying to graduate school. I was fortunate to land at the University of Missouri for my dissertation research, studying hominin and ape postcrania under the mentorship of Carol Ward [chap. 44]. Carol had longtime connections with Meave Leakey, and I was able to participate in some of Meave's fieldwork in East Turkana. I fell in love with Turkana—it is a surreal, beautiful landscape and home to some of the most spectacular hominin fossil discoveries. The specific research questions that I now pursue are, in some ways, cumulatively a stroke of luck and circumstance.

GAME CHANGER

 For the scientific community more broadly, I imagine this must be Raymond Dart's 1924 discovery of the Taung child cranium, which repositioned hominin origins in Africa and documented the fact that big brains evolved later in the game. For me, the 2009 publication by Tim White [chap. 45] and his team of the *Ardipithecus ramidus* skeleton (ARA-VP-6/500)[1] was a game changer for modern paleoanthropology. Although not everyone agrees that *Ardipithecus* was an early hominin (I believe it *was* a hominin based on study of the originals in Addis Ababa and the phylogenetic work of Carrie Mongle), the postcranium provides clues about what was happening during ape and hominin evolution during a critical time period. The Ardi skeleton reveals that it is unlikely that we evolved from an ape that resembles living African apes, which was the paradigm for many years. The scientific reach of Ardi remains to be seen in the coming years, but it has already reshaped ideas about early hominin origins for many scientists.

TIME TRAVEL

 I would travel to the past (hopefully with a camcorder!) during the period of the divergence of the chimpanzee-hominin lineages to see what the last common ancestor looked like. This is the question that keeps me awake at night. We have a fairly good idea about the shape of hominin evolution and the major patterns and trends from *Australopithecus* to *Homo*. There are surprises like *Homo floresiensis* and *Homo naledi*, of course, but we generally can see the big picture. People who bemoan the "paucity of the human fossil record" must not actively work with fossils—there are thousands of hominin specimens! What remains hazy in hominin evolution is the starting point for hominins. Even Ardi does not satisfactorily address this question.

EVOLUTIONARY LESSONS

Regarding our present and future, we could take a lesson or two in sustainability from *Homo erectus*. Our own species, *Homo sapiens*, has been around for 300,000 years or so. We have managed to achieve remarkable feats during this time that differ from those of any species to have ever lived on this planet, but we are now on the precipice of ecosystem collapses and irreversible global warming. *Homo erectus* ventured out of Africa, was successful in a broad range of geographical environments, and had a relatively low impact on the environment and natural resources. Existing from about 1.9 million years ago in eastern and southern Africa until perhaps 30,000 years ago in Asia, *Homo erectus* was also a more successful (longer enduring) species than we have yet been shown to be. Can modern humans modify our diet, lifestyle, and population size to reduce our imprint on the natural world?

SPECIAL?

This may be an unpopular opinion, but yes, I believe humans are special when compared to other animals on this planet. Humans do unusual things: walk bipedally, control fire, and make complex tools requiring planning and spatial understanding. Humans also do bizarre things: play organized sports, build monasteries and monuments, and compete in March Madness. We are a species unlike any other in terms of our cognitive complexity and sociality, but we are also bounded by certain unalienable aspects of our biology as primates. There is much to be learned from other primates, especially the great apes, to better understand our species. The one pitfall that we should actively work to avoid is modeling earlier stages of human evolution based exclusively on modern great apes. Our closest living relatives have also been evolving independently for millions of years, so they are derived for their own behaviors and are not a "window into the primitive."

ADVICE

My advice is specific to budding paleoanthropologists. Undergraduate students should try to get as much exposure to different areas of the natural sciences as possible. In retrospect, I wish I had more practical (field or lab) experience in geology and advanced genetics. Paleoanthropology graduate students should try their hardest to be exposed to fossil material from all over the world and from different time periods, so they can appreciate the big patterns in evolution before getting pigeon-holed with the minutiae. My own scholarship on apes and humans has been enriched by working on fossil primates from the Oligocene to the very recent, interacting with living primates, and dissecting both human and nonhuman animals.

INSPIRING PEOPLE

Tim White [chap. 45], Bill Kimbel [chap. 32], Kay Behrensmeyer [chap. 22], and Christoph Zollikofer [chap. 69]. These individuals impress me with the breadth and depth of their knowledge, so I'd like to hear their unplugged, off-the-cuff perspectives on humans past and future.

29

SONIA HARMAND

onia Harmand is a field archeologist in the Anthropology Department and in the Turkana Basin Institute at Stony Brook University. For more than twenty-five years, she has conducted fieldwork in France, Syria, Djibouti, Ethiopia, and Kenya, where she is the West Turkana archaeological project director. Sonia's primary research focus is the dawn of human technology. She was part of the team that in 2011 published information on the oldest Acheulean stone tools found so far (about 1.7 million years ago; from Kokiselei 4 in West Turkana, Kenya).[1] Soon after that, she and her team discovered the 3.3-million-year-old archaeological site of Lomekwi 3 nearby, which contains the world's oldest stone tools.[2]

YOUR BEGINNINGS

I had no idea what scientific research was until I was in my twenties. During my teenage years, I read a lot about great explorers in Africa such as Henri de Monfreid, Joseph Kessel, and Arthur Rimbaud. I had always been attracted by explorations in Africa, and I thought I would become an ethnologist. When I entered college in France, I discovered primatology, and my readings changed to great women exploring the primate world in Africa: Jane Goodall and Diane Fossey. I wanted to be like them. My early years in college taught me how to become a scientist in the fields of ethnology and geography. When I was twenty-two, I met the archaeologist Hélène Roche in a CNRS [the French National Centre for Scientific Research] lab in Meudon, a western suburb of Paris. This was a turning point in my career. On the door of her office was a picture of Kanzi, the bonobo, knapping stones, and I was very intrigued by it. I pushed through the door, and we talked for many hours. For me, that day was the starting point of my career as an archaeologist. I joined Hélène in the field in Kenya, and she taught me everything.

GAME CHANGER + TIME TRAVEL + DRIVING FACTORS

 We hear a lot about biological evolution, but what fascinates me the most, and has always fascinated me the most, is our behavioral story. How did we develop this special ability of snapping stone, a unique ability that makes us different from the other primates? My team and I found the oldest stone tools known so far, and the most fascinating questions for me remain: Why did we start to make stone tools? How did that happen? If I had a one-shot roundtrip in a time machine, I would love to stand close to the first knapper and share the unique moment when the first flake was stroked from a core, the sound of it, and maybe the sense of contentment felt at that particular moment. That moment changed everything for us and led to our success as a species. Even though it was one of the most important moments in human evolution, producing the first flake was probably an accident. But the necessary cognitive capacities were already in place for understanding the advantages of using a cutting object.

SPECIAL?

Hominin technology is unique! Sure, a wide range of animals *use* tools, including mammals, birds, fish, cephalopods, and insects. Jane Goodall played a huge role in the 1960s by showing tool using in chimpanzees at Gombe, Tanzania.[3] But in most cases, tool-use behaviors in animals are stereotypical behaviors. Only a couple of species can modify a tool (chimpanzees when fishing for termites, for example). Although new research in primate behaviors shows more and more tool-using behaviors, *making* tools and making *stone* tools to act on food and nonfood objects (for example, to make other tools) is unique to our lineage. It is the hallmark of hominid intelligence and was a crucial step in our evolution. Our closest living relatives can only tell us how much hominid technology differed 3.3 million years ago from the cognition and capabilities of extant primates when it comes to tool using and tool making.

ADVICE

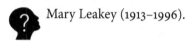 A sweatshirt that I like to wear when I teach has this written on the back: "Go, Fight and Win." I think it synthesizes most of my career experiences as a young archaeologist in Africa. In French we use the expression "sortir des sentiers battus," which can be translated as "go off the beaten tracks." Yes, do that! Find new sites, study with new techniques, go off the tracks both in the field and academically. Think differently, be curious, audacious, and open-minded, but prepare for all sort of challenges.

INSPIRING PEOPLE

Mary Leakey (1913–1996).

30

RALPH HOLLOWAY

R alph Holloway is a paleoneurologist at Columbia University's Department of Anthropology, where he has developed his research since 1964. Ralph studies brain endocasts as a proxy to understand brain evolution in the extinct members of the human lineage. He also studies endocasts of living apes and humans and their brains. As a pioneer in paleoneurology, he was one of the first to investigate the Taung child endocast (a juvenile of *Australopithecus africanus* of about 2.5 million years ago), concluding that although brain size was pretty much like that of living apes its structure was different, perhaps associated with higher cognitive functions, such as language.

YOUR BEGINNINGS

In the 1950s I was majoring in geology at the University of New Mexico and was taking a physical anthropology course dealing with osteology. It wasn't until I came to the University of California, Berkeley in 1960 that I became interested in the brain. I was one of Sherwood Washburn's students, and I had to take a lot of anatomy courses as well as osteology. I was not interested in primate behavior, but in human evolution. Around 1962, I took a course in neuroanatomy from Dr. Marian Diamond, and that was the first time I took a real interest in the brain. And my interest mainly concerned differences in the primate brain. When I finished my dissertation in 1964, I didn't think paleoneurology could cast much light on the evolution of the human brain. My doctoral dissertation was on quantitative aspects of the primate brain, and when I came to Columbia University, I wanted to do research on different primate brains. It wasn't until I went to South Africa in 1969 and worked in Professor Raymond Dart's lab on the Taung endocast that I became hooked on studying the evolution of the brain. In 1971 and 1972, I was back in South Africa and working in Kenya on new fossil discoveries.

GAME CHANGER

 It was Julian Huxley's *Evolution: The Modern Synthesis*.[1] It was like an epiphany, and I felt a great sense of freedom, from religion in particular.

AMAZING FACT

 I am amazed that we haven't slaughtered each other yet and that there remains so much stupidity in our species.

TIME TRAVEL

 I would travel between 3 and 1 million years ago when both bipedal locomotion and brain reorganization, as well as brain size and social organization, were under strong selection pressures, leading to a mosaic of hominin evolutionary changes.

DRIVING FACTORS

 Bipedal locomotion, precision grip of the hand, brain reorganization, and social behavior, leading to self-domestication.

FUTURE EVOLUTION

 I suspect pandemics and violent climatic changes will test humanity's ability to last through the next one hundred years, if that long. Did I mention nuclear war?

EVOLUTIONARY LESSONS

 Evolutionary studies should teach us that things are never static but are always changing, including our own genomic background, behavior, and adaptation to ever-changing climates, food resources, and the like.

SPECIAL?

 Humans are nothing more than an advanced species of ape. I think our brains are special but basically ape, although language and the use of arbitrary symbol systems do make us special.

When I was working at Berkeley, I read a paper about a microcephalic woman with a brain the size of a chimpanzee's who could talk. Ever since that time, I have avoided making relationships between language and the size of the brain. I am sure that the quality of language probably does depend on brain size. But in terms of the evolution of the brain, I believe that the case for language can be made when we find stone tools made to standard patterns, such as the Acheulean hand axes and possibly earlier stone tools going back to maybe 2.5 million years ago. More important than the size of the brain would have been the brain's organization. That is why the Taung skull is so important; it showed reorganization of the cerebral cortex.

RELIGION AND SPIRITUALITY

 Beats me! I doubt it has helped us any.

ADVICE

 Steep yourself in the biological sciences, particularly organic chemistry, and especially genomics, where neurogenomics will cast the most light on our evolution past, present, and future.

INSPIRING PEOPLE

 Raymond Dart.

31

KEVIN D. HUNT

K evin D. Hunt (Indiana University) studies the functional morphology of living apes and other primates as a means of reconstructing the locomotion, posture, and behavioral ecology of early hominins. Kevin founded the Semliki Chimpanzee Project in 1996, based in the Toro-Semliki Wildlife Reserve, Uganda, where he has conducted pioneering primatological fieldwork on the reserve's "savanna chimpanzees." Semliki chimpanzees include more terrestrial bipedalism into their locomotor repertoires, compared to forest chimpanzees. Kevin's extensive research provides key insights into both modern ape biology and the emergence of the human career.

YOUR BEGINNINGS

High school textbooks tend to shy away from controversy, and my high school biology text ran about par for the course. Human evolution was too hot a potato to cover in any meaningful way, which meant evolution in general was left until the last few pages of the text, and human evolution, or rather its possibility, was consigned to the last couple of paragraphs. I was already familiar with the basic concepts of natural selection and knew something about fossil hominins before I read the text, yet when I came to these brief paragraphs, I fussed over them compulsively. The last few sentences discussed Neanderthals, described as disappearing suddenly. It seemed to me that this disappearance and its cause were of earth-shaking importance, and I felt the science of their extinction could hardly be as unsettled as my textbook portrayed: Neanderthal fossils are abundant as are their artifacts; I sensed this inconclusive summary was a dumbed-down, simplified abridgment tailored to the high school audience. Yet when I searched for a more thorough review of the "Neanderthal Problem," I found that "it's a mystery" was a pretty popular synopsis of human paleontology's view of Neanderthals. Human fossils and evolutionary processes were suddenly an obsession, and in short order a life-goal became studying human evolution.

GAME CHANGER

Stern [chap. 92] and Susman's [chap. 42] 1983 publication, "The Locomotor Anatomy of *Australopithecus afarensis*,"[1] simply bowled me over, and I have read it dozens of times. Before I encountered this article, I thought I understood australopiths pretty well. In the image of Lucy on the cover of Don Johanson's *Lucy: The Beginnings of Humankind*,[2] where Lucy's grayish white bones lay conspicuously on a deep red background. This image became seared into my memory: Lucy's skeleton looked altogether human. Likewise, the description of Lucy and other early australopith fossils recorded feature after feature paralleling the human condition—and virtually none that resembled apes. When I had examined australopith casts—before reading Stern and Susman—I saw all the features that shouted "human" (the valgus femur, the prominent lateral lip on the knee, the short, bowl-like pelvis, the broad sacrum) and none of the chimpanzee-like features. Stern and Susman's exhaustive element-by-element description, wonderfully illustrated, was a revelation. It completely altered my view of australopiths. Rather than human-like from the neck down, I realized that Lucy was quite chimpanzee-like from the crown of her head to her first lumbar vertebra, and quite (but not completely!) humanlike from L1 [the first lumbar vertebra] down. Here and there apelike features could be seen wherever they interfered little with bipedal locomotion, features such as long, curved toes, short legs, smaller joint surfaces below the waist, a more plantar-flexed set of the ankle, and a convex external tibial platforms. Australopiths are chimeric, I now feel certain, a mix of human, chimpanzee, and unique features.

AMAZING FACT

The steady, relentless ratcheting up of brain size. When I observe the masterful competence of chimpanzees in their natural environment as they forage, socialize, and problem-solve, I can hardly imagine any organism negotiating this environment better than they do. Why did *we*, hominins, need and, what's more, how could we possibly *afford* a brain three times the size of chimpanzees? Brains are expensive to grow, expensive to power, and heavy to lug around. A big brain slows maturation, leaving its owner a slow starter in the mating game, forced to delay reproduction until later in life, seemingly at a severe disadvantage compared to smaller-brained rivals. Brain-punishing phenomena lurk everywhere in nature. A drought might pick off a calorie-wasting large-brained individual, while a more economical hominin might make it through. Competitors are many, each waiting to edge out a big-brained rival who has delayed reproduction and who wastes time gathering the extra calories needed to power that brain. Yet amidst these disadvantages we have taken on three times the brain burden of the astonishingly smart apes. It's a surprise that the balance has so consistently fallen on the side of brain growth and for so many millennia—and only so in our lineage.

TIME TRAVEL

 I would visit the first bipeds. Australopiths have chimpanzee-like brain volumes and male-female body weight differences that are gorilla-like, with males something like twice the mass of females, far in excess of the contrast between the sexes in humans or chimpanzees.[3] Among living primates, this level of body size dimorphism is found among species where the primary social group is a single male bonded to several unrelated females.[4] These early hominins have chimpanzee-like arms and ribcages but humanlike pelves. What were they doing with that strange, chimeric body? Living species with australopith upper body anatomy harvest fruit while hanging by one arm. Only humans have a humanlike pelvis, but we're pretty sure we understand the functional anatomy of bipedalism, and they were almost certainly bipeds. Presumably they were both bipeds and arm-hangers, but that's not something we see in any living primate, perhaps siamangs and orangutans come close. I have speculated on how they melded these two behaviors, but I would dearly love to see the actual beings in action.

Their oddness goes beyond mere morphology. Australopiths have robust faces, heavier than those of chimpanzees and even gorillas. Foods chimpanzees harvest bipedally are found in the drier areas of their habitat, and the foods are fibrous and tough. A population confined to a dry habitat where foods are tough and harvested best with a bipedal posture might have selected for bipedalism. This model fits with some—but not all—reconstructions of the early hominin habitat. In even a quick visit with early bipeds, an observer could see their habitat, social behavior, diet, and locomotion. There's still more information in their faces and jaws. Years ago, Kay and Hylander[5] showed years ago that fruit size is correlated with incisor size. Australopiths have small incisors, so they were likely eating smaller fruits than chimps, and small fruits are more typical of the small trees found in dry habitats, suggesting a link between dry habitats and hominin origins.

Australopiths have robust, ventrally curved digits—even more curved than chimps—and like chimpanzees they have distinct flexor sheath ridges, anchors for the thick tendons typical of primates with a powerful grip. Why would an ape have arm-hanging traits like these without having the long fingers of apes? They must have been hanging from smaller branches. Where do you find smaller branches? In short trees, the kind found in dry habitats. Small trees also fit with the fact that australopiths have arm-hanging traits but lack vertical climbing features: You don't get into small trees with hand-over-hand climbing, you just leap up to the lowest branch. That's why goats can get into trees. Again, even a few hours with australopiths would confirm or refute this small-tree hypothesis.

DRIVING FACTORS

To a great extant, I buy Coppens's "East Side Story"[6]—the apes that became hominins were trapped in the dry habitats of (mostly) east and south Africa. They evolved heavy faces to process the tough foods typical of dry habitats and they evolved bipedal anatomy to gather those fruits while standing on the ground. Well, while partly arm-hanging and partly standing, using their raised arms to stabilize a bipedal posture. The chimpanzees I study in the dry habitat at Semliki [the area just east the Semliki River, the natural border between the Democratic Republic of the Congo and Uganda] have heavier faces and more humanlike legs than other chimpanzees.

The dry habitat food-gathering perspective I outlined above makes a very clear prediction when it comes to early fossils that some believe are hominins, such as *Sahelanthropus* and *Ardipithecus*. These two fossils have more lightly built faces than australopiths, more like chimpanzees. Either my hypothesis—that robust faces and bipedalism among hominins are linked adaptations—is wrong, or these apes were not bipeds. *Orrorin* is a more difficult fossil to interpret. Its femur resembles that of Miocene apes as much as australopiths, but the open angle of the femoral neck is not like that of australopiths. This suggests to me that its knee was not valgus, a condition seen among adult humans (but not babies!); *Orrorin* must have been more bowlegged, like chimpanzees. Its femoral neck is long, like that of australopiths, but that open femoral neck troubles me. Still, it is only one individual. If further finds show it is not a biped, then either the 6- to 4-million-year-old bipeds are hiding from us, which is not that unlikely, or *Sahelanthropus*, *Orrorin*, and *Ardipithecus* are hominins and bipedalism arose *within* our lineage around 4 million years ago, not at its beginning around 6 million years ago.

FUTURE EVOLUTION

I believe humans will look more or less the same for many hundreds of thousands of years because we now have the capacity to curate our genome. We may have ethical qualms when it comes to messing around with our DNA, but I believe we will put those aside in incremental steps and decide to keep ourselves pretty much the way we are today. After some dithering, I expect we will purge genetic diseases from our genome. Otherwise, in short order, I predict that we will take command of our biology to more or less halt evolution.

EVOLUTIONARY LESSONS

Many current health problems, both physical and mental, are a consequence of our history. We evolved as hunter-gatherers for 2 million years. For millions of years before that we were apelike fruit foragers. Other than nibbling around the edges to acquire the ability to drink milk and tolerate diseases that come with high-population densities, we live, think, and breath like hunter-gatherers. This is illustrated in our propensity to succumb to lifestyle diseases such as stroke and heart disease; in the way we make decisions about mate choice, child bearing, and child rearing; and in our fascination with animals and plants—E. O. Wilson's "biophilia." Our current frailties come from living in bodies that are still those of hunter-gatherers. After much head scratching among economists concerning the human tendency to make irrational financial choices, the importance of considering evolved psychology to understand economic decisions has emerged as a subdiscipline of economics. In 2017 Richard Thaler received a Nobel prize for his work in this area. Because humans evolved their instincts and cognitive abilities as hunter-gatherers, we face a mismatch between our evolved cognitive tools and our current built and social environment. Public policy rarely—possibly *never*—takes into account that humans make choices using cognitive tools that evolved for hunting and gathering and for social maneuvering in a society of only a few hundred people. We make irrational decisions about which college to attend, how to save for retirement, and whether to change jobs because our brains are designed to meet ancient rather than contemporary cognitive challenges. It is virtually axiomatic that any new public policy initiative will be followed by a cascade of unanticipated and undesirable consequences. Consideration of evolutionary psychology and mismatch theory could help us craft more effective public policy, not to mention help each of us make better personal decisions. There is a place for deep-time scholars in designing the built world and the rules that govern it.

SPECIAL?

This is a sticky question, and perhaps a little ambiguous. Every species is special, but I assume you mean "do humans deserve some special consideration other animals do not?" If I didn't believe that more self-aware, more cognitively advanced beings were more special, I wouldn't eat fish rather than pigs. In keeping with my biases, I would never eat a parrot, but I do eat chicken. Our intelligence makes us special, certainly, but that very statement also lands us on the slippery slope of "are smarter people more special?" At times I am certain I don't believe that—but then someone frames a scenario that makes a monkey out of me. During some catastrophe, if I had to choose between saving a severely cognitively handicapped human—an anencephalous person, for example—and a healthy wild chimpanzee, I would choose the chimpanzee. But then, I believe chimpanzees should have human rights too. So perhaps I would be more likely to say that more self-aware animals (that is, "animals with a more sophisticated self-concept") are special.

But perhaps you mean "are humans as a species *more distinct* from their close relatives than any other species?" Are humans uniquely unique? Well, our brains are uniquely neuroned-up, but having said that, I immediately think that orangutans are uniquely unique in that their hips are uniquely flexible; gibbons are uniquely competent at brachiation. Our brains are off the chart, and that is certainly important, so if the measure of specialness is intelligence and dominance over other species, we are special.

RELIGION AND SPIRITUALITY

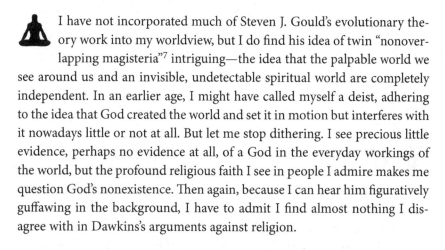 I have not incorporated much of Steven J. Gould's evolutionary theory work into my worldview, but I do find his idea of twin "nonoverlapping magisteria"[7] intriguing—the idea that the palpable world we see around us and an invisible, undetectable spiritual world are completely independent. In an earlier age, I might have called myself a deist, adhering to the idea that God created the world and set it in motion but interferes with it nowadays little or not at all. But let me stop dithering. I see precious little evidence, perhaps no evidence at all, of a God in the everyday workings of the world, but the profound religious faith I see in people I admire makes me question God's nonexistence. Then again, because I can hear him figuratively guffawing in the background, I have to admit I find almost nothing I disagree with in Dawkins's arguments against religion.

ADVICE

 Travel widely; summon the courage to talk to the leaders in your field when you actually have something to ask them; listen attentively; change your mind now and then; and study other animals.

INSPIRING PEOPLE

 Darwin, of course. His name will be offered up so many times, so I'll offer as an alternative, Georges-Louis Leclerc, Comte de Buffon.

32

WILLIAM "BILL" KIMBEL

William "Bill" Kimbel was the Virginia M. Ullman Professor of Natural History and the Environment at Arizona University, the former Institute of Human Origins director (2008–2021), and an elected fellow of the American Association for the Advancement of Science. Bill combined theoretical, lab-based, and fieldwork research in the evolution of the human lineage, particularly during the Plio-Pleistocene of Africa. At Hadar (Ethiopia), his decades-long fieldwork efforts recovered the most extensive fossil collection of *Australopithecus afarensis*. He was also well-known for his sharp mind when applying theoretical principles of evolutionary and systematic theory to current paleoanthropological discussions. In addition, he was a beloved teacher and mentor to many in the field. On April 17, 2022, Bill passed away after fighting cancer for more than three years.

YOUR BEGINNINGS

In 1973, as a university sophomore, I took a course on human evolution from Don Johanson. But the seed had been planted by a high school teacher in 1971–72, who taught an "elective" class in anthropology, which led me to summer excavations at Valley Forge State (now National) Park.

GAME CHANGER

 "Lucy."[1] My proximity to the newly discovered Hadar hominin fossils (then in the United States on loan from the Ethiopian government), first as an undergraduate and then as a graduate student, pretty much set my course in paleoanthropological research. In addition, my first (but by no means only) reading of Franz Weidenreich's "The Skull of *Sinanthropus pekinensis*"[2] taught me the importance of close observation of morphology and the value of being able to communicate it in writing.

TIME TRAVEL

 I would like to spend a week or two in close observation of *Ardipithecus ramidus* in the wild. Second choice would be *Australopithecus boisei*. *Ardipithecus* is morphologically and adaptively transitional, and I would like to see how that translates into observable behavior in a natural setting. *Australopithecus boisei* is so uniquely specialized in its morphology that I would want to know whether and how that translates into functional-adaptive characteristics (diet, feeding behavior, etc.). So the answer is similar in both cases, I am motivated by curiosity about whether or not our "commonsense" inferences from morphology to behavior in the past are sensible.

At issue is the notion of "driving forces" placing our ancestors "on the road leading to the human lineage." The question itself is set up to place universal *modern human* traits (e.g., bipedality) as defining of *hominin* origin(s). I would separate origin ("on the road"), which is cladistic, from adaptation ("driving forces"), which may have no bearing on the origin event itself—they are conceptually, and potentially temporally, distinct. For example, consider (facultative) bipedality and nonhoning canines, both of which appear very early (indeed, contemporaneously) in the fossil record. The great majority of explanatory hypotheses about hominin origins focus on terrestrial bipedality. Very few focus on the canines, which have at least as many implications for behavior as locomotion. (Some theories cite both: Jolly[3] [chap. 77] and Lovejoy,[4] for example.) Why is this the case? Could it not be the case *cladistically* that the earliest hominins were quadrupeds with reduced, nonhoning canines? And that bipedality on the ground was a subsequent development, following the origins event by hundreds of thousands (or even more) years? Or perhaps *none* of the characters that happen to preserve in the fossil record have *anything* to do with the ecological and behavioral parameters of populations involved in the cladistic event! All of this reduces to my earlier comment that some of this information may be unknowable—and therefore, almost by default, we tend to examine the question from the modern perspective, projecting our traits back in time and asking how and why they were important to, indeed defining of, our ultimate origin.

DRIVING FACTORS

 This is unknown and perhaps unknowable. To date facultative bipedality, a reorganized central cranial base, and small, non-honing canine teeth are all candidate changes at or near the base of the hominin clade. Which, if any, among them was "the main driving factor" cannot yet be identified with any confidence. Our historical bias has been to claim bipedality as the major factor, but this is post hoc. There is no reason to think this was the one factor that *initially* set us apart from other great apes.

EVOLUTIONARY LESSONS

The study of human evolution teaches us that we are inextricably linked to the natural world, not separate from it, that the human lineage is but one (minor) twig among the millions of species (living and extinct) on the tree of life, and that all human beings are born equal by virtue of our uniquely close genealogical ties to one another relative to any other of those millions of species. These realizations have had numerous ethical implications throughout history that most world leaders have failed to comprehend.

INSPIRING PEOPLE

Darwin, of course.

33

FREDRICK "KYALO" MANTHI

F redrick "Kyalo" Manthi is a senior research scientist and director of Antiquities, Sites, and Monuments at the National Museums of Kenya. Since 1986, Kyalo has been involved in multiple archeological, paleontological, and geological expeditions in different parts of Kenya. He is the leader of the West Turkana Paleontology Project, centered on Kanapoi, the type site of *Australopithecus anamensis*,[1] as well Lomekwi and Natodomeri fossil sites dated respectively at around 3.4 million and 200,000 years ago. In addition, he reconstructs the paleoenvironments of fossil sites through the study of their small mammals. Kyalo is also committed to local outreach and education; he is the founder and chairman of the Prehistory Club of Kenya, whose mission is to educate young people about Kenya's prehistoric heritage.

YOUR BEGINNINGS

My passion for research and particularly prehistory research began when I was in primary school in rural Makueni, about 120 kilometers east of Nairobi, Kenya. My father worked at Olduvai Gorge with Dr. Mary Leakey, and he brought me books on prehistory and geology. When reading these books, I was amazed at how researchers unravel and piece together information about life forms that lived millions of years ago. When I joined high school, my interest in prehistory research blossomed. I continued to read and learn more about prehistory and human evolution, and by this point in time it was apparent that I would pursue a career in prehistory research.

GAME CHANGER

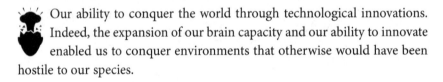 The discovery of human fossils more than 2.5 million years old, especially species attributed to the genus *Australopithecus*, changed the way most people perceived human evolution. For a long time, much of what we knew about human evolution was derived from fossils attributed to the genus *Homo*, especially *H. erectus*, recovered in both Africa and Asia. Because stone tool making was for a long time largely associated with members of the genus *Homo*, the discovery of stone tools associated with earlier humans, especially australopithecines, was to me a game changer in the way we viewed our own evolution. Certainly, the discovery of the australopithecines, especially *Australopithecus anamensis*,[2] provoked my interest in the kinds of environments in which these early humans evolved.

The proposed ancestor-descendant relationship between *A. anamensis* and *A. afarensis* is interesting to me because it bridges a gap in the fossil record that had not been filled until *A. anamensis* was discovered. I am also intrigued by the mesiodistally longer maxillary canine crowns of *Australopithecus anamensis*, which resemble those of the earlier *Ardipithecus ramidus*.

AMAZING FACT

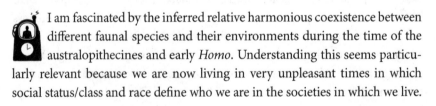 Our ability to conquer the world through technological innovations. Indeed, the expansion of our brain capacity and our ability to innovate enabled us to conquer environments that otherwise would have been hostile to our species.

TIME TRAVEL

I am fascinated by the inferred relative harmonious coexistence between different faunal species and their environments during the time of the australopithecines and early *Homo*. Understanding this seems particularly relevant because we are now living in very unpleasant times in which social status/class and race define who we are in the societies in which we live.

DRIVING FACTORS

 I believe the inferred environmental changes in eastern Africa during the Late Miocene, which included retreat of the rain forests and opening up other environments, helped set an ape apart that began the journey to humanity. Key among the basic adaptations developed during these early stages of our evolution, about 6–7 million years ago, was "primitive" bipedalism, as evidenced in the fossil remains of two taxa: namely, *Sahelanthropus tchadensis* and *Orrorin tugenensis*. Indeed, bipedalism and a suite of skeletal changes that followed—not only in the legs and pelvis but also in the vertebral column, feet and ankles, and skull—were critical for the journey to humanity. It is conceivable that bipedalism was favorable because, among other things, it freed the hands for reaching and carrying food, saved energy during locomotion, and provided an enhanced field of vision. These features are believed to have been advantageous for thriving in the new savanna and woodland environment created as a result of the retreat of the rain forests in the eastern African Rift Valley.

FUTURE EVOLUTION

A very interesting question! Unlike our ancient ancestors whose existence and survival was significantly influenced by prevailing environmental conditions, the progressive technological advances of modern-day humans have, to a large extent, tamed the natural environment so we can live and survive relatively well in different environments. However, these technological innovations and the lifestyles that accompany them have created new health hazards to the same human populations and are polluting the environment in which we live. Indeed, the effects of technological innovations and political and economic competition between nations across the world have introduced novel diseases and exacerbated universal phenomena such as global warming. These effects have the potential to significantly shape evolutionary trajectories and the existence of the human species. It is difficult to predict how humans will look 100, 100,000, or 1 million years from now. I hope we will not have destroyed ourselves through novel diseases, some caused by lifestyle choices and others caused by pollution of the environments in which we live. At least I will not be there to witness this!

EVOLUTIONARY LESSONS

Evolutionary studies document the stages that led humans to become the "intelligent" species we are today. We are the end product of a long evolutionary journey that took place over millions of years and was full of challenges. Our ancient ancestors endured and conquered many difficult challenges, and as a species we have the potential to conquer new challenges that come our way now and in the future. Human evolutionary studies have also shown that all humanity across the world share a common ancestry in Africa. Indeed, regardless of the inherent differences in skin color between human populations, an aspect that has been imperially reinforced over the years to advance the culturally constructed concept of race, we are All One People who share a common ancestry in Africa and should live in harmony with one another.

SPECIAL?

I think the word "special" is subjective! However, our intelligence does enable us to deal with complex situations, and this sets us apart from all other animals, including our closest living relatives, the apes. Genetic studies have shown that humans and chimpanzees share about 98.6 percent of our total nucleotide sequence in common. Despite this very strong genetic similarity, many traits set us apart from chimpanzees. Our high level of intelligence and habitual bipedalism are two traits that make us a special species.

RELIGION AND SPIRITUALITY

Some religious groups across the world take issue with the scientific explanation of how humans originated. In my view, there should not be any conflict between evolution/science and religion. Science reports that humans evolved over millions of years, and we find a similar time line in the Bible in 2 Peter 3:8: "But do not forget this one thing, dear friends: With the Lord a day is like a thousand years, and a thousand years are like a day." The Bible also states that God created man (humans) last, which corresponds with scientific thought: the journey to humanity began during the Mio-Pliocene boundary, about 7 to 6 million years ago, after all major groups of nonhuman animals had appeared.

Many religious denominations have come to accept that biological evolution has produced the diversity of living things over billions of years of Earth's history. Here are some thoughts from religious groups and leaders who seem to appreciate evolution:

"There is no contradiction between an evolutionary theory of human origins and the doctrine of God as Creator." (General Assembly of the Presbyterian Church)

"Students' ignorance about evolution will seriously undermine their understanding of the world and the natural laws governing it." (Central Conference of American Rabbis)

"New findings lead us toward the recognition of evolution as more than a hypothesis. In fact it is remarkable that this theory has had progressively greater influence on the spirit of researchers, following a series of discoveries in different scholarly disciplines. The convergence in the results of these independent studies—which was neither planned or sought—constitutes in itself a significant argument in favour of the theory." (Pope John Paul II, Message to the Pontifical Academy of Sciences, October 22, 1996)

In my view, it is relatively easy to hold both scientific and religious views on the origin of life and humanity. These views should complement each other and give meaning and understanding to the many situations we encounter in our interactions with one another and the environments in which we live.

ADVICE

Humans are innately curious about our past and what the future holds for us. Understanding our past as a species, including our long evolutionary journey, is important because it will enable us to appreciate and deal with the different situations we encounter now and in the future.

INSPIRING PEOPLE

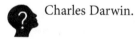 Charles Darwin.

34

MARY MARZKE

ary Marzke, a professor of anthropology at Arizona State University (where she was based since 1978) and a fellow of the American Association for the Advancement of Science, was a beloved teacher. A world leader on human evolution, Mary's research focused on the relationships among precision grips, tool behaviors, and hand morphology in extinct members of the human lineage. Her multidisciplinary approach included dissections, electromyography, kinematic analysis, and novel 3D quantitative analyses of complex bony morphologies. Mary passed away September 3, 2020, after a twenty-year battle with multiple myeloma.

YOUR BEGINNINGS

In junior high school, my family became good friends with a physical anthropologist at the University of California, Berkeley (Prof. Theodore McCown) and his family, and we often went camping and skiing together. I heard firsthand about his field trips and paleoanthropology research, especially at Mt. Carmel, and I asked many questions. By the time I was in high school, I was committed to a future university major in anthropology with a focus on fossil hominin research.

GAME CHANGER

In 1958 Professor Washburn moved to the University of California, Berkeley, where I was an undergraduate. He had recently written papers on new approaches to the study of physical anthropology that strongly influenced our views about the human evolutionary process.[1] Conversations with him encouraged my interest in studying the comparative functional analysis of living and fossil hominin morphology, which I was able to do with him on my committee.

TIME TRAVEL

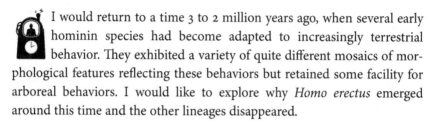 I would return to a time 3 to 2 million years ago, when several early hominin species had become adapted to increasingly terrestrial behavior. They exhibited a variety of quite different mosaics of morphological features reflecting these behaviors but retained some facility for arboreal behaviors. I would like to explore why *Homo erectus* emerged around this time and the other lineages disappeared.

EVOLUTIONARY LESSONS

Human evolutionary studies are fundamental to an understanding of our morphology and behavior, both aspects distinctive to humans and those shared with other primates. From these studies, we learn about the adaptability of our human ancestors to changing environments but also about the kinds of constraints that have channeled evolutionary trajectories of hominin species and are likely to be among the interacting factors in future change.

SPECIAL?

Humans have a unique evolutionary trajectory, but this is true for all animals and plants and does not make us special. Apes and other primates help us to elucidate the details of our early trajectory by revealing functions of morphology and behavior that they shared with us. Comparative studies increasingly show that humans are unique, not so much in individual features as in the relative development and patterns of interactions of the shared features. Early concerns about human biological adaptability to rapid cultural change have been substantially modified by a new understanding of multiple sources of intergenerational genetic similarities and of the interaction of gene and culture evolution. However, pressures ranging from warfare with innumerable causes to widespread population movements, climate changes, disease threats, and global economic stress seem to be challenging human biological, behavioral, and cultural adaptability. This raises the question of whether the special aspect of human evolution may prove to be its relatively short duration.

35

EMMA MBUA

E mma Mbua is the group leader of the Paleoanthropology Section in
the Department of Earth Sciences at the National Museums of Kenya
(Nairobi), where she's been based since 1979. In 2005, she cofounded
the East African Association of Palaeoanthropology and Palaeontology,
whose mission is to strengthen prehistoric research in the region and to
unite scholars from different backgrounds. Emma's research interests revolve
around Pliocene human evolution and adaptations and paleoenvironmental
reconstructions. She's currently directing fieldwork at the Kantis fossil site on
the outskirts of Nairobi, where she found the first undisputed remains of
Australopithecus afarensis in Kenya.

YOUR BEGINNINGS

My interest in paleoanthropological research began in the early 1980s
upon my arrival at the Department of Paleontology, at the National
Museums of Kenya (NMK). I was recruited as a curator of hominins,
and my duties entailed facilitation of local and international scholars in
accessing hominin specimens for study. I was supervised by Dr. Meave
Leakey, who was the head of paleontology, and Dr. Richard Leakey [chap. 54]
was the chief executive of NMK. My arrival at the museum coincided with a
period of great debates and varying opinions regarding early human evolution, particularly issues touching on the taxonomy and mode of evolution.
Also during this time, the paleontology department was frequented by
renown paleoanthropologists, paleontologists, archeologists, and geologists
with some famous names, such as Dr Alan Walker, Prof. Bernard Wood
[chap. 46], Dr, Michael Day, Dr. Glynn Isaac, Prof. Jack Harris, Dr. Tim

White [chap. 45], Dr. Peter Andrews [chap. 2], Dr. Pat Shipman, Dr. Kay Behremeyer [chap. 22], and by numerous students who are distinguished scholars in their respective disciplines today. These international affiliates came to Nairobi yearly to study the many early human fossils housed at NMK. Whenever there was a congregation of scholars in Nairobi, special seminars with guest speakers were organized by the department on various topics on human evolution. It was the scientific discussions and interactions with visiting scholars that raised my interest and fascination with paleoanthropology, and at that moment I knew I was in the right field.

AMAZING FACT

I am amazed and fascinated by the fact early human groups had the ability to adapt and survive for many years despite many environmental and predator bottlenecks. From the many years of human evolution studies, there is now growing evidence that at the time of divergence from the putative common ancestor (about 6 million years ago) our ancestors could have occupied forested woodlands, probably possessed the ability to climb trees, and also walked upright on two legs intermittently when on the ground. The forested environments gave way to open savanna grasslands in the Middle Pliocene in eastern Africa, and emerging new threats to early humans necessitated new strategies for survival (e.g., habitual upright walking and freeing of the hands for other chores). Some side branches on the human evolutionary tree died, but our own lineage (*Homo* sp.) was able to adapt to the grassland environment, devising new strategies to enable their survival (e.g., embracing culture items such as stone tools) in mitigation of environmental forces.

TIME TRAVEL

If I had a one-shot roundtrip in a time machine, I would travel to the Pliocene period (3.8 to 3.0 million years ago) in Africa when the forests were giving way to the savanna. This was an exciting time, and I could see the great diversification of early hominins into various species. Research suggests the existence of a variety of early hominin species that probably overlapped in eastern Africa within a similar environment in the Pliocene. *Australopithecus afarensis* is widely known in the Pliocene and is currently known from sites in Ethiopia, Kenya, and Tanzania between 3.8

and 3.3 million years ago.[1] Several other species of Pliocene hominins are known in eastern Africa: *Kenyanthropus platyops* from west Turkana in Kenya[2] and *Australopithecus deyiremeda* from Waronso-Mille.[3] *Australopithecus bahrelghazali*,[4] another Pliocene hominin, is known from central Africa in Chad, about 2,500 kilometers west of the Great Valley, demonstrating a wider range for the Pliocene hominins. Although the species of Pliocene hominins that gave rise to early *Homo* is not yet clear, the Pliocene is an interesting time period that may provide crucial information about early human adaptive strategies in changing environments.

FUTURE EVOLUTION

 Environmental conditions and cultural innovations will probably be the drivers for change in our species 100,000 years in the future. The current increase in greenhouse gases associated with industrialization (continued use of fossil fuel and other pollutants to run our industries) is having a negative impact on conditions for species survival, including our own. It is imperative to minimize the greenhouse effect to protect all animal species in the future. In addition, culture and innovation, when applied in consideration of environmental preservation, could play a major role in mitigating adverse effects on survival of our species. Our ancestors, the early *Homo*, embraced cultural innovations, particularly making stone tools about 2.8 million years ago, and this culture diversified and evolved into many varieties during the emergence of modern humans in later prehistoric time. Stone tools were used as weapons and for cutting, pounding, and even digging for edible plants, enabling the survival of our ancestors. The population of the world has greatly increased, and we will require abundant technological innovations and resources for survival. As the nations of the world clamor for development through diverse technologies, it is imperative to entrench laws and guidelines to protect the environment to ensure our species' future survival. Ultimately, our body shape and size will depend on how our species treats the environment.

RELIGION AND SPIRITUALITY

The acrimony that arises whenever human evolution and religious beliefs are discussed together is unnecessary. Understanding these two concepts requires that we exercise patience and tolerance.

Religious beliefs advocate for the creation of living things through the divine hand, whereas human evolution promotes a biological process driven by natural selection, as well as cultural influences in the later stages of this evolution. Being a scientist in the field and having grown up in Kenya where almost every family attends church and reads the Bible, I am fascinated by human evolution studies and also tolerant of religious beliefs. Human evolution is supported by a wealth of evidence in the form of fossilized remains and scientific facts emanating from the fossils housed in African museums, including in my own in Nairobi. I have been surrounded by a wealth of scientific information on human evolution, but I have learned the teachings of Christianity, which mostly mold spiritual being and character. As a result, I see biological change through natural selection and other biological factors as main drivers for evolution, but I am open-minded to divine interventions in this process. If nature alone is responsible for life, there probably would be many mistakes in life forms.

INSPIRING PEOPLE

 The late Dr. Alan Walker!

36

ROBYN PICKERING

Robyn Pickering is a senior lecturer in the Department of Geological Sciences and at the Human Evolution Research Institute in the faculty of science at the University of Cape Town, in Cape Town, South Africa. As a geologist, she seeks to understand where and when early human ancestors evolved. In the past decade, Robyn has developed novel technical approaches to gathering key information from rocks associated with the fossil remains of early members of the human lineage (especially at sites in the "Cradle of Humankind," South Africa). She also studies flowstones, stalagmites, calcretes, and tufas to extract information about the environmental conditions during the time of our distant relatives.

YOUR BEGINNINGS

I have been fascinating with rocks my whole life. As a little girl, I have vivid memories of rambling around the Melville koppies ["koppie" is a small rocky hill, an Afrikaans word used ubiquitously in South Africa] in Johannesburg with my father when I was four or five years old, marveling at the beautifully preserved ripple rocks and pestering him to explain how they were formed—again and again. I loved biology at high school, and it was something of a revelation to me to discover fossils, and human fossils in particular, which I thought were the ultimate combination of biology and geology. I poured over Don Johanson and Blake Edgar's book *From Lucy to Language*[1] and dared to dream about being a paleoanthropologist myself. I then read all of Richard Leakey's [chap. 54] books and enrolled at the University of the Witwatersrand to do a general science degree in geology and archaeology. However, it was not until I had almost finished my

PhD that I really committed to the idea of being a researcher—despite these very strong, early inclinations!

GAME CHANGER

 This is a little obscure, but stay with me: in 1998 Richards et al.[2] wrote a landmark paper about using the uranium-lead technique to date speleothems (cave carbonates). This paper made me want to do a PhD, my dream PhD, looking again at the stratigraphy of Sterkfontein, South Africa, and opening the possibility, for the first time, of directly dating these deposits with a well-established, reliable method. Having a chronology for the South African caves that dates the deposits themselves, and that does not rely on faunal comparisons with east Africa, is something of a game changer in itself.

The South African hominin sites have a very different geological setting from those in eastern Africa—caves vs. river and lake sediments exposed through the tectonic activity of the East African Rift system. We don't have active volcanoes in South Africa, and without the interbedded volcanic tuff layers so famously used to date the hominin deposits in East Africa, the South African deposits were seen as being undatable. I was taught this as an undergraduate at Wits [University of the Witwatersrand, Johannesburg]. U-Pb [uranium-lead] dating is a well-established chronometer, but it is usually applied to small resistant minerals, such as zirchon. Cave carbonates, usually in the form of stalagmites, are also well established as archives of paleoclimate information. The Richards's paper was one of the first intersections of these two fields and served as an enormous inspiration to me. I did my PhD research on continuing to develop the U-Pb technique so it would be applicable to the South African cave deposits, which are particularly low in uranium. I have not changed research direction very much since those early PhD days, and finally, after over a decade of work, in 2019 we published a comprehensive, precise, and direct chronology for the South African cave sites and their associated hominin and other fauna based on U-Pb dating of the cave carbonates.[3]

AMAZING FACT

I really love hominin fossils. I find them incredible and feel in such awe of the distance in time between when these ancestors of ours were alive and how remarkable it is that this one, this individual, was preserved, fossilized, and is sitting in front of me now. The isotopic studies on dietary reconstructions are probably my favorites. I find it amazing that the isotopes (carbon and strontium) preserved in hominin teeth can expose these otherwise secret insights into how these individuals and groups lived, what they ate, where they grew up, how long the babies breastfed, and so forth.

TIME TRAVEL

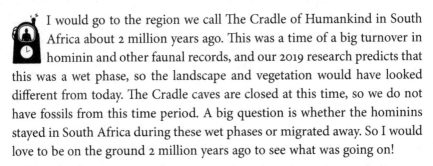 I would go to the region we call The Cradle of Humankind in South Africa about 2 million years ago. This was a time of a big turnover in hominin and other faunal records, and our 2019 research predicts that this was a wet phase, so the landscape and vegetation would have looked different from today. The Cradle caves are closed at this time, so we do not have fossils from this time period. A big question is whether the hominins stayed in South Africa during these wet phases or migrated away. So I would love to be on the ground 2 million years ago to see what was going on!

EVOLUTIONARY LESSONS

The more we discover about the human fossil record, the louder and clearer the story of diversity stands out. As an undergraduate student, I was taught that human evolution was a very linear this species replacing this species story. Today we recognize that a lot of diversity and even interbreeding went on in our human past. This can only be helpful in informing and framing current inclusivity and diversity discussions that are taking place not only in society at large but within academia itself.

ADVICE

This is a competitive field, with a lot of big personalities; it is not easy to slice out a piece of the pie for yourself. Find something you are truly fascinated by and stick with it; that is my best advice based on my own experience. Do not be afraid to try something new either. I was directly told that my PhD was pointless, that we already knew how old the South African caves were. This was not pleasant to hear, but if anything, it made me more determined to continue with my work and to be successful. So be brave, be bold, and be persistent.

INSPIRING PEOPLE

I feel loyal to the South Africans here, and although he was born in Scotland, I would love to sit down to tea with Robert Broom and chat about his single-minded quest to find early human ancestors. I have had the great privilege of many cups of tea with the wonderful, generous, and very funny Bob Brain; he'd be delighted to discuss these questions, and I'd love to hear his answers. I also greatly admire Mary Leakey, an exceptional scientist but also a woman and a mother in this field, so she is on my list too.

37

J. MICHAEL "MIKE" PLAVCAN

J. Michael "Mike" Plavcan is the vice-chair of the Department of Anthropology and the graduate studies director at the University of Arkansas. He is also a fellow of the American Association for the Advancement of Science. Mike is an expert in primate sexual dimorphism and associated sociosexual behavior, and in implementing 3D digitizing and analytical methods to study living and fossil primate morphologies. Mike is also engaged in fieldwork in Kenya as a principal member of the West Turkana Paleontology Project, working with Fredrick Kyalo Manthi [chap. 33] and Carol Ward [chap. 44]. This project is crucial to understanding the origins of the australopiths as well as modern humans.

YOUR BEGINNINGS

I was curious about nature, and I grew up playing in nature. I remember collecting butterflies in a summer field, prying rocks from a clear cold stream by a well house to look for crawfish, photographing birds, studying books on nature, and in high school reading Stephen J. Gould's columns and books. I was fascinated by paleontology, and in college I got hooked on comparative anatomy. Here's how my career in paleontology and primatology began. I was in a comparative anatomy class, and one afternoon when we were in the lab I heard Dr. John Lundburgh talking to a student next to me about paleontology. The student wasn't particularly interested, but I said to him, "I'm interested in paleontology." He looked at me and asked what sort, and I told him I liked primate evolution. He told me to call Rich Kay in the Anatomy Department, and he gave me his number. As soon as the lab was over, I nervously dialed Rich's number. When he answered the phone, I introduced myself, told him that Dr. Lundburgh had told me to call him, and he invited me over. And at that point my career began.

GAME CHANGER

If I have to point to a single discovery, it would be the Taung child.[1] The history of paleoanthropology, even today, is one in which the fossil record dictates the evolution of models and ideas about human evolution itself. Most recently genetic discoveries have led the community to rethink our ideas about the course and nature of human evolution. Such data were truly unexpected and have led to a tremendous excitement at the new courses and understanding we have about human evolution. Of course, new fossils, such as *Paranthropus boisei*, *Homo floresiensis*, and the Denisovan material, also force us to rethink human evolution. Even so, I think the biggest game changer of them all was Taung.

The big questions in human evolution have always been and continue to be driven by the obvious things that set us apart from other species—culture, big brains, short faces, and walking on two legs. We are arguably a narcissistic species and certainly are most proud of our big brains and culture. It is no surprise that these obvious human traits we value so highly were seen as the fundamental targets of human evolution. It was thought that early humans would develop large brains and culture first, and the evolution of upright walking and changes to the face would follow. This was the concept that underpinned the acceptance of Piltdown—after all, the key to success of a good con is that the "mark" is ready and willing to believe a story.

Taung flipped our entire understanding of human evolution. When finally accepted, it forced paleoanthropologists to accept that hominins first were defined by changes in locomotion and diet, and that the evolution of large brains and culture only came later. No discovery since has so fundamentally changed the paradigm of paleoanthropology. As for influencing my own career, the discovery provides an enormously entertaining way to introduce students to the nature of hypothesis testing and model building. It also reminds me periodically that when data falsify our models, no matter how cherished, we need to drop them and move on.

AMAZING FACT

 Nothing challenges our sense of self like the knowledge that multiple species of *Homo* coexisted. It is an easy statement of fact to make, but it is a stunning statement about human evolution that cuts to the core of our place in nature. We see ourselves as unique in the world. Our religions place humanity at the center of attention of God, and for most people humanity stands not just as the pinnacle of the natural world but separate from and above nature. We see ourselves as beyond nature: we have stepped outside of a "niche" and are not just invulnerable to the vicissitudes and forces of nature but actually control (and destroy) nature. Most people believe in some sort of soul—a metaphysical expression of self that is not formed to fit into nature but transcends the physical world itself. Given all of that, how could there be another species of human? What would it be like to encounter another animal that is like us, yet different? Would they think the same, act the same, share our beliefs? Clearly they must have been different, but if living today, how would people view their metaphysical state? I suspect people would act toward another species as they act toward outside groups today—with hostility and denigration (dehumanization of people we don't like is one of the fundamental tools used to define "us versus them"). Regardless of how we might imagine such a state, the fact of coexistence with other humans cuts to the core of the definition of what it means to be human. It reminds me that the nature of modern humans—physical, mental, emotional, and even spiritual—may be unique today, but it is not the only outcome in human evolution. This fact should be the last blow to any vestiges of the scala naturae that we hold.

TIME TRAVEL

About 6 million years ago because that is when bipedalism originated. I would love to see how our ancestors walked before they were committed bipeds.

DRIVING FACTORS

That is the biggest question in paleoanthropology. We really have no idea, but it is becoming apparent that, like the rest of evolution, small changes started our ancestors down a path that diverged and created something truly weird and different.

FUTURE EVOLUTION

Humans will largely look the same as we do now physically, but no one really knows. Certainly in a hundred years nothing will be perceptibly different. In a million years—who knows? The fossil record tells us that species can change rapidly, but sometimes they stay pretty much the same for millions of years. Perhaps a better question is whether humans will be here at all in a million years. The fossil record also teaches us that everything becomes extinct, but somehow we have trouble accepting that fact. Much like we all know that we will die, but it is hard to wrap our head around that reality. The extinction of humanity runs against the grain of our self-congratulatory notion that we are above nature.

EVOLUTIONARY LESSONS

Human evolutionary studies offer a dose of sobering reality to our opinion of ourselves. I am amused when people denigrate studies of human evolution as being useless information. Nothing creates more controversy than the statement that humans evolved from apes. If evolution occurred for all organisms except humans, it would not garner the hostility and attention that it gets in the public domain. The moment you say that humans evolve, people react because it places humanity squarely within the natural world. Human evolution tells us that we are animals and that the forces of nature have shaped us.

I sometimes tell students this: If you strip yourself naked and stand before the mirror and think about your natural behavior (the urge to eat, sleep, mate, form pair-bonds, compete, and so on), you look and act like an animal because you are one. This statement elicits a reaction from some students who are shocked and refuse to believe that we are "just animals." But we are animals, and understanding that creates a conflict between the idea that we are self-directed and our free will is the determinant of whether we are gay or law-abiding or kind or angry versus the idea that much of our behavior is biologically influenced. We hear about that conflict every day on the news, and it is a major concern for people in their daily lives. People really do care about human evolution, and that fact is shown by the effort people exert in trying to prove that human evolution is wrong. The fact that we changed over time to become what and who we are forces us to accept our place in nature.

SPECIAL?

When compared to other animals, of course we are special. We have changed the world and are the only species that can appreciate that impact on nature and then knowingly correct it. One of the most important things about studying other animals is to define what makes us unique versus what does not. We can only understand what makes us unique by understanding that which does not.

RELIGION AND SPIRITUALITY

You cannot falsify God, so saying that science disproves religion is silly and false. But to those whose faith relies on a dogma of special creation, evolution is lethal. To others of faith, the study of human evolution is merely the revelation of how God created us. The "debate" about human evolution, especially in the United States, is a great illustration of this. Human evolution clearly falsifies a literal interpretation of Genesis. For many people, this is a deal killer, creating a clear black-and-white dichotomy of truth—either humans evolved and there is no God or the biblical account is true and humans did not evolve regardless of any evidence people claim proves otherwise. The interesting part is why this dichotomy occurs at all. All Christian denominations accept the fundamental precept of atonement. This is the idea that Christ died for our sins, and the basis of the concept of sin is that humans are fundamentally sinful as part of our nature. In the literal interpretation of the Bible, our sinful nature is the product of the fall, when Adam and Eve ate the forbidden fruit. In this case, the actual act of Adam and Eve is fundamental to the nature of Christ—no forbidden fruit, no sin, and therefore no reason for Christ. If evolution is true, then the biblical account of creation and the origin of sin is not true, and therefore the fundamental tenet of Christianity is not true. If the creation story is not literally true, the fear is that the rest of the Bible might not be true, then faith becomes a matter of personal interpretation, and the Word of God is nothing more than a compilation of human stories. If you believe in that view of Christianity, evolution is fundamentally incompatible with the way you see the world. Therefore, it cannot be true, even if a bunch of scientists say the evidence proves the biblical story is false.

Another way of looking at this is to see truth in the Bible as something that transcends the text. The easiest example of this is the parables in the

Gospels. Jesus taught most effectively with stories. The stories of the Good Samaritan or the Prodigal Son are not literally true. Indeed, whether they are true or not is irrelevant. Instead, these stories convey a truth—in the first case, that God judges us based on our actions and not on our ethnicity or religion; and in the second case, that God forgives and welcomes sinners to heaven just as He welcomes the faithful. By extrapolation, the creation story then carries a higher truth: Genesis 1 is a statement that God is the creator, and Genesis 2 is a statement that humans have a sinful nature and rebel against the will of God. If this is true, then evolution is irrelevant and stands only as the means of creation.

ADVICE

 Form you own opinions based on evidence, and remember that you make mistakes just like the rest of us. Don't cherish your ideas to the point where you are not willing to let go of them.

INSPIRING PEOPLE

 Aristotle.

38

KAYE REED

Kaye Reed is a president's professor at Arizona State University's School of Human Evolution and Social Change and a research associate at its Institute of Human Origins. Kaye analyzes the mammalian fauna from Plio-Pleistocene localities in Africa to understand changes through time in the community, the structure, and the paleohabitats of extinct members of the human lineage. Kaye's current main field project focuses on the disappearance of *Australopithecus afarensis* and the appearance of the genus *Homo* in the lower Awash Valley in the Afar region of Ethiopia. At the Ledi-Geraru site (about 2.8 million years ago), her team has identified the oldest fossil of the human genus.[1]

YOUR BEGINNINGS

I wasn't one of those people who knew what I wanted to do at age six. In fact, I had returned to college to get a bachelor's degree at the age of thirty-four, majoring in English—because, what else? I had to take three social sciences classes as a requirement to graduate, and in the quarter system at Portland State University that was a full year. I chose anthropology. After I completed the Introduction to Physical Anthropology, I decided to change my major to anthropology—but definitely the biological aspect. I had a great mentor in Marc Feldesman, when I picked up an assignment in his office he asked me, "What are you doing with your life?" I said, without thinking, "I am going to graduate school." And he said, "Sit down; let's talk." I had never thought about doing research as a career until that moment. Marc suggested I apply to Stony Brook, which I did, and I have been researching mammal communities ever since—past and present. And whether the focus is on modern primates or fossil hominins, I love every minute of it.

TIME TRAVEL

 I would go to the Plio-Pleistocene transition (roughly) because I want to see what *Paranthropus* species were eating, how they lived, and their preferred habitat. I would want to contrast this with the earliest *Homo* species, so if I have only one shot, I would have to already know a place where both species were sharing the landscape! But I also want to see all of the other species that shared the landscape with these hominins to see if our reconstructions are accurate. I might have a hard time coming back!

DRIVING FACTORS

A lot of people think that bipedalism is the key. I can't think of a good reason to become bipedal, except *maybe* walking between resource patches if one started out as a suspensory ape. Is that enough selective pressure? We ignore the hand in early *Australopithecus* species—it does differ from apes—and I wonder what its purpose was at that time. What action or adaptation required a better manipulator than continuing on with the hand of the ancestral ape? I think a change in the function of the hand through time that resulted in the hand of *Homo* would have been a major factor on that road.

EVOLUTIONARY LESSONS

Two things are important. First, in *Homo sapiens* today we can trace the origination of many structures in our body to other organisms from the past. We no longer use these structures in their original form, but we can see evolution working through millions of years—from a notochord to us, for example. When I teach evolution this way, students grasp how a long slow process could create many new species, including humans. I think this helps people understand a difficult concept. The second idea that will inform the future is how the interplay of climate and our lineage can help predict what might happen if habitats become warmer, drier, etc. We can see this in the extinction (probably) of *Australopithecus afarensis*—this species reached some critical threshold and could no longer survive in its current form. This may or may not happen to us; we need to listen to science about climate change.

SPECIAL?

Yes, humans are special. As my colleague Kim Hill likes to say, no other animal has been able to get to the moon and all that implies about learning and technological advancement. That's on one level. On another level, we are just the current end point in the long history of evolution—that is, just another mammal. Studying other primates (and birds, whales, etc.) shows us that some of the ideas we have about what makes us unique may not be accurate. Certainly, studying apes (who have also survived through 5 million years or so of evolution from a common ancestor) is helpful for explaining the path we might have taken to get where we are as far as having some cognitive abilities and cooperating with others. In the long run, I don't think studying primates can elucidate anything other than that we are unique. We would get closer to understanding when and how we became unique by studying the fossil record of our lineage—or developing a time machine!

39

BRIGITTE SENUT

B rigitte Senut is a professor at the French National Museum of Natural History (Paris, France). She studies fossil mammals and the origins of modern great apes and humans from their Miocene ancestors. Since the 1980s, Brigitte has conducted paleontological surveys in Kenya, Uganda, South Africa, and less explored parts of Africa, such as Angola, Botswana, and Namibia. Together with Martin Pickford [chap. 18] and Glenn Conroy [chap. 8], Brigitte discovered the first fossil ape in subequatorial (and western) Africa (*Otavipithecus namibiensis*; Namibia, about 13 million years ago).[1] She is also the codiscoverer (with Pickford and their team) of *Orrorin tugenensis* (Kenya; about 6 million years ago), one of the earliest members of the human lineage.[2]

YOUR BEGINNINGS

It was not really a moment but more a succession of events in my life. When I was a little girl, I remember collecting small fossil shells in my parents' garden. They were the witnesses of a very old sea that covered the Paris Basin millions of years ago. I have always been interested in nature, and rocks in particular, and I was fascinated by volcanoes, caves, and mountains. In those days, these fields were more for men than for women, and I considered them more as a hobby than a job. However, I joined a scientific club for teenagers because I wanted to know more about these past worlds. During my last years in high school, I was taught by several professors in geology and natural sciences who were passionate and transmitted their passion to students. At the same time, major discoveries were being made in the Omo Valley in Ethiopia by Arambourg and Coppens and their team, and I found that captivating. When I registered in the University of Sciences (Pierre and Marie Curie University) in Paris, it was evident that it would be in geology (paleontology being part of geology in those days). This is where I met a fantastic professor in general geology and sedimentology, Prof. C. Pomerol. He was a true naturalist who taught us to read landscapes, to distinguish the particular metallic sound of phonolites under the blows of the hammer, and to sniff the fossil! Frequent field trips provided us with live training, and we could experience the geological element in its grandeur, a contact that definitively gives meaning to the life of the field geologist. I was enjoying the paleontology courses but was aware of a lot of misogyny from some lecturers or professors, so I didn't think I would be able to have a career in paleontology. However, even if it was just a hobby, it was clearly a field in which I wanted to be involved. To make sure I could make a living, I registered in applied geology and geochemistry together with paleontology. During my postgraduate studies, Prof. Y. Coppens gave a lecture on australopithecines (it was 1976, just after the discovery of Lucy). During his conference, he described the skulls, teeth, and tools but did not say too much about the rest of the skeleton other than to announce that these early hominids were bipedal creatures. In those days, few scholars were interested in locomotion. If the fossil hominid was a biped, the type of bipedality was not really investigated. I was puzzled. I went to see him, and after a long meeting in his office at the Musée de l'Homme in Paris, he agreed to supervise my master's in paleontology on the upper limb of early hominids. This was the beginning of long years of cooperation! He was one of the few professors who did not consider gender to be a problem, and he gave me the opportunity to realize my dreams: studying the past in the lab but, even more, doing fieldwork, which is the most exciting part of the job and also the most challenging part. You can misinterpret a section or a fossil, but the field never lies to you: it yields all its value to you.

GAME CHANGER

The discovery of Lucy in itself did not affect my way of looking at our own evolution, but it changed the way I practiced science. Lucy's discovery in 1974 in Ethiopia was a major breakthrough.[3] It really influenced my career, and I was happy to study it for my PhD. The results I obtained concerning systematics and locomotion were new and challenged the ideas of some leaders of paleoanthropology. I was highly criticized by some of them! But a special moment is still clear in my mind. I was at the meeting "Old Bones: The Hard Evidence," held at the American Museum of Natural History in New York in 1983. I was invited to present the results of Coppens's team on *Australopithecus afarensis*. This meeting was a major event in the world of human paleontology, and it was possible to access a lot of original specimens from diverse parts of the world concerning our heritage. It was really impressive. Most leaders of the discipline were there, and it was an important test for me as a young female researcher. At the end of the session, during the questions, one colleague stood up, pointed at me, and shouted: "You are a heretic!" There were no comments, it was just a statement. I was quite perturbed, but as we left for the lunch break, I was approached by Prof. Charles Oxnard who said: "Brigitte, please join the club!"

For some scholars, science is like a religion: you must agree with the majority. But science is not a vote; it is a matter of discussions, of sharing evidence. If 90 percent of a scientific audience supports an idea, it doesn't mean that the remaining 10 percent are stupid! Heresy has no place in science, and since then I have refused to join teams where lobbying was more important than thinking and debating. It is unfortunate that this has become more and more frequent in the evaluation of scientific papers. If your ideas don't support the results of the reviewers (or of the editorial committee), in most cases it is clear that the article won't be published! This creates a club in which there is no debate, and in the end science becomes uniform and does not take into account the diversity of thought. This will lead to the impoverishment and possibly even the destruction of science. I tell my students that they have to think for themselves, and if they can properly support their ideas, they have to follow them. This is what we call freedom of thought.

Coming back to my own field experience, a major advance was made when scientists realized that the Miocene epoch was key to reconstructing our past. To understand how early hominids emerged (in terms of anatomy, environment, etc.), it was evident that we needed to search in older stratigraphic levels, and Miocene was the answer. I remember a colleague telling me in the early 1990s that it was stupid and a waste of time to search in the Miocene, it was too old. It was clear that our ancestors lived in the Pliocene and that hominoid evolution took place in East Africa. Soon we found the first subequatorial hominoid, *Otavipithecus namibiensis* from the Middle Miocene of Namibia, and then I was lucky to find a hominoid upper half molar in the Lower Miocene at Ryskop in South Africa.[4] These discoveries not only indicated that hominoids had a Pan-African distribution, but also opened a much wider area for fieldwork and a much broader view of their diversity. The discovery of *Orrorin* in the Upper Miocene of Kenya was a major break-through. For the first time in our history, we had found a biped 6 million years old, much older than the early to Middle Pliocene dichotomy between apes and humans predicted by most of the international establishment! As a result, more fieldwork is done now in the Miocene of the Old World.

AMAZING FACT

Not only the adaptability to such varied environments on Earth, but also the ability to cross borders and to dream that one day we will be able to populate other planets in the universe. The capacity to push the boundaries of adaptations and adaptability.

TIME TRAVEL

Definitely the Neogene to test my results on the hominoids: Were the reconstructions of the paleobiodiversity, paleoenvironments, and the biology of our ancestors at large close to what we described? It would be a bonus to travel with stopovers in the Miocene and Pliocene epochs to see how our ancestors were shaped and migrated through time.

DRIVING FACTORS

This is a difficult and complex question. I am inclined to think that the adaptability of members of our lineages was a benefit for their expansion. They were probably less restricted and more opportunistic.

FUTURE EVOLUTION

At the moment this is more fiction than reality. Basic direct social contact may be erased by machine-connected communications. We may still be here in a hundred years but be so connected that we lose our feelings and spend too much time in a virtual world as our neighboring environment slowly but surely disappears. The image that comes to my mind when talking about our future is the book *Brave New*

World by Aldous Huxley.[5] Do you really think our species will still be here in 100,000 years (maybe) or 1 million years (I doubt it)? The genus *Homo* has survived almost 4 million years, which is not too bad for a species. Considering the demographic issues on our planet, is it time to give up our place to another taxon?

EVOLUTIONARY LESSONS

There are several aspects to this question. The study of the past is important for human societies, in a cultural sense, because it roots us all in deep time. You know where you belong as a living being; this is also what roots you in your society and makes you what you are as a human being and as part of a group. It stabilizes the group. This is a never-ending quest, and I think it is probably one of the major aspects of our cognition. Culture is important in a society: a society without culture is doomed to extinction. Our future is definitely based on our past.

Another aspect is that studying the past teaches us that nothing can be taken for granted in life. The survival of our species is directly linked to natural forces on the planet. Understanding the past, the environments, the Earth's pulses, and the distribution of fossil species should make us more respectful of nature and give us the knowledge to survive the impact of natural events. We can learn from the past.

SPECIAL?

This question is not the right one to ask. Humans are not special, they are different. All animals have evolved through time over longer or shorter periods, being shaped through time with similarities and differences. The biodiversity of today expresses all these variations accumulated or lost through geological time. Compared to other animals, humans have developed features that are more evolved than those in other mammals, but other mammals have their own specific traits. Referring to primates, or more specifically to hominoids, it is clear that humans have developed a form of bipedalism. But we are far from being experts in knuckle-walking as practiced by chimpanzees and gorillas, for example. In both contexts, these locomotion styles are derived and are special to each species. Genetics teaches us that we share almost 99 percent of coding genes with chimpanzees, but this means nothing. The main difference is in the last

1 percent, which means that chimpanzees are not humans and humans are not chimpanzees.

RELIGION AND SPIRITUALITY

 Human evolution is a scientific issue, it is based on facts; religion is part of our cultural package. There is no link between the two.

ADVICE

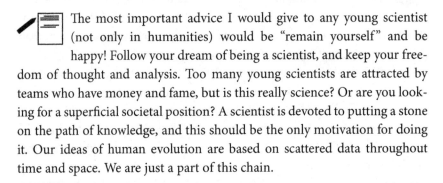 The most important advice I would give to any young scientist (not only in humanities) would be "remain yourself" and be happy! Follow your dream of being a scientist, and keep your freedom of thought and analysis. Too many young scientists are attracted by teams who have money and fame, but is this really science? Or are you looking for a superficial societal position? A scientist is devoted to putting a stone on the path of knowledge, and this should be the only motivation for doing it. Our ideas of human evolution are based on scattered data throughout time and space. We are just a part of this chain.

INSPIRING PEOPLE

 Jean-Baptiste de Lamarck.

40

RICHARD "RICH" SMITH

Richard "Rich" Smith is a professor of biological anthropology and the Ralph E. Morrow Distinguished University Professor at Washington University in St. Louis. Rich is interested in methodological issues associated with making inferences concerning human origins and the biology of human ancestors; in particular, how and why inferential biological judgments and the essential consideration of uncertainty are often reduced to "objective" statistical statements. In this work, he applies insights from several disciplines, including cognitive research on heuristics and biases, confirmation bias and wishful thinking, along with discussions from the philosophy of science on issues such as causation, uniqueness, historical explanation, underdetermination, and contingency.

GAME CHANGER

I think the game changer for paleoanthropology, and for me, began with the great works of the modern synthesis: Dobzhansky's *Genetics and the Origin of Species*,[1] Mayr's *Systematics and the Origin of Species*,[2] and Simpson's *Tempo and Mode in Evolution*.[3] It then took others, most notably Sherry Washburn, to explain how to do human evolutionary biology within this framework.[4]

My first encounter with biological anthropology was the opportunity to work with Howard Bailit, who did his PhD with W. W. Howells. He had possession of the dental cast collection from the Harvard Solomon Islands Project[5] and was examining patterns of geographic microdifferentiation. When I approached Howard about working with him, I was at the University of Connecticut mostly doing other things, and my background was entirely clinical. I had never had a course in evolutionary biology or biological anthropology. He gave me those three books to read and told me to come back when I had finished them. They opened the door to a world I did not know, and three years in his lab put me on track for a career in paleoanthropology.

AMAZING FACT

From what I understand about the evolutionary patterns of large, long-lived, mobile terrestrial mammals, the pace of change in the hominin lineage over the past 4 million years is remarkable and is unlike that of other mammals from similar times and environments. It is difficult to understand the mechanisms for these changes. I tend to think of changes within a lineage as mostly associated with microevolutionary fine-tuning, and large changes as mostly associated with speciation events. Although the identified species in the hominin lineage are probably overly split, there nevertheless seems to have been an extraordinary number of speciation events in a lineage in which one would anticipate significant gene flow. The number of ways modern humans are different from our direct ancestor only 4 million years ago perhaps qualifies as amazing. At the other end of the spectrum, why so many extinction events? These complex species were not merely steps on the way to us. They were successful species adapted to their own environments. Why did they, over such a wide geographical range and with a diversity of adaptations, all disappear?

TIME TRAVEL

My interest in hominins is very much associated with the species that are different from modern humans and great apes. I am intrigued by the difficulty of understanding animals for which there are no modern analogues, and I am mostly a methodologist, interested in understanding how to make scientifically sound inferences about the past. Once hominin ancestors are variations on a large-brained striding biped, the difficulties, and therefore my interest, are reduced. I would want to spend time between 2.5 and 2.0 million years ago, when this transition from decidedly non-human like to more humanlike appears to have taken place.

DRIVING FACTORS

I don't think we can know, and I don't think we are doing science when we ask this question. This was a unique sequence of historical events. It is not subject to experimentation, and it is exactly what the term "contingent" means: a complex set of interactions in a highly path-dependent process. Occasionally, a story we tell about this can be shown to have false components. More often, equally compatible stories can proliferate independently of each other, and ad hoc additions can adjust to any new evidence.

FUTURE EVOLUTION

For the short term, 100 years from now, I can't see how much change will be detectable, if it has occurred. The human population is too large, too mobile, inhabiting too many environments, and subject to too much technological intervention for there to be any predictable effects of natural selection. As for 100,000 years, I'm pessimistic about life on Earth in general. We tend to think of climate change in terms of higher sea levels, more frequent extreme weather events, and the decimation of species diversity. This is climate change as a linear effect. Humans are poor at understanding nonlinear effects—multiplicative (2-4-8-16-32) or even exponential—and I have little doubt that these are occurring now for processes that lead to threshold effects: no change is apparent until the process literally falls off a cliff. The climate change taking place today because of human activities is not part of the normal cycle of climate changes that have gone on for billions of years. The bodies nearest to us that are closer to the Sun, further from the Sun, and the same distance from the Sun—Venus, Mars, and our moon—do not have atmospheres that sustain life. The delicate balances that result in our atmosphere are fragile. We are destroying it. I would be surprised if 100,000 years from now Earth has an atmosphere that sustains life.

EVOLUTIONARY LESSONS

In teaching human evolution to large lecture classes for over sixteen years, my greatest hope was that I was giving students an appreciation for the continuity of life on Earth in both time and space. We are continuous with other living species and with our species in the past and in the future; we occupy but a moment in time. Human evolutionary studies give us a perspective on our place in the world and on our place in the history of life.

SPECIAL?

We are special in the sense that elephants are special, giraffes are special, hummingbirds are special, and aye-ayes are special. We are special in that we do not follow a generalized pattern that can also be observed in a number of closely related species. Otherwise, we are special only in thinking of ourselves as special.

RELIGION AND SPIRITUALITY

From the religious point of view, if one allows for the idea that evolution is God's mechanism for creation, then one can believe in some form of a deity and accept the facts of evolution. Other religious beliefs—heaven, hell, the Earth is a few thousand years old, or there is someone up there who hears and acts on our prayers—are entirely incompatible with the facts of human evolution.

Should a scientist acknowledge that there are things about the universe that are beyond the reach of human perception, human understanding, and scientific investigation? I think so. Our senses and our brains are limited. I am confident that there are things I do not understand and cannot know. Religion, as commonly understood, is not part of my life, but neither is a dogmatic insistence that there is nothing more because we cannot conceive of it. That position would be entirely unscientific because it is based on belief rather than on evidence. The fact that religion is universal in human societies is an interesting scientific question for cognitive psychology. I found Pascal Boyer's *Religion Explained: The Evolutionary Origins of Religious Thought*[6] particularly interesting on this topic.

INSPIRING PEOPLE

 I would like to bring W. E. Le Gros Clark, William Gregory, Adolph Schultz, and Sherry Washburn into the present time and invite them to dinner to discuss what they think of the story of human evolution today.

41

DAVID STRAIT

David Strait is a professor in the Department of Anthropology at Washington University in St. Louis. He is interested in understanding how and why the various species of early hominins diversified. Much of his work relates to the study of hominin taxonomy and evolutionary phylogenetics (cladistics). He is also interested in diet and feeding evolution in fossil humans using engineering, experimental, comparative, and ecological methods. David complements his lab research with paleoanthropological fieldwork to recover new fossils.

YOUR BEGINNINGS

I can identify with clarity the moment I decided to become a paleoanthropologist. As background, both of my parents were college professors, so I went to college thinking that I might find something to study and then become a professor myself. I had been good in math and science in high school but wound up dabbling in philosophy my freshman year. By the beginning of my sophomore year, it was clear that I had no aptitude in philosophy, and I was casting around for something else to study. My roommate was an anthropology major (he is now a vertebrate paleontologist), so I decided to take the introductory anthropology class as well as a human origins class with him. I vividly remember the first lab that I attended. A number of casts were laid out on the table, and I picked Sts 5 ("Mrs. Ples"). I thought to myself, "This is cool." I had grown up in New York City and always enjoyed seeing the dinosaurs at the American Museum of Natural History, but I had never considered a career studying evolution. Shortly thereafter I arranged to meet with the anthropology graduate student who was a resident in my dormitory, Dan Lieberman [chap. 55], now a paleoanthropologist himself. I told him that I thought I wanted to be a paleoanthropologist, but I wanted to try fieldwork before I made my decision. He arranged for me to meet with his advisor (Ofer Bar-Yosef), who in turn arranged for me to participate on a paleolithic archaeology excavation that summer in France. That trip confirmed that I wanted to study human evolution—and I met my wife in a bar in Paris. . . .

GAME CHANGER

Historically the obvious answer is the discovery of the Taung child and the accompanying description in 1925 of *Australopithecus africanus*. However, this happened far before my time, so I can't really say that it had much of a direct impact on me. The most impactful discovery that has influenced me was the discovery of the Black Skull and its subsequent recognition as *Paranthropus aethiopicus*. Adding to this, the paper that made the significance clear was not the original description by Walker et al. in 1986[1] but a subsequent paper by Kimbel et al. in 1988[2] explaining its phylogenetic significance [read about coauthors of these studies Bill Kimbel (chap. 32) and Richard Leakey (chap. 54)]. This discovery made clear that patterns of early hominin evolutionary relationships were extremely complex and that any hominin phylogeny had to include huge amounts of homoplasy (i.e., parallel, convergent, and reverse evolution). Basically, all of the work on hominin phylogeny since that point in time has been dealing with the implications of this discovery. As an aside, although new discoveries are often presented as being game changers, they rarely are. Most new discoveries provide complementary information. The discovery of the Black Skull radically changed our understanding of early human evolution.

AMAZING FACT

I am astonished that modern humans possessing body forms adapted for living in warm environments dispersed into and colonized Europe during the worst part of the last ice age (i.e., the last glacial maximum).

TIME TRAVEL

 I would like to go to central Africa during the Late Pliocene to see if australopiths were living in the rain forest.

DRIVING FACTORS

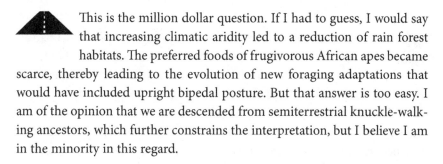 This is the million dollar question. If I had to guess, I would say that increasing climatic aridity led to a reduction of rain forest habitats. The preferred foods of frugivorous African apes became scarce, thereby leading to the evolution of new foraging adaptations that would have included upright bipedal posture. But that answer is too easy. I am of the opinion that we are descended from semiterrestrial knuckle-walking ancestors, which further constrains the interpretation, but I believe I am in the minority in this regard.

FUTURE EVOLUTION

Well, in a million years, our species will probably be extinct. But in the nearer term, I predict that our anatomical and physiological evolution will be characterized by increasing variation related to the fact that selection against deleterious conditions will have been largely removed (think, for example, about how many people have poor eyesight).

EVOLUTIONARY LESSONS

 The important lessons derived from human evolutionary studies are:

(a) Human are animals and are subject to the same ecological and evolutionary forces as other animals.
(b) Humans have adapted over long stretches of time to natural oscillations of climate change, but the change that is occurring now is "unnatural" (i.e., caused by us) and will be catastrophic if we don't act to correct it right now.
(c) Race has no basis in biology but is a social construct. It is nonetheless real because racism is real, and as a result race can, paradoxically, have profound biological effects (on health, life expectancy, etc.).

SPECIAL?

Humans are special only in the sense that our particular suite of derived adaptations (particularly, manual dexterity, tool use, brain size, and language) have led to profound emergent properties, such as our ability to profoundly modify the environment to allow explosive population growth.

RELIGION AND SPIRITUALITY

 My father was a minister. My brother is born again. We've never had an argument about human evolution. Faith and science are (or should be) completely different realms of the human experience. I don't know any scientists who spend their time thinking about how to "disprove" religion. And I think that an inability to incorporate facts about the natural world into a belief system is the sign of a weak theology. But five hundred years ago, the idea that the Sun was the center of the solar system was heretical. I predict that society will eventually get over the perceived conflict between faith and evolution.

ADVICE

Be persistent and passionate.

INSPIRING PEOPLE

 I would very much like to know what Darwin would think about our current state of knowledge of human evolution. In retrospect, he got so much right using such limited evidence.

42

RANDALL "RANDY" SUSMAN

Randall "Randy" Susman is a professor emeritus of anatomical sciences at Stony Brook University School of Medicine. He studies the origins of the human lineage and the subsequent behavioral and anatomical changes during human evolution. Regarding the former, Randy studies the origins of terrestrial bipedalism through comparative studies of modern apes' and humans' functional morphology and the ecology and behavior of African apes. He was a pioneer in field studies of bonobos in the Lomako Forest of the Democratic Republic of the Congo in the 1970s and 1980s, and this work is presented in his book *The Pygmy Chimpanzee*.[1] Regarding the latter, he has investigated the evolution of ape and human positional behavior and tool behavior in early hominins in East and South Africa.

YOUR BEGINNINGS

I have spent a long time thinking about human evolution or, as Richard Klein refers to it, the "Human Career."[2] From the 1970s to the present time, a wealth of early hominin and other fossils have been collected in the field, and lab studies have been enhanced by technology and data acquisition and processing. We now know a great deal about the maintenance behavior of key primates (chimpanzees, bonobos, and gorillas) under natural conditions, which helps us understand the "function" term in the functional/morphological equation (e.g., Lauder, 1981).[3]

I conducted research in comparative primate anatomy as an undergraduate at the University of California, Davis in 1969, and I also dissected a gibbon and a baby chimp in a comparative anatomy course taught by Warren Kinzey. As an undergraduate, I studied mammal locomotion with Milton Hildebrand, anatomy with Warren Kinzey, primate behavior with Donald

Lindburg, and archeology with Martin Baumhoff. Kinzey suggested I go to graduate school at the University of Chicago. In my senior year I started reading the papers by John Napier, a surgeon who had started to study primates and who understood comparative primate anatomy. He led the studies of the hominid postcrania that the Leakeys were recovering from Olduvai Gorge. The Olduvai fossils sampled an opportune interval of the geological record, and the deposits, particularly Beds I and II, were datable in absolute terms. By the time I got to graduate school in the fall of 1970, exciting discoveries were being reported in *Nature*, *Science*, and other specialized journals just about every week. Russell Tuttle had studied knuckle-walking, but he did no morphometrics save for accumulating muscle weight data. In the 1970s, Tuttle began doing hard-wired EMG [electromyogram] on juvenile gorillas at Yerkes Primate Center, and Napier was pioneering telemetered EMG at Chicago when I got there.

GAME CHANGER

The sum of the advances in primate behavior in the 1960s and 1970s (e.g., Schaller 1963, Lawick-Goodall 1968),[4] the fossil discoveries in East and South Africa, and Napier's laboratory studies of comparative primate anatomy allowed functional interpretations of fossil morphology for the first time.[5] The "game" was definitely changing for people who acquired domain knowledge in three complementary areas: (1) behavior (particularly in the field but also occasionally in the lab), (2) fossils, and (3) comparative anatomy. Lots of people were attracted to the fossils, but some proceeded with a one-sided bracket, namely, humans.

AMAZING FACT

How faithful the continuum tends to be in a variety of different, extant, anthropoid morphologies to forms we find in aspects of the earliest hominins. Other things about human evolution also amaze me.

DRIVING FACTORS

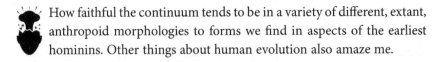

A major impetus for the advent of the hominid career may have been the desiccation and shrinking of the expansive Miocene forests throughout Central Africa. A hominid precursor *traveled* quadrupedally for greater distances and was occasionally a biped on the ground and was also bipedal on vertical trunks and along boughs in the trees. A major adaptation to life on the ground for early hominins was, as with most primates past and present, increasing body size.

SPECIAL?

Before the work of Jane Goodall, chimpanzees were vegetarian, non-tool-using apes. Humans engaged in tool-using behavior, ate meat, and had a definitive social organization. After Goodall's work at Gombe, chimpanzees were known to be hunters, and they were tool users that even made tools. Thus, after observing free-ranging apes under natural conditions, the very definition of what was "human" changed. As for the question of humans being special, the differences between humans and our closest primate relatives are continuous ones.

INSPIRING PEOPLE

The only question I *would* ask is to the person working at Olduvai Gorge Bed I, FLK-NN Level 3 who in 1960 excavated the OH 35 leg. This person cleaned the fossil, then picked it up and ran 200 yards or so to show it to Mary Leakey. Why did this person not return the fossil to the OH7/OH8 floor where the OH hand and foot were found?

I don't have any actual knowledge that OH 35 was "mis-located" after it was recovered, so I base my belief on the association of OH 35 on "circumstantial evidence." It is highly unlikely that the OH 8 foot and OH 35 leg both have (1) postmortem damage in the same location at the fibular malleolus, (2) have the same patina, (3) are both described by Mary Leakey in Olduvai Gorge, Volume 3, as coming from a "gray-green Marley clay with nodular limestone inclusions," (4) articulate well at the talocrural joint (despite what has been said by someone in a paper that attempted to judge the fit between the talar trochlea and the distal tibia-fibula), and (5) the very humanlike cruro-pedal index, which matches the index and morphology of the foot and leg in humans. We may never know the real provenience of OH 35, but I think it's at least possible to consider different options. I certainly wouldn't want to cast aspersions on Mary Leakey and her field team, who did a masterful job of excavating Olduvai Gorge. The Leakey's work at Olduvai was remarkable given the era in which they worked, and they identified and recovered everything from skulls and teeth to distal phalanges. Thanks to them for their attention to all the postcrania!

43

PETER UNGAR

P eter Ungar is distinguished professor and director of the University of Arkansas's environmental dynamics program, a fellow of the American Association for the Advancement of Science, and a member of the Johns Hopkins Society of Scholars. Peter works primarily on the role of diet in human evolution. His research includes field observations of apes and other primates, the study of modern hunter-gatherers in Tanzania, investigation of fossils, and developing new methodological techniques to analyze teeth morphologies to extract clues about diet. He is also the author of popular books on teeth and human origins, such as *Evolution's Bite*.[1]

YOUR BEGINNINGS

 I was about six years old and on a visit to the American Museum of Natural History in first grade—the dinosaur exhibit.

GAME CHANGER

 I can't point to one study, but Darwin and Wallace[2] and others effectively moved us from being apart from nature to being part of nature.

AMAZING FACT

 There were something like half a dozen sentient (as far as we can tell) upright walking species of hominin roaming the planet in the later part of the Pleistocene.

TIME TRAVEL

 Around 2 million years ago. It'd be amazing to see *Homo habilis* and *Paranthropus boisei* walking around the savannas of eastern Africa.

DRIVING FACTORS

 I don't think we can answer this, but I'd say it probably relates to a unique solution to the problem of changing environmental conditions (biotic and abiotic). I hesitate to use "road leading to"; that's not the way evolution works in my mind.

FUTURE EVOLUTION

 I don't think we can answer this question.

EVOLUTIONARY LESSONS

 We are all children of Africa. We all share a common heritage. Knowing this is profoundly important as we try to figure out how to share this pale blue dot together in peace.

SPECIAL?

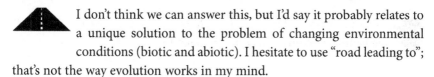 Of course, we're special—no one else is, to the best of our knowledge, contemplating this question. Studying other primates can help set the baseline for how we differ—albeit the more we learn, the more the line blurs.

RELIGION AND SPIRITUALITY

 I don't think human evolution has any bearing on spirituality. Clearly, though, it makes any literal interpretation of "creation" currently in vogue a difficult pill to swallow.

ADVICE

 Think hard and long before you choose this path. The number of emerging young scholars seems to be increasing more rapidly than the job market can handle. If you're bright and hard working enough to make it, you're likely to find a much simpler and more lucrative path to a fulfilling life. However, if you are committed to paleoanthropology and understand that success is far from guaranteed given the vagaries of the job market, I would not dissuade you. Just have a plan B.

INSPIRING PEOPLE

 Aristotle. I would love to see the look on his face when told that all life is descended from a common ancestor.

44

CAROL WARD

C arol Ward is a curator's professor in the Pathology and Anatomical Sciences Department at the University of Missouri's School of Medicine and a Fellow of the American Association for the Advancement of Science. Carol studies ape and human evolution through the fossil record. Using biomechanical approaches and novel 3D analytical tools, Carol reconstructs the locomotor and feeding behavior of fossil members of the human lineage and Miocene apes. In Kenya, she coleads the West Turkana Paleo Project, with Kyalo Manthi [chap. 33] and Mike Plavcan [chap. 37], searching for new evidence of the early stages of human evolution and their environment.

YOUR BEGINNINGS

I remember what inspired me to study paleoanthropology as a career. The summer after my freshman year in college I was working at an amusement park, a job that involved operating rides but also cleaning up after people who became sick on the rides. I decided that job wasn't for me and that I would take a summer course instead. I wanted something unusual because it was the summer. A friend suggested "biological anthropology," a subject I had not heard of that my friend told me was "cave men and stuff," which I thought sounded cool. I took the class and fell in love—in love with the fossils and the puzzle of figuring out how to use them to learn about human evolution. I have been studying paleoanthropology ever since.

I was only eighteen years old, and the class was a short intensive summer course at the University of Michigan. Milford Wolpoff [chap. 68] was the professor. He walked into class the first day, said his name and that he was a paleoanthropologist (he wrote that on the board), then turned to us and said: "The Earth is a giant disk riding on the back of a huge turtle, and the stars are painted on a canopy above us." We then spent two long class periods arguing about it. It took me a good twenty years to realize that the point he was making was about models and hypotheses and testability. It was actually brilliant, and the class was great.

TIME TRAVEL

 This is a tough one—there are many fascinating time periods from which to choose. Each time I think about answering this question, I choose a different "when." I guess, ultimately, I cannot help wanting to head back to just over 4 million years ago in eastern Africa. This time period marked the appearance of some of the earliest committed bipeds and the beginnings of the tough-chewing adaptation on which later australopiths capitalized. What were these earliest australopiths like? What did they spend their days doing? What did they eat, how did they get their food, and how did they get there? What sort of social lives did they have? How did they interact with other animals? The best stories start at the beginning, and for me this is the beginning that interests me the most.

DRIVING FACTORS

I think it was a suite of events that occurred at the end of the Miocene. The world was cooling and drying, and forests were receding and becoming more spread out. Prior to this time, most apes were forest dwellers, spending most of their time in trees eating the fruits that grew in those trees. As the climate changed, more open habitats appeared. Hominin ancestors capitalized on living not just in forests but in these new habitats as well, changing their jaws and teeth to be able to process not just fruits but seeds, tubers, nuts, and more. They already were intelligent and ingenious animals like apes are today, but this allowed them to be flexible and adaptable. Eventually they specialized in moving about on the ground to travel about as well. This combination of adaptations allowed them to flourish, whereas the ancestors of today's African apes remained tied to a fruit-based diet and forest habitats, which were shrinking. Today hominins inhabit the entire globe, but only a few apes have survived. It was a successful strategy.

FUTURE EVOLUTION

There is no way to answer this question because in the process of evolution what happened to happen, happened. Evolution is shaped by natural selection, in which some environmental pressures caused some individuals to produce more successful offspring and grand-offspring than others, passing on those traits in their genome that provided them with success. Over thousands of generations, these beneficial traits spread and the species changed, and what causes selection on our lineage has changed over time. But thanks to human cultural innovations such as nutrition, medicine, assistive technologies, and more, which have improved at an exponential rate over the past few thousand years, few hundred years, and especially the past few decades, our environments and our very Earth are continuing to change dramatically. Factors that caused selection in the past—such as disease resistance, whether a woman's pelvis was large enough to give birth to large-brained babies, an inability to cope in harsh climates, and many more—are no longer arbiters of differential reproductive success among people. As a consequence, we have no real way of knowing what factors are causing natural selection in our species, so we cannot predict the long-term changes our species will experience. The only way to know would be using a time machine that could leap tens or hundreds of thousands of years into the future. Give me one of these, and I will be able to tell you what happened.

SPECIAL?

All species are special compared to any others. Still, humans have no parallel. The evolutionary biologist Richard Alexander referred to humans as a "uniquely unique" species.[1] Not only do we move about the world on two feet like no other living animal, but we have relatively little body hair but ever-growing hair on our heads, unparalleled manual dexterity, and language and complex culture on which we rely for every aspect of our lives. We live in incredibly large and complex social groups, inhabit all reaches of the globe, and have widely variable social systems and ways of life that all involve strong and variable pair-bonds, friendships among non-kin, and reciprocal social relationships. Our brains are capable of solving incredible challenges; we can engage in mental time travel, complex social negotiations, and problem solving. We have a theory of mind and are capable of studying ourselves. And so much more. No other species has ever, or likely will ever, share this unprecedented complex of features and abilities. By studying our closest living relatives, we can explore how we are similar to and different from each and use this information to reconstruct what our common ancestors would have been like. In fact, our understanding of how humans evolved would not be as rich without the diligent work of primatologists who bring us insights into these similarities and differences. In addition, the wide variety of living primates form a set of natural experiments to study the rules of life and how and why evolution shapes species, including our own. Living primates let us investigate aspects of biology invisible to the fossil record that are integral parts of species' biology. Combining multiple lines of evidence about primate evolution provides a rich network of inferences we can use to reconstruct human evolutionary history.

45

TIM WHITE

Tim White is a professor emeritus at Berkeley, the former director of its Human Evolution Research Center, and distinguished chair in life and physical sciences. He is also a fellow of the American Association for the Advancement of Science and a Member of the U.S. National Academy of Sciences. He codirects the Middle Awash research project in Ethiopia, where he has been instrumental in discovering and describing several new species of the human lineage, including *Australopithecus afarensis*,[1] *A. garhi*,[2] and *Ardipithecus ramidus*.[3] Tim has mentored several paleoanthropologists who subsequently became prominent figures in the field, including Berhane Asfaw [chap. 21], Yohannes Haile-Selassie [chap. 27], and Gen Suwa.

YOUR BEGINNINGS

 I don't remember, I would have been too young.

GAME CHANGER

 On the Origin of Species,[4] whose claim that "there is grandeur in this view of life" has been amply demonstrated—and whose humble hope that "light will be thrown" on our origins continues to inspire.

AMAZING FACT

 The fact that we are such bizarre mammals—bizarre in our anatomy, physiology, behavior, biogeography, locomotion, etc.

TIME TRAVEL

 Unthinkably limited. I would hijack the machine and reprogram it, setting it to stop four times every 25,000 years, over the past 10 million years, at carefully chosen habitable locations of Africa, Europe, Anatolia, and Asia. Those four hundred stops would barely be enough to get even the broadest sense of what was happening. Alas, the current paleoanthropological record pales in comparison to the knowledge that would come from such an exercise of imagination. But at least the knowledge of our past that we have gained so far is based on actual evidence rather than science fiction.

DRIVING FACTORS

 Natural selection operating in myriad ways that we do not understand.

FUTURE EVOLUTION

 Demographics will drive human evolution in the future. In a hundred years, we will look the same. For longer timescales, it's unpredictable.

EVOLUTIONARY LESSONS

 Resources are finite.

SPECIAL?

 Every life form is, or was, special (at least at some level). Unless the word *special* is defined, the second question can't be answered.

RELIGION AND SPIRITUALITY

 In relation to human evolution: cognitive dissonance.

TIM WHITE

ADVICE

 Be broadly inquisitive, self-critical, persistent, and evidence-bound.

INSPIRING PEOPLE

 Samuel Clemens [you might know him by his pen name, Mark Twain].

46

BERNARD WOOD

Bernard Wood is the university professor of human origins at The George Washington University. Bernard studies how to recognize different species of the human fossil record and how to figure out their phylogenetic relationships. Over his fifty-plus-year career, his research has covered large and minor aspects of morphological variation and evolution, from ankles, to teeth, to muscles in living primates and humans. In addition to an extensive research publication record, he authored several textbooks, including *Human Evolution: A Very Short Introduction*,[1] and is the editor of the *Wiley-Blackwell Encyclopedia of Human Evolution*.[2]

YOUR BEGINNINGS

There was no eureka moment. I did my first research project when I was a sixteen-year-old schoolboy; I looked at how soil pH influenced the numbers and types of earthworms on the scarp face of the Cotswold Hills in England. I decided to try to become a doctor, and in those days you needed biology, physics, and chemistry to get into medical school. I was much better at biology than at either physics or chemistry, but I also liked history and geography. I enjoyed human anatomy so much that I decided to be a surgeon, so I took an extra degree course that focused on the anatomical sciences. Michael Day, who was the supervisor for my research project, suggested I look at the OH 8 talus, and later he arranged for me to go to Kenya where I had the good fortune to spend six weeks with Richard Leakey [chap. 54] and his team exploring what was then known as East Rudolf. Richard took a punt on a callow inexperienced researcher and asked me to join the team that described and interpreted the fossil hominins that were being recovered "thick and fast" from East Rudolf. Surgery is one of those things you should do *only* if you cannot contemplate doing anything else. Eventually paleoanthropology won out over surgery. Although fieldwork is essential for paleoanthropology, it was never my passion; what excited me was the challenge of making sense of the evidence discovered by others.

GAME CHANGER

I don't think there are game-changing discoveries; if there are, it would mean that we don't understand the "game" as well as we should. Occasionally I read an elegant (i.e., well-written and well-argued) paper I wish I had written myself, but frankly, not often. The book that introduced me to human evolution was Theodosius Dobzhansky's *Mankind Evolving*.[3] Two of my heroes, W. W. (Bill) Howells and Wilfrid Le Gros Clark, wrote books, monographs, and research papers that also influenced me. Bill Howells's *Mankind in the Making*[4] and his later book *Getting Here*[5] are elegance personified. Although Le Gros Clark was on record as being skeptical of many of the claims made by Raymond Dart and Robert Broom about the australopith discoveries that had been made in South Africa, in 1946 he went to see the evidence for himself. In October 1947, not long after his return—via East Africa—he published a comprehensive and insightful review of the fossil evidence in which he admitted that his initial assessment had been mistaken.[6] It is seldom cited, but it should be. My reprint of that paper—"borrowed" from Michael Day's reprint collection—is a treasured possession.

AMAZING FACT

Not one fact, but several. First, apparently modern humans are more closely related to chimpanzees and bonobos than the latter are to gorillas. That means that "relatedness" cannot be neatly mapped onto, or inferred from, function and behavior. Second, a great deal has happened in the modern human lineage since we split from the hypothetical common ancestor of chimpanzees/bonobos and modern humans some 8 to 6 million years ago. Unless our interpretations are badly awry, our ancestors lived alongside creatures so intriguing and weird (e.g., *Paranthropus boisei*) that we struggle to find living primates that are functional comparators. Third, not that long ago we shared the planet with potentially a handful of other taxa. It is difficult to point to any other mammal group that, in so short a time, has undergone as many grade shifts as the modern human clade has done.

TIME TRAVEL

 I would travel around Africa about 2 million years ago to regions *other* than those where we find fossil evidence of early hominins. I want to see where else hominins were living, what they looked like "in the flesh," how they were behaving, and how they were interacting with each other and with their environment.

DRIVING FACTORS

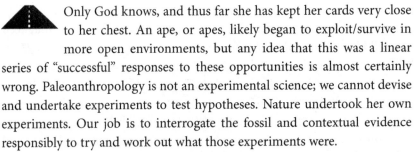 Only God knows, and thus far she has kept her cards very close to her chest. An ape, or apes, likely began to exploit/survive in more open environments, but any idea that this was a linear series of "successful" responses to these opportunities is almost certainly wrong. Paleoanthropology is not an experimental science; we cannot devise and undertake experiments to test hypotheses. Nature undertook her own experiments. Our job is to interrogate the fossil and contextual evidence responsibly to try and work out what those experiments were.

Paleoanthropology, for me, is making sense of any data that my preconceptions—which always have to be reviewed on the basis of the data—suggest might be relevant to reducing our ignorance about what happened between the hypothetical common ancestor of modern humans and chimpanzees and bonobos and the present day. This does not mean that my interest is confined to the hominin clade. There may be extinct clades that are more closely related to modern humans than chimpanzees and bonobos, and interpreting the hominin clade may—and almost certainly should—mean understanding what is going on in other parts of the higher primate/primate Tree of Life, and also the parts of the Tree of Life that involve other mammals.

EVOLUTIONARY LESSONS

It is not clear to me what investigating our evolutionary history can teach us. If I thought it would encourage us to be humble, then it would have some benefit, but otherwise I am skeptical that it has any utility. There may be a few times when understanding the evolutionary context of the human condition helped develop therapies, but my guess is that these are few and far between. Understanding our evolutionary history is intrinsically no more important than understanding the evolutionary history of frogs or butterflies or snakes. The more we know and value the natural world, the more likely we are to change our behavior in ways that increase the chance that we will not destroy the world around us. It is salutary that a hominin species like *Paranthropus boisei*, which was almost certainly not one of our ancestors, lasted about a million years before disappearing from the fossil record. Modern humans have been around between a quarter and a third of that time.

SPECIAL?

Yes, and no. We are special in the sense that we are the only ape with the means to investigate and debate our evolutionary history. But in other respects, our morphology and behavior are reassuringly "apelike." The investigation of the aspects of our phenotype that make us special is particularly challenging. It is the classic N=1 [sample size of one] problem. If there are no relevant comparators, then the comparative method is of little help. Human evolution research is hampered by the lack of any extant comparator between us and chimpanzees and bonobos. The origins of spoken language or complex tool manufacture or art or of anything that makes modern humans distinctive would be a much more tractable problem if Neanderthals were still around!

RELIGION AND SPIRITUALITY

You can make religion compatible with human evolution easily if you are prepared to stray from biblical literalism. If I believed that an all-powerful God was in control of the world, then I would consider natural selection and genetics a very smart way to enable animals and plants to adapt to changes in the environment. If organisms were not malleable, then *any* change in the environment that departs from the conditions for which those organisms were "designed" would threaten their survival.

ADVICE

The older I get, the less inclined I am to give advice, but here goes. Resist the temptation to think that the evidence available to you is more significant than it is. Consider *all* the possible interpretations of that evidence, not just the one you are currently wedded to. Instead of thinking of all the reasons your interpretation is the correct one, think of all the flaws in your argument. Remember that we are unlikely to have all, or even a modest percentage, of the relevant evidence, so we should avoid coming up with comprehensive explanations for complex problems. Finally, we are in the business of reducing our ignorance about our evolutionary history. We are not trying to be "right," just less "wrong."

PART III

Becoming Human

ANSWERS FROM THOSE WITH VAST
EXPERTISE ON MORE RECENT ASPECTS OF
HUMAN EVOLUTION

47

EUDALD CARBONELL

Eudald Carbonell is the former chair of prehistory at the University "Rovira i Virgili" (Spain). In 2004, he founded (and directed until 2015) the Catalan Institute of Human Paleoecology and Social Evolution. He's also the vice president and general director of the Atapuerca Foundation, created to support paleoanthropological work at the famous Atapuerca sites in northern Spain, a fieldwork project he has codirected since 1991. The sites have yielded the largest human paleontological record in Europe, especially of early Neanderthals,[1] and remains of *Homo antecessor* (about 1 million years ago).[2] The research of Eudald's team has earned multiple prestigious awards (e.g., Príncipe de Asturias Prize), and as an author and philosopher of science, he has published multiple essays and popular books.

YOUR BEGINNINGS

I unconsciously decided at age five that I wanted to do research on archeology or paleontology. A few months before then, in 1958, I began to collect marine fossils in the town of my grandparents in Santa María de Besora, in the province of Barcelona. This is where my love for evolution was born, which later turned into a passion for human evolution that continues to this day.

GAME CHANGER

The Time-Life books had a strong impact on me; they helped me consolidate my interests more in line with archeology than paleontology. Until then, I had been fond of minerals, ancient fossils, and clandestine excavations of prehistoric materials in the caves of the place where I was studying in the Catalan Pyrenees. Amateur and excited.

AMAZING FACT

What surprises me most about human evolution is its uniqueness, our socialization capabilities, and the great discoveries made by our genus. I mean all the species of the genus *Homo*, including our own. I mean the tools, language, fire, art, and beliefs. It amazes me how the increase in sociability through these factors makes us special as primates.

TIME TRAVEL

If I could move back in time, I would like to visit the African Pliocene, more than 3 million years ago, and observe how the first stone tool was consciously made. I would like to see how the gesture of production of stone tools is repeated until artifacts become socialized. If I could travel to the future, I would like to experience the beginning of our lineage on Mars.

DRIVING FACTORS

The ability to integrate the sequence of acquisitions has turned this integration into exponential growth, which is the current evolutionary result of our lineage. With accumulated knowledge, we humanized faster than the time it took for the entire sequence of hominization.

FUTURE EVOLUTION

In a hundred years, we will probably be dehumanizing ourselves by forced marches. We will find ourselves at the beginning of posthumanization, and also transhumanization. In 100,000 years, there will be many different lineages, and the Sapiens species will be an interstellar memory. Forms of transhuman consciousness will have evolved in immeasurable ways. Time and space will be fused in these interstellar consciousnesses. In a million years, we will most likely not exist, and the system will have lost our memory.

EVOLUTIONARY LESSONS

The study of the past, to which I have dedicated my life, will be useless if our species, *Homo sapiens*, does not decide how it wants to evolve in the future. Without a vision for the future, the system memory is useless. I have come to realize that without a critical consciousness of the species, neither knowledge of the past nor the ability to think about the present is useful.

SPECIAL?

Humans are not special, but we are unique compared to other primates. Evolution has endowed us with skills and attitudes that other primates do not have. Our large brain, our communication skills, our ability to learn throughout life, our ability to know and think, and our consciousness are phenomena that our first cousins do not share—on a qualitative or quantitative level.

RELIGION AND SPIRITUALITY

Rationality and spirituality are two sides of the same coin, typical of a brain that learns from both its animal and human behaviors. A scientist must always keep rationality and logic ahead of his personal spirituality. The spiritual should not be collectivized. Collective individuality is preferable.

ADVICE

Study, analyze, investigate. Humanity is probably the most complicated path of studies carried out by our species, both in life and earth sciences as well as in social sciences. Because we study ourselves, the objective and the subjective are linked by incredibly strong electromagnetic forces. Surely knowing who we are will make us better specimens, but our human consciousness will probably disappear when we have reached this goal. We must study each other. I think of young people, who do not stop getting to know each other—research is the most humane thing that exists.

INSPIRATIONAL PEOPLE

Two humans have had more influence than many others who have influenced me, apart from my grandmother; they are Aristotle and Darwin. I would like to have a meeting with them outside of our singular time space.

48

MANUEL DOMÍNGUEZ-RODRIGO

M anuel Domínguez-Rodrigo is a professor in the History and Philosophy Department of Alcalá University (Spain) and a visiting professor at Rice University. Manuel also codirects the Institute of Evolution in Africa and the Olduvai Paleoanthropology and Paleoecology Project. During more than thirty years in the field, he has researched human origins in Ethiopia, Kenya, Tanzania, and South Africa. Manuel is interested in reconstructing early human behavior from archeological and paleontological records and in the emergence of the genus *Homo*. His critical approach makes his work controversial at times, but it is always thought-provoking.

YOUR BEGINNINGS

When I was ten years old, I wanted to understand life, my own existence. The need to answer the deepest philosophical question that humans have had since the dawn of time came from deep within me.

GAME CHANGER

My interest was to understand how we became humans, and this has much to do with "humanization" as complementary to "hominization" (which is more focused on the physical, that is, our biological evolution). I was initially interested in two popular books that were presented to me as a freshman (*Lucy*, by Johanson and Edey,[1] and *The Sex Contract*, by H. Fisher[2]). However, the game changer for me was Glynn Isaac's posthumous publication of *The Archaeology of Human Origins*.[3]

AMAZING FACT

Self-awareness, our ability to combine the most selfish behaviors with the most solidarity attitudes, empathy, our concept of time, how humanness is composed of a web of interdependent behaviors from reproductive to dietary to social. I am fascinated by how similar we are to other primates in some features and how different in others. I am most interested in the differences, those are the ones that make us human.

TIME TRAVEL

Slightly before 2 million years ago, when behaviors were evolving and leaving material traces in the form of archaeological sites. More specifically, 1.84 million years ago, to test whether everything I am interpreting from the Olduvai sites is very far from the reality.

DRIVING FACTORS

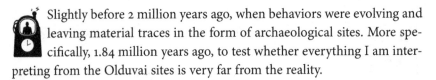

Ecology in its broadest sense: environmental variability and food availability, interspecies competition, intraspecies competition, and predation. Everything else ultimately relates to these basic principles. More specifically, all the ecological features that make up the structure of the African savanna biome are driving factors. They are unique, and no other biome reproduces them. Such a unique set of selective factors led to our evolution.

FUTURE EVOLUTION

Not playing psychic here, but I guess our future will be very intertwined with robotic and artificial intelligence technology. Our selective environment will be predominantly artificial.

EVOLUTIONARY LESSONS

How we were. Are we inexorably restricted to pair-bonding, cultural monogamy, and social reproductive structures? Did we evolve to exercise? What did we evolve to eat? How did our particular physiology evolve? How was the human mind shaped by natural selection? Evolution can help us make the right choices for our health (if we understand how our anatomy and physiology evolved) and also help us understand the essential human social and reproductive strategies. For instance, I am currently fascinated by being able to trace our social reproductive behavior archaeologically and finding no trace of it beyond 50,000 years ago.

Modern human foragers universally adopt a social reproductive structure based on individual reproductive units (i.e., families or households). Each unit uses the camp space independently and generates discrete clusters of materials that are archaeologically traceable. This was initially noticed by Yellen in the 1970s.[4] This particular type of "ethnoarchaeological record" can be documented archaeologically up to 50,000 years ago. However, in sites earlier than that age, we have not yet found a spatial distribution of archaeological materials like those we document today ethnoarchaeologically. After discarding other taphonomic processes, this suggests that the social reproductive organization of hominins prior to that age was different from that documented today in our species. For example, the 2-million-year-old sites from Olduvai are very different both in spatial distribution and content of materials from what we can observe among modern foragers.

SPECIAL?

 We are indeed different, not necessarily special.

RELIGION AND SPIRITUALITY

Saint Thomas said science and faith were incompatible. Saint Agustin said they complemented each other. My current position is with the former. One is a rational attempt to understand reality, the other is an irrational need to survive reality. However, both could be adaptive.

ADVICE

Switch to data mining or business if you are in time; those are more productive . . . just kidding. This is vocational. Do not expect to be wealthy or have an easy life. However, few things will enrich you as much inside as letting yourself be transported to your own past.

INSPIRING PEOPLE

 Einstein, the last true genius.

49

DEAN FALK

D ean Falk is a distinguished research professor and the Hale G. Smith Professor of Anthropology at Florida State University. Her research focuses on human brain evolution; the emergence of language, art, and music; and the history and philosophy of science. Dean's research on the *Homo floresiensis* endocast suggests that, despite having a brain the size of a chimpanzee's, the organization of its cerebral cortex was advanced. Dean has published numerous scientific books and articles, including some that focus on the role of women and children during evolution. She is currently writing a book tentatively titled *Basket Weaving 101: Hominin Evolution During the Botanic Age* (University of Toronto Press, forthcoming).

GAME CHANGER

The use of medical imaging technology (fMRI, functional magnetic resonance imaging; CT, computed tomography) to study the anatomy and functioning of extant and fossil hominin brains is a game changer. CT provides the technology for producing "virtual endocasts" that reflect the morphology of the cerebral cortices reproduced from crania of fossil hominins; fMRI permits the formation of hypotheses about how the brains of our early ancestors functioned. These methods allowed me to describe the cerebral cortices of, and speculate about the functions of, the brains of australopiths,[1] early *Homo, Homo floresiensis* ("Hobbit"),[2] and Albert Einstein.[3]

TIME TRAVEL

 I would travel 5,500 years into the future because that's approximately how long ago reading was invented (as far as we know). I'd like to see where current information technology (e.g., computers, AI, etc.) takes us fast-forwarding the same amount of time into the future.

FUTURE EVOLUTION

Technology will shape future human evolution (and epigenetics will speed things up). In 100 years, the frequency of high-functioning autism (HFA) will have increased significantly. If our species still exists in 100,000 years, HFA may have become the "new normal." And 1 million years from now?—I suspect *H. sapiens* will have become extinct because of WMD (weapons of mass destruction).

The HFA idea is developed in my most recent book, *Geeks, Genes, and the Evolution of Asperger Syndrome*,"[4] which discusses the increasing frequency of autism from the point of view of microevolution and Simon Baron-Cohen's findings regarding places like Silicon Valley.

SPECIAL?

Yes. No other species (on Earth) has *grammatical* language that permits the symbolic expression and receptive understanding of an infinite number of ideas. Comparative studies of the brains of humans and apes illuminate the nature of the evolved distributed neurological networks that facilitate the ontogenetic acquisition of language (and other advanced cognitive abilities, e.g., music, mathematics, reading) in the former.

RELIGION AND SPIRITUALITY

 They are not compatible with evolutionary thought; except to say that spiritual/religious views are interesting products of human nervous systems.

50

KATERINA HARVATI

K aterina Harvati is the paleoanthropology director at the University of Tubingen's Senckenberg Center for Human Evolution and Paleoecology. She is an expert on Neanderthals and early *Homo sapiens*. Katerina leads one of the most productive research teams whose research is featured in the most prestigious scientific journals and has earned her numerous distinctions (e.g., the Leibniz Award). Katerina also codirects the DFG Centre for Advanced Studies, "Words, Bones, Genes, Tools: Tracking Linguistic, Cultural, and Biological Trajectories of the Human Past," which aims to establish the theoretical foundations for a new cross-disciplinary field of biocultural coevolution.

YOUR BEGINNINGS

Coming back from the Koobi Fora field school in the summer of 1993, I realized that this was what I needed to do with my life. Fieldwork remains my favorite part of our work—although working with fossils is a close match.

GAME CHANGER

Throughout human evolution there were multiple species present, living at the same time, sometimes also at the same place. It is as if there were multiple experiments in humankind, of which we are only one. It also amazes me that we are now the only ones left. What does this say about us and our ancestors?

TIME TRAVEL

 Easy: Europe, about 50,000 years ago. I would like to meet the Neanderthals and experience their culture firsthand, see what this alternative humanity really looked like and how they lived. What would it be like to meet them? Would we be able to communicate? Would it be a hostile or a friendly encounter? Would they be more humanlike than expected, or completely alien?

EVOLUTIONARY LESSONS

The most important things human evolutionary studies can offer us are perspective and humility. Through paleoanthropology and related disciplines, we come to realize the enormity of geological time and the true place of humanity in the context of the evolution of life on our planet. Studying the past of our own genus and species, we also realize that things we take for granted in our everyday life (for instance, the relatively stable and mild conditions of the Holocene that enable our current lifestyles) are in reality always in flux and can change rapidly, with consequences for human life—including extinction of local populations or even past human species.

51

YOUSUKE KAIFU

Yousuke Kaifu is a professor at the University Museum at the University of Tokyo, Japan. Yousuke's work focuses on reconstructing the history of humanity (especially in Asia) through the study of the fossil record. He focuses on the emergence and diversification of different *Homo erectus* populations, modern human migrations during the Pleistocene, and the origins of the Japanese people. As a hands-on researcher, he led a project reenacting the voyage of humans from Taiwan to Japan 30,000 years ago. They used no smartphones, no watches, and no maps.

AMAZING FACT

The discovery of *Homo floresiensis*, reported in 2004,[1] has upended the definition the genus *Homo*. This species had a bizarre combination of *Homo*-like gracile cranio-dental morphology and *Australopithecus*-like small brain and body size, and it had survived on an isolated island of Indonesia until recently (about 50,000 years ago). This markedly contrasts with the known general trends of body and brain size increases that occurred on the continental *Homo* populations during the last more than 2 million years. Later discoveries further informed us that such "non-mainstream" evolution of *Homo* occurred in other regions at the continental margin (*Homo naledi* from South Africa, first reported in 2015,[2] and *Homo luzonensis* from the Philippines, reported in 2019).[3]

TIME TRAVEL

 I definitely would travel to western Pacific areas of 50,000–30,000 years ago to ask the first Australians, Filipinos, and Japanese Islanders: "Why did you decide to take the risk of crossing the open ocean to have a new life on the unknown offshore islands?" Motivation is something for which we have no evidence to use in scientific research (fossil bones, archaeological remains, DNA, etc.).

FUTURE EVOLUTION

We can think about that, but this may not be our duty, nor our monopoly. One of the missions of evolutionary anthropologists is to provide useful materials for other people who can think about these questions.

EVOLUTIONARY LESSONS

We learn the situations of each nation, population, and individual through history. Likewise, human evolutionary studies tell us about our current situation. For example, we all recognize that substantial biological, cultural, and linguistic diversities are present in contemporary humans and our societies. But to learn how these were shaped, we must study our deep past on an evolutionary scale.

SPECIAL?

Of course, yes. More than thirty years ago, Robert Foley argued that we are special only in the way that each other animal is.[4] This is an important view, but our uniqueness is outstanding in some ways. In particular, *Homo sapiens* are distributed all over the world, but we maintain genetic homogeneity within the level of a single biological species. This strangeness is highlighted when we compare *Homo sapiens* with the situation of other extant primates, which are diverse (more than two hundred species) but are largely restricted to the tropical and subtropical areas of the world.

RELIGION AND SPIRITUALITY

Spiritual and religious views of the world are also a human product. We can study how and why such views have emerged. I am not sure if that will lead to compatibility, but this is really an intriguing question.

52

RICHARD KLEIN

Richard Klein is the Anne T. and Robert M. Bass Professor of Anthropology and Biology at Stanford University. He is also a member of the American Academy of Arts and Sciences and the National Academy of Sciences. Richard studies the evolution of human behavior through archeological and fossil records. His classic book, *The Human Career*,[1] brings together both aspects like no other. Richard's research focuses on understanding the behavioral changes that enabled *Homo sapiens* to spread to Eurasia around 50,000 years ago, eventually replacing other human species (e.g., Neanderthals).

GAME CHANGER

I've long focused on modern human origins, and like many others I was deeply influenced by the 1987 article "Mitochondrial DNA and Human Evolution."[2] The researchers showed that *Homo sapiens* evolved exclusively in Africa, which ended a persistent argument that the Neanderthals and other nonmodern Eurasians had been significantly involved. In 1997, the authors of "Neandertal DNA Sequences and the Origin of Modern Humans"[3] confirmed the core conclusion when they produced the first Neanderthal mtDNA sequence. Subsequent ancient DNA studies have reconfirmed this, although they have also demonstrated that nonmodern Eurasians and modern humans of African origin occasionally interbred. In 2018, David Reich[4] summarized the genetic evidence for later human evolution, especially evidence that he was instrumental in developing for interbreeding. I think Reich's book will prove to be the most influential book on paleoanthropology in decades. It has deeply influenced my thinking and writing.

TIME TRAVEL

 I've often thought it would be instructive to visit a half-million-year-old Acheulean site to check my inference, long after the fact, that the people were not proficient big-game hunters. This conclusion is based on my research at the Duinefontein 2 site [near Cape Town, in South Africa; about 270,000 years old] and at other comparable open air sites.[5] The rarity of butchery marks suggests that Acheulean people had little to do with the accumulated carcasses.

DRIVING FACTORS

I assume that reduction in equatorial forest cover in response to a cooler, drier climate in the later Miocene between 10 and 6.5 million years ago favored ground living in the ape species that produced the hominin lineage/clade. A decrease in the amount of energy necessary to move from place to place on the ground could have favored bipedalism.

The savanna hypothesis does not require that early hominins adapted to life in the open savanna, only that hominin-style bipedalism was energetically less costly than ape-style quadrupedalism for walking between increasingly dispersed food patches.[6] The patches could have been stands of trees that provided both food and refuge. Interpreted in this way, the savanna hypothesis depends on the well-documented climate-driven trend for grasses to expand versus trees and bushes at low latitudes between 8 and 3 million years ago.[7] It cannot be falsified by indications for woodland or forest, such as those arguably associated with *Ardipithecus ramidus* in the Middle Awash, Ethiopia, or with *Australopithecus africanus* at Makapansgat and Sterkfontein Caves, South Africa. These indications are strictly local, and they reveal only the setting in which individuals of a species died or in which their bones accumulated, not the range of settings to which a species could have been adapted or the particular setting it preferred. *Australopithecus* underscores the point, because East African sites indicate that it often died in more open settings than those implied at Makapansgat and Sterkfontein.[8] The variety of known settings is compatible with the savanna hypothesis.

FUTURE EVOLUTION

Cultural evolution has vastly outpaced physiological evolution over the past 50,000 years or so. I think this trend will continue, which means that if humanity survives another 100,000 or 1 million years people will look pretty much like they do today.

EVOLUTIONARY LESSONS

I don't think human evolutionary studies have significant practical implications, present or future. They're worthwhile for the same reason that a good play is. Some people enjoy plays and some don't, but almost no one thinks they can help solve present-day social, economic, or health problems. If paleoanthropology is to survive long term, it must convince a sizable portion of the educated public that it produces results that are both interesting and valid.

53

CARLES LALUEZA-FOX

C arles Lalueza-Fox is the director of the Museum of Natural Sciences of Barcelona and the principal investigator at the Institute of Evolutionary Biology's Paleogenomics Lab (Barcelona, Spain). His research centers on studying the structure, function, and organization of ancestral genomes, especially in Neanderthals. Through ancient DNA (aDNA), Carles and his team address questions involving broader aspects of human evolution, population dynamics, diversity, adaptive processes, and ancient human migrations. In 2014, his team published a study sequencing the first European Mesolithic genome.[1] They discovered that African versions of pigmentation genes determined his skin color but that the individual had blue eyes.

YOUR BEGINNINGS

In the slightly overcrowded home of my infancy, there were many books about history and the past. I remember that Time-Life books *Early Man* by Clark Howell[2] and *Evolution* by Ruth Moore[3] greatly impressed me, as did *Gods, Graves, and Scholars* by C. W. Ceram.[4] Later, in my third year of an (until then) unappealing biology degree, the combination of a teacher who explained all zoology themes under the light of evolution and yet another book—*Lucy: The Beginnings of Humankind* by Donald Johanson and Maitland Edey[5]—convinced me that I would like to study the past from the perspective of evolution.

AMAZING FACT

There are parallels between recent history and human evolution in the sense that many of these processes were probably influenced by contingency. Of course, we just see the final product of the process and, therefore, have the feeling that this was all inevitable. This happens in general with the perception of evolution; we are amazed by the adaptations achieved because we are blind to the countless failed attempts—yet some of them will likely be found in the fossil record. In a more general context, we will need to integrate the enormous complexity of all evolutionary processes—not just the human one—that are being uncovered by genetics. Whenever we have data, we see admixture events in the past among evolutionary lineages, yet these lineages finally diverged and became different species or populations. We have to build a view in which we can integrate all of the fossil and genetic evidence without avoiding the underlying complexity of the process. We will need a strong computational approach to do this.

SPECIAL?

Humans are not special when compared to other animals; for instance, general evolutionary rules such as the decrease of body size in island isolates has also applied to humans in the distant past. Of course, aspects of humans such as our complex symbolic capacities—we can write books, for instance—make us notoriously special in the natural world. The genetic study of great apes so far has not helped us to understand a great deal about our own evolutionary singularities; it is disappointing that we still don't understand the mechanisms for traits such as the bipedalism. I think this is due to two factors: the long time elapsed since the great apes diverged from the hominin lineage, and the fact that many differential traits are probably due to regulatory changes affecting complex networks of genes—and not just changes in the coding sequence of genes. The good news is that both aspects are scientifically approachable; we can now retrieve ancient genomes that are informative at least of the last hundreds of thousands of years of evolution, and we can also work out with functional studies some of the genetic changes that make a human a human and an ape an ape. But the work ahead is enormous.

SPIRITUALITY AND RELIGION

If we think some books are "sacred" and have been directly written by God, with not a single typo, this belief is incompatible, not only with human evolution but also with democracy, tolerance, justice, equality, and many other aspects of the modern world. If we are talking about personal forms of conduct outlined in books such as the Bible, I find this perfectly compatible with evolution, including human evolution.

ADVICE

We tend to think we know (almost) everything now, but in fact the unknown is infinite; every time we discover and establish some scientific fact, the unknown grows, sometimes unnoticed, sometimes in directions we did not even know existed. Therefore, there will be plenty of room for research in the future. We are people fighting the unknown, and we are people fighting ignorance; both fights are tough and, in many ways, crucial for the future of humankind, and this is a task open to everyone.

INSPIRING PEOPLE

Western society was created by overlaying different views of the world—Greek philosophy, Judaism, humanism, and the enlightenment—and I would like to talk with a representative of each of them, perhaps Plato, Jesus, Michel de Montaigne, and Voltaire.

54

RICHARD LEAKEY

Richard Leakey was a Kenyan-born paleoanthropologist, conservationist, and politician. Son of legendary Mary and Louis Leakey, his work as a paleoanthropologist involved fundamental discoveries and many years serving as the director of the National Museums of Kenya. Work at Koobi Fora began after a chance landing in the area in the late 1960s. With the team of incredible Kenyan fossil hunters led by the late Kamoya Kimeu, he made incredible discoveries in the areas surrounding modern-day Lake Turkana. The "Turkana Boy," the skeleton of a young and very complete *Homo erectus* individual around 1.6 million years old, is probably the most remarkable.[1] After surviving a plane crash in 1993, and multiple health problems over the years, Richard passed away on January 2, 2022, soon after his seventy-seventh birthday.

YOUR BEGINNINGS

I grew up with parents [Mary and Louis Leakey] who were doing research, although I was obviously not involved in writing papers. I had been in high school but didn't enjoy it. I was a high school dropout. Initially my experience grew from going out to the badlands looking for fossils and then showing them to my parents. They collected them, described them, and wrote papers on them. No, I didn't want to do it myself. Many years later I happened to find (in Lake Baringo, in the Chemeron Basin, Kenya) a complete skeleton of a Pliocene baboon-type animal called *Paracolobus*. It was complete from the head to the toes and to the end of the tail, and it was beautifully preserved. I excavated it, and then, well, I thought nobody at the time was writing about colobines. There was no competition for the topic, so

I got some books on anatomy and described the bones. I worked from *Gray's Anatomy*[2] originally, and I produced a fairly detailed description. I talked to a few people, including Clark Howell, and they said I should publish it. So my first paper was based on my discovery and my description of one of the most complete fossil pre-colobines ever found.[3] That was probably my first research experience, and I found it quite rewarding.

At the time, I was a young man. I had a job, I had a child, I knew nothing about anatomy, and I had just started to work at Lake Turkana. I decided I needed to learn a lot about anatomy, and bones and muscles, to be useful. Meave, my current wife (she wasn't my wife then), was looking after small primates with my father outside Nairobi. Because I was in and out of the museum looking after things my father wanted me to look after, I asked if she had studied monkey anatomy, especially *Colobus*, and I asked if she perhaps could take me on as a student and take me through some dissections. She was a busy young woman, and she wasn't hugely enthusiastic. However, I learned that she particularly wanted (she told other people) a perfused male and a perfused female of *Theropithecus*, the Ethiopian gelada. As director of Kenya's museum, I was able to use my office to request the Ethiopians to send me a couple of these animals that were shot crop raiding and had been perfused in Addis in the anatomy department for a friend. I rang her one day and said: "I have two perfused baboons ready to be dissected on the condition that I do the dissection with you from a to z." And so, for the next six months I dissected two baboons with my [now] wife. I learned what I could about which muscle went where and what pulled on what, etc., and then she and I decided to get married anyway. From then on I enjoyed doing research and participating in it.

GAME CHANGER

From my early teens to when I left home in my early twenties, my parents frequently had many of the great names as visitors: Clark Howell, the young Coppens, Gaylord Simpson, etc. As a result, I was exposed to theories and ideas, and I became inspired with the new discoveries. Not so much a game changer, but an awakening.

The discoveries of Olduvai, particularly "Zinj" by my mother,[4] were also important in my formation from the early years because I realized that one single skull could make such a difference in the world's opinion. I keep reminding myself that we didn't know how old Bed I was [the oldest sedimentary level of Olduvai] when OH 5 was discovered. We thought it was probably only half a million years old, and when Louis, Evernden, and Curtis published their paper, we found out that it was clearly very old, 1.8 million years old.[5] That was a game changer.

AMAZING FACT

It is amazing that evolution has produced such a stupid creature. We all spend hours writing and talking about the big brain and how this distinguishes us from everything else. But if you look at the way we run the world, and how we run our lives, we are not at all smart. That seems to me to be one of the most astonishing facts about human evolution.

TIME TRAVEL

I think I'd go back about 3 million years to the savanna around the Rift Valley. I'd like to observe the different ecosystems and see how large-bodied apes were adapting to them in terms of bipedalism and the diversification of food resources.

For the future, I would go very far into the future, and I don't think any humans would be there. We will be gone. I would like to see the last people trying to survive.

DRIVING FACTORS

To me the most plausible explanation is the relationship between intellectual cognitive skills to make things that didn't exist before (very simple tools) and being able to talk about what to make, which is imagination. Once you have imagination and speech, God is the limit.

FUTURE EVOLUTION

I don't think there will be humans around in a million years. In 100,000 years, I would be very surprised to see them. In the next few hundred years, I think we will see greater assimilation and further domestication of *Homo sapiens*. We will be seeing human "breeds" rather than "races." There will be breeds than run faster, breeds that swim faster,

I don't know. I don't think much about the future because I don't have any faith in it.

EVOLUTIONARY LESSONS

There is a pejorative white view of Africa: being a "black continent," being a "backward continent," being a "primitive continent." First of all, research (e.g., archaeology, anthropology, paleoanthropology) has demonstrated very clearly that color is a meaningless concept. Second, however you want to play the game, now we can go "under the skin" and look at genetics. We can hardly afford to be prejudiced against color when not so far back those "pale skins" were "dark skins." I think human evolutionary studies in the future will level the significance of "difference" and provide an opportunity for better relationships. Also, the relationship between "us," our evolution, the resources we use, and the climate is enormous. This is becoming very clear. It is absolutely evident that previous climate changes have led to extinctions, not necessarily of humans, but it will certainly lead to the extinction of humans if we don't do something about it. Learning that the past was dynamic and that the future will probably be equally dynamic and that we were not put here as a "super being" by a fake God will help a lot.

SPECIAL?

I don't think we are special at all. We have expanded our repertoire of behavior and response through language and culture, but fundamentally we are the same. I see no justification for suggesting that we are special, although "spiritualists" would say, "well, we have a soul that the others don't." Clearly that can't be the case given what we know about evolution: When did the soul come in? What is this soul? I think there is a lot of science fiction in this game. In much of Africa, the "Brits" came along and introduced Christianity and this wonderful God, and then said God had a wonderful child with blue eyes and blond hair and that he was a *mzungu*. How do you expect people to believe such claptrap? The only thing that makes us special is our stupidity. We have at our disposal means to access so much more, to do so much more. With artificial intelligence and technology, we could solve all sort of issues. The problem lies in the fundamentals: A troop of baboons lives happily in a valley, and yet there are also politics all through them; they know what they can and cannot do. Watch the House of Commons and the U.S. Senate, and it's all primate behavior!

RELIGION AND SPIRITUALITY

I think Darwin was the first to say that you can't really believe in a God that's the benefactor of humanity. Given how utterly stupid we are, and how we treat the people that don't share our view, it can't be a God like that. We have a "spiritual" component that has been kidnapped by people and used as another element to explain what science at the moment doesn't explain. I have no problem with people having some sort of spiritual guidance on how to behave. However, there should be a separation between spirituality and scientific explanation. Both have coexisted so far. For example, good sciences came after Islam (early mathematicians in the Mediterranean were Muslim). Many great scientists today are also spiritual. I am not because I don't see any need for it. Here, in America, some teachers say that "you have to learn about evolution, but don't believe it. You only need to learn it to get your grade, but you don't have to believe it to guide your life." Evolution is not a story, there are a series of hard facts that are connected and that can be reasoned. Religion and spirituality have come out of our human imagination.

ADVICE

Be willing to look at things rationally and recognize that there is some very bad news, and we need to talk about it. We always try to dress things up to seem less serious than they are. For example, climate change: if all the ice in the world melts, the water will flow over coastal cities, and these cities will be under water. This is not imagination; this can be measured. Will the world have the resources to relocate the seaports and major civilizations? And, if so, to where? This is not "fake news." We don't have the "luxury of the past"; there were twenty-four tornados and hurricanes in Arkansas this year [2019] as opposed to twelve in past years, and also huge storms in Southeast Asia and Africa. It wasn't like that in historic times, but it has happened before in prehistoric times. Sea levels rose 20–30 meters, lakes overflowed, rivers dried up, etc. It has all happened before, and geology shows us that. What wasn't there then was people. We are the most recent hominin species to be affected by this situation.

I hope that people studying human evolution come at it from a biological, ecological, as well as an anthropological perspective. I hope they see us not as special entities but as organisms that breath and die, that live and have progeny. We depend on the same things we needed hundreds of thousands of years ago (for example, food!). Recognizing the interconnectedness of all life is absolutely fundamental to any new human origins studies.

INSPIRING PEOPLE

I still have enormous admiration and respect for Charles Darwin. I think he had a vision that was larger than that of many others. Maybe I am wrong. He also had time, but we don't have time anymore. In any case, I think he would be the one who would have the most insight.

DANIEL "DAN" LIEBERMAN

D
aniel "Dan" Lieberman is the Edwin M. Lerner II Professor of Biological Sciences at Harvard University. Dan's research seeks to
understand how and why the human body is the way it is and how
that evolutionary story is relevant to health in modern humans. Dan combines experimental biomechanics and physiology, paleontology, and comparative anatomy. He also runs fieldwork projects in Kenya and Mexico. In
addition to numerous articles, he has published several books, including *The
Evolution of the Human Head*,[1] *The Story of the Human Body*,[2] and *Exercised*.[3]
He is an avid runner.

YOUR BEGINNINGS

My love of research began when I was a freshman in college and inadvertently asked David Pilbeam [chap. 19], my professor, why several
researchers had estimated the cranial capacity of OH 7 (the type specimen of *Homo habilis*) so differently despite using similar methods. Pilbeam
set me to work measuring the endocranial volumes and parietal dimensions
of chimpanzees and humans to see if I could come up with a single regres-
sion that would estimate both species accurately. When I couldn't, he invited
me to collaborate on a solution to the problem, and that ultimately led to my
first coauthored publication.[4] I was hooked! I then did a senior honors thesis
evaluating the probability that two fossils, KNM-ER 1470 and KNM-ER 1813,
could belong to a single species, and that was even more exciting and also led
to a publication.[5] But I didn't finally commit to doing research as a career
until I spent a postgraduate, pre-PhD year at Cambridge University studying
phylogenetic systematics, visiting Africa, and thinking about all sorts of

questions in human evolutionary biology. That year was so much fun that I couldn't imagine doing anything else.

GAME CHANGER

Many studies have inspired and changed the way I think, but one that especially transformed my thinking was Nesse and Williams's *Why We Get Sick*.[6] That book opened my eyes to how we can use evolutionary theory and data to address problems of health and disease. For all I loved about the book, it raised more questions than it answered. I was struck by how little information it incorporated on what actually happened in human evolution: how little we knew about many diseases, how little it addressed musculoskeletal diseases, and how much it focused on explaining why we got sick without proposing evolutionary-informed solutions to treating and preventing disease.

AMAZING FACT

Everyone knows how uniquely smart, communicative, cooperative, long-lived, and reproductively successful humans are, but I still find it astonishing just how spectacular average everyday humans are at endurance athleticism, especially long-distance running. Even middle-aged professors can outrun horses in a marathon. (Yes, I once ran the "Man vs. Horse" marathon in Prescott Arizona; and despite a mediocre time, I was faster than all but thirteen of the fifty-one horses I raced against.)

TIME TRAVEL

Although it would be useful to travel back 6 or 7 million years to see what was going on around the time hominins diverged from chimpanzees, I'd most want to visit Africa 2–2.5 million years ago and observe early species of the genus *Homo*, including *H. erectus*. In many ways, the emergence of the human genus—however it occurred—was the most momentous anatomical, physiological, and behavioral transformation in human evolution. But we still don't know what really happened or what these early ancestors were like. What kind of hunter-gatherers were they? How dimorphic were they? What kind of groups did they live in? How did they hunt, talk, cooperate, prepare their food, grow up?

DRIVING FACTORS

We will never know for sure, but I'll bet it was climate change. I think the most likely scenario is that fractionating forests required our generally chimpanzee-like ancestors in less forested habitats to travel farther and to rely more heavily on fallback foods. In such conditions, selection favored those who were better at traveling efficiently on two legs and chewing mechanically demanding foods (hence, the two signature sets of derived features of the earliest hominins: bipedalism and postcanine tooth enlargement and thickening).

FUTURE EVOLUTION

I've often sidestepped this common question. Although all good scientific hypotheses make predictions, they aren't really science unless we can test them. That said, perhaps the fact that people so often ask this question suggests that I need to take it more seriously. If forced to guess, I'd predict that as cultural evolution continues to accelerate it will have ever-more powerful effects on reproductive success, causing new kinds of co-evolution. For example, if current energetic trends persist, maybe we will become better adapted to handle persistent physical inactivity and obesity. I have little confidence, however, in this prediction.

EVOLUTIONARY LESSONS

Much as I enjoy learning about human evolution just for the sheer pleasure and interest of figuring how we got to be the way we are, I am increasingly passionate about how that information is relevant to solving the problems we currently face. Humans today confront numerous serious challenges including (among many) the rapid global rise in chronic diseases. We need the lens of evolution to address why we are so prone to obesity, heart disease, osteoarthritis, cancer, type 2 diabetes, and autoimmune diseases. These insights will help us better prevent and treat them. I also believe we need to understand how the way we typically treat these diseases creates its own evolutionary dynamic, which I have suggested we call "dysevolution." Briefly, dysevolution occurs when we treat the symptoms of mismatched diseases rather than their causes, thus allowing or even making these diseases more prevalent and severe.

SPECIAL?

I dislike human essentialism, but to paraphrase George Orwell: while all species are special, some are more special than others. One of the central questions of human evolutionary biology is to understand how humans got to be the very special creatures that we are, and there is no way to answer that question without understanding what happened in human evolution. That means studying our closest relatives to reconstruct our last common ancestors and using the powerful tools of comparative biology to test evolutionary and biological hypotheses.

RELIGION AND SPIRITUALITY

As a scientist who was raised an atheist, I struggle to comprehend what it is like to be religious. In addition, it seems to me that the greatest challenge natural selection theory poses to any religious explanation of humanity is that evolution happens on its own without any agency. That said, I am struck by how many intelligent, thoughtful, educated people—some of them scientists—are also religious. Clearly, there is great value to studying the biological and evolutionary roots of religion and spirituality.

ADVICE

Good science starts with asking good questions. Pick good questions that matter not only to you but also to others.

56

BIENVENIDO MARTÍNEZ-NAVARRO

ienvenido Martínez-Navarro is a paleontologist and ICREA research professor at the Catalan Institute of Human Paleoecology and Social Evolution (IPHES-CERCA) and in the Prehistory Department at the University Rovira i Virgili. For more than thirty years, Bienvenido has studied the Quaternary mammals from Europe, Asia, and Africa. His focus is on understanding the ecological scenario in which the earlier members of the genus *Homo* evolved and dispersed throughout the world. His research efforts include understanding the systematics, phylogeny, biochronology, paleobiogeography, and ecology of large fossil mammals (e.g., hyenas, saber-toothed tigers, hunting dogs, bears, elephants) and the potential ecological relationships among them and with extinct humans. In *El Sapiens Asesino,*[1] Bienvenido explains his perspective on the extinction of the Neanderthals.

YOUR BEGINNINGS

I wanted to be a geologist, but in 1982 when I was eighteen years old, I was invited by Dr. Josep Gibert to participate in the excavation at the site of Venta Micena in Orce (southern Spain), the most complete, extensive, and best-preserved Pleistocene paleontological site in Europe. When I started to dig at that marvelous site, I decided to be a paleontologist.

GAME CHANGER

 When I came back to Terrassa (Barcelona), where I was living at the time, from Orce in 1982, I read *Lucy: The Beginnings of Humankind.*[2] I was really impressed with the discovery of *Australopithecus afarensis*

and the work in the Afar region, and my passion for human paleontology and for Africa began then.

AMAZING FACT

Researchers have invested a lot of time and energy going deep and explaining how important several advances in human evolution have been for our lineage. For me, the most important human advance was the change from a mostly vegetarian diet to a diet with a high percentage of animal food (proteins and fats). This change was only possible because humans started to make tools (stone knives and hammers) with their hands to process the carcasses of large mammals. The physiology of our ancestors changed, and their intelligence increased, as was explained by Aiello and Wheeler in 1995.[3] This was the beginning of increased sociality and cooperation on our way to becoming modern humans.

TIME TRAVEL

Although this is not my research subject, I would like to go to the Upper Paleolithic times, especially to the Magdalenian. In 1997 I was invited by my friend Alain Turq to visit the original cave of Lascaux (now the site is closed, and it is only possible to see the copy, which is exhibited close to the original cave). That was the most beautiful artwork I have seen in my life. I am still asking myself how the artist was able, in that time, to make these paintings, distributing and taking advantage of the three-dimensional space using paleolithic techniques.

EVOLUTIONARY LESSONS

First of all, human evolutionary studies tell us that we are another species in the community of mammals. We are not an isolated species without a connection with the natural world; we are part of it. Second, to save ourselves, we must protect the natural world. And to survive in the future means that we have to help the planet's biodiversity survive.

Of course, studies on human evolution show us the progress of our lineage until now. I am not sure that we can make conclusions about how and what we will be in the future.

SPECIAL?

Well, I will say that we are different rather than special when we compare ourselves to other animals. Our skills to create things and our social capabilities make our species unique among all the species that populate or have populated the planet. Of course, the study of other species, including large social carnivores, and especially apes and other primates, helps us see and explain their behavior and how human behavior evolved. Our increased capabilities and sociality evolved from ancestors who were scavengers and hunters and ate substantial quantities of meat and fat.

RELIGION AND SPIRITUALITY

Independent of our education in religious or secular families and schools, I think all of us have been dominated by dogmas since we were born. Nobody is free of dogmatic ideas, regardless of whether they are religious or not. Of course, if someone literally interprets the Bible or the Quran without any critical spirit, that person cannot do research. But if the person separates religious beliefs and has an open mind to evaluate all the possibilities that science offers, why not?

The scientific community, including human evolution researchers, is full of outstanding scientists who have religious beliefs.

ADVICE

 The only thing I will say is to keep an open mind, focusing on your research without any dogma. Only serious data can support good ideas and interpretations. Humans are not the center of the world.

INSPIRING PEOPLE

 I would like to ask these questions to the late Prof. F. Clark Howell.

57

BRIANA POBINER

B riana Pobiner is a paleoanthropologist and educator at the Smithsonian's National Museum of Natural History (Human Origins Program). To understand the evolution of the human diet (focused on meat-eating), she has done archeological fieldwork in Kenya, Tanzania, South Africa, and Indonesia, and she studies additional fossil collections in Europe and the United States. Briana also studies evolution education; this research has been funded by the National Science Foundation. In 2021, she received the American Association of Physical Anthropologists and Leakey Foundation Communication and Outreach Award and a National Center for Science Education Friend of Darwin Award.

YOUR BEGINNINGS

It was a series of moments more than a single moment. I began my first year of undergraduate studies as a prospective English major, but when I stepped into my first anthropology class at Bryn Mawr College, I was entranced by human evolution and prehistory, mainly due to the amazing teaching and mentorship of Janet Monge. The second moment was my first night in the field in my human origins field school in South Africa run by Lee Berger at Makapansgat. We ate stew made with mopane worms for dinner and slept in tents; Lee made sure we all knew what sounds a leopard made in case one tromped through our campsite at night. I was already enthralled by fieldwork and knew that I wanted to include it in my future, but when we later went to Kruger National Park and saw a lioness whose mouth was dripping with blood from eating from a fresh kill, I wondered what my reaction would have been if I was a hungry ancient hominin. Would

I run in the other direction? Or would I follow her? My interest in studying hominin scavenging was born. The expert faunal analysts at the field school could pick up even small fragments of fossil bones and teeth and tell us what bone they were from (and sometimes what species), and I thought, "I want to be able to do that!" So I focused my studies on learning to identify and analyze animal fossils. When I started to find fossils myself, that feeling of reaching through time to touch the past was it. I was hooked. On the last night of the field school, I told Lee that I now knew I wanted to pursue paleo-anthropology as a career, and I asked for advice on a next step. He recommended that I gently pester people until someone gave me an opportunity. And he was right.

AMAZING FACT

I find it amazing that for most of human evolutionary history there was more than one species of hominin walking the Earth. Our species is mighty in number but has very little genetic diversity. It's mind-blowing to me to imagine that there would have been two, three, or even four similar species around at the same time, and sometimes in the same place. We even know they sometimes interacted—ancient genomics indicates that many of them interbred. When considering potential mates, how different was too different?

TIME TRAVEL

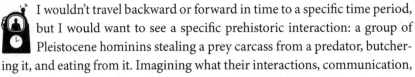

I wouldn't travel backward or forward in time to a specific time period, but I would want to see a specific prehistoric interaction: a group of Pleistocene hominins stealing a prey carcass from a predator, butchering it, and eating from it. Imagining what their interactions, communication, planning, and thinking would have been like is fascinating to me. And then, of course, I'd want to collect the leftover bones to study.

RELIGION AND SPIRITUALITY

As someone whose job entails a lot of science communication activities with varied audiences, I'm invested in making sure that human evolution is as nonthreatening as possible to religious communities, especially given the prevalence of religion in the United States. I'm interested in helping people *add* a scientific perspective to their worldview, not replace their religious faith with one; these perspectives do not need to be in conflict. Everyone comes to their own understanding of the science-religion interface, and I find the view advocated by Stephen Jay Gould called NOMA, or "nonoverlapping magisteria,"[1] most comfortable for me. In this view, these two ways of knowing the world don't conflict because they address different domains: science deals with observable evidence in the natural world, and faith deals with purpose, values, and meaning.

ADVICE

Find the questions that really excite you—and equally important, find the kind of work you could do all day without getting bored.

As a zooarchaeologist and taphonomist, I could spend all day trying to identify what bone a little scrap of broken fossil is from, or studying bone surfaces with a magnifying lens to see what stories the traces on them tell about that bone's history.

58

MARCIA PONCE DE LEÓN

Marcia Ponce de León is a senior lecturer at the University of Zurich's Anthropological Institute and Museum. She is one of the pioneers in the field of virtual paleontology. Through advanced imaging and morphometric techniques, her extensive research has provided numerous insights into the interrelationships between the brain and pelvis, and development during ape and human evolution. A person of multiple passions, Marcia combines research with teaching, curating the primate collections in Zurich, reading books (from poetry to scientific essays), playing piano, and traveling the world.

YOUR BEGINNINGS

It was always important for me to get to the bottom of things—to unravel in order to understand; to look for the causes in order to explain the effects—in other words, to explore. Research is simply a continuum in my life. I had and still have sparks of enlightenment, alone as well as in company, but I cannot pinpoint a decisive moment.

GAME CHANGER

As an avid reader it is difficult for me to name a specific discovery, study, or work that made the real difference. Delving into the works of Hans Spemann (developmental biologist, 1869–1941) and Adolph Schultz (evolutionary developmental primatologist and anthropologist, 1891–1976) was transcendental for my scientific development as a paleoanthropologist.

AMAZING FACT

 The structural plasticity of the hominid brain, in general, and its mechanical plasticity during the human birth process, in particular.

TIME TRAVEL

 I dream of a trip to the Miocene, when Eurasia was full of different great ape species living in a warm and green environment. On the way back to the present, I wish to stop over in the time between 1.7 and 1.5 million years ago when humans evolved their modern brains.

DRIVING FACTORS

 Nothing more than contingency.

FUTURE EVOLUTION

Technoculture will shape human evolution as it is doing now, but we do not know in which direction because it has its own evolution. What humans will be in 100,000 years depends on the care we take of our environment and resources. In those distant times, humans will probably look the same as today, apart from secular or millennial fluctuations. I tend to believe that our species will no longer exist in 1 million years. Either we will have become extinct (as our cousins, the Neanderthals and Denisovans), or we will no longer be recognizable as *H. sapiens*. We might have lost our beloved spherical brains and pointed chins.

EVOLUTIONARY LESSONS

Certainly one of the most important—but somewhat overlooked—aspects is that migration is a driving force of human evolution. Acknowledging this as a fact and teaching it to a broad audience, rather than treating it as an anomaly, may be helpful for public awareness and evidence-based global politics during the coming decades.

SPECIAL?

 We are unique in several respects. But are we special? Every species is special by definition.

ADVICE

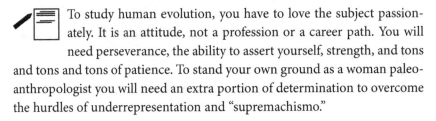 To study human evolution, you have to love the subject passionately. It is an attitude, not a profession or a career path. You will need perseverance, the ability to assert yourself, strength, and tons and tons and tons of patience. To stand your own ground as a woman paleoanthropologist you will need an extra portion of determination to overcome the hurdles of underrepresentation and "supremachismo."

INSPIRING PEOPLE

 Giordano Bruno (1548–1600), an Italian Dominican priest, cosmologist, philosopher, poet, mathematician, and heretic. He challenged each and every dogma, was repeatedly excommunicated, and ended up on the pyre.

59

MARY PRENDERGAST

M ary Prendergast is an associate professor of anthropology at Rice University. She's an archeologist studying how the human and animal worlds are intertwined. Her fieldwork centers in eastern Africa, where pastoralism has been central to the lives of humans over millennia. She seeks to understand the challenges ancient herders faced in new environments and how their relationships with hunters, gatherers, and fishers shaped economies, social worlds, and local ecologies. Her more recent work focuses on research ethics and developing meaningful collaborations with museum curators and community groups. This has led to key ancient DNA studies of the spread of herding and farming in eastern Africa and the genetic history of African foragers.

YOUR BEGINNINGS

Actually, I don't remember; and I always like to tell my students this: it's okay to not know what you want to do, take your time figuring things out. Not everyone in this field has known their calling since age seven, although that happens, and I'm so happy for people who get to live out their childhood dream! (If I did, I would be a large-animal veterinarian—close enough for someone who studies animal bones.) I had a long and messy journey coming in and out of archaeology before I decided to stay. A big part of that decision was realizing what a privilege it is to work with people who are also curious about the past, to share an intellectual journey, to slowly expand the bubble of collective human knowledge, and to talk about what we've learned with people outside the field. Those social and communicative aspects of research are why I love working in teams and teaching, which are two big parts of the typical archaeologist's job.

GAME CHANGER

I don't think it's an overstatement to say that the ability to extract and sequence ancient DNA from archaeological skeletal remains is a game changer for studying our past; most famously, perhaps, in documenting Neanderthal and Denisovan ancestry in some people living today,[1] and in demonstrating that hominins we previously put in separate categorical boxes sometimes had children together.[2] These findings call into question not only species designations but also the use of terms like "archaic" and "modern," building on conversations we were already having based on archaeology and fossils. Ancient DNA is also transforming how people collaborate across disciplines—genetics, archaeology, biological anthropology, linguistics, cultural anthropology, and history—and it is accelerating long-held conversations about research ethics and how to balance the desire to generate new knowledge with respect for both living and deceased people. In Africa, ancient DNA research is scaling up rapidly, from the first ancient genome published in 2015 to more than one hundred and counting today. These ancient people's genomes are revealing things we cannot fully get at in the archaeological and fossil records, such as Pleistocene demography or how groups of people moved across the continent and mixed (by having kids together) over recent millennia. For example, genetic research is helping resolve debates about the extent to which different kinds of artifacts (pottery or stone tools) can be correlated to groups of people—the old "pots-and-peoples" idea—and about when, where, and how frequently our bodies adapted to new dietary changes such as milk consumption, which we know from archaeology began at least 7,000 years ago in parts of Africa.

AMAZING FACT

I am constantly amazed that we are the only hominin species left on this planet (even if some of us carry traces of other hominins in our DNA) and that this is a totally anomalous blip in time compared to the last 6 million years.

TIME TRAVEL

I understand the appeal of fact-checking our interpretations by traveling back in time, but I would love to go only about fifty years into the future to see how we have transformed our field into what I hope will be a more just, human-centered, and sophisticated study of the past. Then I would bring those lessons back here to improve our present (since it's a roundtrip journey!). Archaeology and human origins research—like other aspects of anthropology—developed largely under colonial regimes and within racist and sexist intellectual frameworks that persist in some ways today. There's increasing recognition of that, and we're now having usefully uncomfortable conversations about how to address our past, train more diverse generations of researchers, and support new lines of research. Will we be successful? What will we have failed to do? That's what I'd like to learn.

EVOLUTIONARY LESSONS

It's been interesting to watch the concept of the "Anthropocene" develop. There's plenty of debate about this term and when, where, and how it can be pinpointed in geological records, but archaeology helps us understand that humans have been shaping their environments for a long time in ways often durable and detectable, even back to the Late Pleistocene. Gone are the old views of "pristine" hunter-gatherers and overly simplistic megafaunal extinction models too. Whether through cutting and burning vegetation, establishing camps, hunting and fishing, or transporting plants, animals, or pathogens, people have been shaping their environments for a long time with varied intensities across time and space. There are plenty of real and serious examples of us harming our planet that have long-term consequences, but some impacts can be positive. For example, both ancient and present-day pastoralists and their livestock benefit local ecologies by creating "nutrient hotspots" for grass growth, upending some assumptions that have informed land management policies. If we can arrive at a more sophisticated understanding of human impacts on the planet in the past, it will help us have more informed conversations about protecting our planet (both its people and its resources) in the present and the future.

ADVICE

This may sound obvious, but never forget you are studying people. When we get caught up in our specialized bubbles, sometimes we talk about archaeological entities (Oldowan!) or transform these into groups (the Natufians!); use language like specimens, samples, and data points; or think about human-environment interactions in very abstract ways. Fundamentally, though, we're talking about people. They may have been dissimilar to present-day humans, or quite similar, depending on when and where in the world you are working. But it's important to remember that even very ancient hominins would have been like us in certain ways: thinking, feeling, probably loving, and not always acting in ways that are rational or predicable based on our imperfect models. Although it might be difficult or impossible to get at these issues using scientific evidence, by keeping them in the front of your mind you'll remember why you're doing this work.

60

LORENZO ROOK

Lorenzo Rook is a professor of vertebrate paleontology in the Department of Earth Sciences at the University of Florence (Italy) and former president of the Italian Paleontological Society. His research includes fieldwork in Neogene and Quaternary paleontological sites and reconstruction of the evolutionary history and functional morphology of carnivores and primates. One of his remarkable field discoveries is the *Homo erectus* from Buia (Eritrea; about 1 million years ago).[1] He is also a key member of the team working at the early *Homo* site at Dmanisi (Georgia; about 1.8 million years ago).

YOUR BEGINNINGS

As a student in geological sciences, I was fascinated by how my professor in paleontology, Danilo Torre (1928–2014), was stimulating my curiosity and interest in a discipline located at the interface between biological and geological sciences. His lectures fascinated not only me but all the students in his paleontology classes, both in the geological sciences and natural sciences and in undergraduate and graduate courses for thirty years. Thanks to his profound knowledge of philosophy and natural sciences, his eloquence as an orator, and his ability to popularize any subject, each one of his lectures and talks was a real stimulus to go further into the topic.

GAME CHANGER

 The human fossil record discovered at Dmanisi (Georgia, Caucasus) is a wonderful example of how an "unexpected" discovery (both in terms of geological age and geographical location) forces the scientific community to change their "model" of thinking about human evolution. The skepticism that often follows discoveries made by research groups from countries outside the canonical "elite" did not spare Georgian colleagues when, in 1991, the first mandible of the oldest hominid outside Africa was found at Dmanisi.[2] Thanks to the extraordinary discoveries in Georgia, the ruling model was shown to have a large flaw. What were those hominids doing there some 2,000 kilometers north of the Rift Valley? Very similar forms of hominids were well known from the sites in South and East Africa, and it was believed that they remained in that geographical distribution and were limited to the African continent, at least until later human forms had developed a technology that enabled them to cross into different environments and colonize Europe, about 1.5 million years later (the hypothesis known as the "short chronology" of human expansion). Nowadays, in addition to having yielded fossil material from several individuals,[3] Dmanisi offers paleontologists and paleoanthropologists a unique opportunity to study a sample of fossil populations, and I am proud to be part of the international working group coordinated by Dr. David Lordkipanidze of the Georgia National Museum.

AMAZING FACT

The study of human evolution (in its broader sense) has changed dramatically in recent decades. The availability of new and more advanced tools to conduct integrated analyses and the extensive use of new techniques (dating, computational depth, noninvasive ways to study internal structures, virtual reality, etc.) and the growth of the empirical base and the convergence of an impressive mass of different data (paleontological, archaeological, genetics, paleo-biogeographic, paleoecological, etc.) are composing a completely different story. This forces us to shift from a linear model of human evolution to a phylogenetic picture that looks much more like a huge intricate bush, in which our species represents only a very recent branch that originated 200,000 years ago. At many points on the bush, we can observe temporal overlaps between different species that have coexisted in the same period and often shared the same geographical areas. We have been alone, as a human species, for only a few thousand years. Furthermore, humans continue to evolve, and today evolution no longer occurs only on the biological level but also on cultural and technological levels.

TIME TRAVEL

 I would like to travel in the central Mediterranean region between 8 and 5 million years ago. The endemic insular faunas of the emerged lands that characterized the area prior to the final emergence of the Apennine mountain chain (and definition of the physiography of the Italian peninsula) were inhabited by a number of odd animals, such as the five-horned ruminant *Hoplitomeryx* (in the Apulian area) and the enigmatic ape *Oreopithecus* (in the Tuscan-Sardinian area). I would be especially fascinated to see the latter. Since its first description in 1865 by Paul Gervais, it has held the status of the most debated fossil ape—both in term of phylogenetic relationships and postural/locomotor behavior interpretation. Having the opportunity to see it in life would be really great!

ADVICE

 If you are really and deeply interested in working in the research field concerning the human past (or future), you have to work hard and remember this: There is no luckier person than the one who can make a job out of a passion.

INSPIRING PEOPLE

Leonardo da Vinci, a genius and a universal talent of the Renaissance, an inventor, an artist, and a scientist. He fully embodied the spirit of his era, bringing to it the greatest forms of expression in the most diverse fields of art and knowledge. He was also a great precursor of geology studies and among the first to realize that fossils were remnants of living organisms. Leonardo also thoroughly explored the human body. He was fascinated by the machine represented by the human body, which he considered far more perfect than anything created by man.

61

ANTONIO ROSAS

Antonio Rosas is a research professor in the Paleobiology Department at the National Museum of Natural Sciences (Spain). He studies the paleobiology of fossil humans, including their life history. A former member of the Atapuerca research team, since 2003 Antonio has been the head of anthropological studies at the Neanderthal site "El Sidrón" (Asturias, northern Spain) and a member of the Neanderthal Genome Project. Among his major discoveries is a study of a juvenile Neanderthal from El Sidrón that indicates, in general, that Neanderthals developed at the same pace as modern humans.[1] His team is currently developing paleoanthropological research in Equatorial Guinea.

YOUR BEGINNINGS

I was ten years old when I saw some bones of a late prehistoric skeleton in a cardboard box. That was the moment! The history is as follows: My family used to spend the summer holidays in a small village in southern Spain. In 1971, just after our arrival at the village of my brother, I went to visit some of our local friends—sons of the mayor, who was an amateur archeologist. A few days earlier he had received the remains of a prehistoric burial that someone had discovered by accident. When I saw those bones in our house, I was in love. And a voice coming from the bottom of my soul said, "I want to spend my life studying these things." And here I am.

AMAZING FACT

 "Deep morphology" has always been my great passion. That is, understanding the biological forces that shape "biological form," most especially how time is enclosed in the form. For many years, the processes underlying the evolution of the peculiar Neanderthal phenotype was my uppermost concern. Currently, questions raised around reconstruction of the last common ancestor we share with chimpanzees have, somehow, substituted this scientific unease.

SPECIAL?

Of course (and perhaps unfortunately for the planet) we are special. Evolution has produced special "things." We humans are biological entities with evolutionary roots similar to those of every other species, but evolution has given rise to a singular product. The Anthropocene epoch best illustrates the impact humans have had on Earth. And that impact is a way to visualize humans as special creatures.

RELIGION AND SPIRITUALITY

 I'm not a religious person, or better said, I do not believe that the forces of the universe are dictated by deities. However, I think it's absolutely compatible to have a spiritual view of the world and be a scientist. There are many examples in history. However, explicative myths of the origin, such as the one of Adam and Eve, are outside science.

ADVICE

Vocation is the most important part of our profession and is the engine moving our work and thinking. Always keep your passion alive. This is the key.

INSPIRING PEOPLE

 I would like to know the opinion of a good paleoanthropologist who lives in the future a hundred years from now. That would be fascinating!

62

CHRIS RUFF

C hris Ruff is a professor emeritus and former director of the Center for Functional Anatomy and Evolution at the Johns Hopkins University School of Medicine. He studies the evolution of the primate skeletal system, with an emphasis on human bipedality and changes in postcranial robusticity in relation to behavior and biocultural innovations. His research attempts to elucidate the relationships between brain and body size and technological sophistication and physical activity patterns. He has also served the field as editor-in-chief of the *Yearbook of Physical Anthropology* and the *American Journal of Physical Anthropology*, and as associate editor of the *Journal of Human Evolution*.

YOUR BEGINNINGS

I had no single epiphanous experience when it came to deciding on research as an occupation. Throughout high school and early college, I had been debating whether to pursue a career in science or the humanities, in particular, music composition. Perhaps the most influential conversation I can remember having during that time period was with my father, who started out as a professional violinist but later switched to teaching history, although he continued to perform in a professional orchestra part time throughout his life. He suggested that music could be a great "hobby" and enjoyed more if you weren't simultaneously using it as a means of financial support! That sounded reasonable, and I haven't regretted my subsequent decision. My introduction to empirical research began with the opportunity to radiograph archaeological skeletal material in my senior year of college (looking for transverse lines of growth arrest, which were there).

This was my first experience looking "inside" of bones, and it inspired me to keep looking further into internal skeletal structures as a graduate student and beyond.

GAME CHANGER

"The Gait of *Australopithecus*," published in 1973 by Lovejoy, Heiple, and Burstein,[1] was a game changer at the time, and probably the most influential study in my early career. The authors applied orthopaedic biomechanics principles to australopith hip specimens, concluding that in all functionally significant ways they were similar to those of modern humans, with differences in morphology attributable mainly to obstetric factors (i.e., the smaller brain sizes of australopiths). Although I don't necessarily agree with all of their conclusions now, the paper was critical in focusing paleoanthropologists' attention on insights gained from the clinical literature. It also gave new meaning to the "obstetric dilemma" first posed by Washburn[2]—the trade-off between gait mechanics and brain size—which has been a central (although much debated) paradigm in human evolutionary studies.

AMAZING FACT

The feedback between biological and cultural evolution over the past million-plus years, leading to an accelerating increase in brain size and a decrease in face (masticatory apparatus) size, is really remarkable and unprecedented.

TIME TRAVEL

Thirty or forty years ago I probably would have selected some point in the future, but now I'm actually afraid of what I might find given recent developments around the world (global warming, environmental degradation, ever-more technologically sophisticated methods of social control). Historically there are many periods, places, and people that I'd love to get a peak at, but from a more anthropological perspective I think being an observer in East Africa about 3 million years ago in an area populated by hominins would probably be the most fascinating. How did they actually move? How would they interact (if at all) with a far-distant cousin—like a modern chimpanzee, or something more "human"? What did it mean to be "human" at that time?

FUTURE EVOLUTION

In the near future, if we avoid a number of potential catastrophes, I would like to think that the breakdown of ethnic/geographic barriers between human populations would continue and accelerate. Ethnic divisions continue to play a major role in many conflicts today, but it's also true that there are more opportunities than ever for different populations to intermix, culturally and biologically. Eventually this may lead to fewer conflicts and greater cooperation, although I'm not terribly optimistic given our past history. In the far future, if we make it, I would expect continuing gracilization of the human skeleton—not to the dwarfed bodies and light-bulb-shaped heads of Martian invader fantasies but to smaller faces and lighter bones. We've already seen these trends twice—at the transition from *Australopithecus* to *Homo* and at the transition from foraging to food production during the Holocene—both as a result of reduced mechanical loadings on the skeleton (masticatory and locomotor). Given the increased pace of technological development and the off-loading of physical tasks characteristic of our present societies, I expect that we're in for a third round. I'm not sure brain size would increase—we already have enough brainpower to design machines that can do most of our work for us if new and cleaner power supplies for them can be developed.

EVOLUTIONARY LESSONS

Perhaps the most important lesson is that we're subject to the same evolutionary processes and constraints as other organisms. That could teach us some humility, which would be good for other species but also for our own self-preservation. We still depend on interactions with the rest of the natural world, no matter how much we may shield ourselves from it.

RELIGION AND SPIRITUALITY

 I once asked a European colleague why creationism and the general science-religion conflict in the United States did not seem to be major issues in Europe, even in European regions known for religious observance. His answer was that science addresses the "what and how," whereas religion addresses the "why." Because they answer different questions, they're not in conflict. If more people realized this, I think we could avoid unnecessary and counterproductive arguments that are detrimental to both sides.

ADVICE

 Humans are fundamentally no different than other animals, so approach human evolution the same way you would with any other species—with scientific rigor and no preconceptions.

INSPIRING PEOPLE

It would be interesting to hear the thoughts of Loren Eiseley, a philosopher, naturalist, and anthropologist, whose book *The Immense Journey*[3] was inspiring to me as a student because it combined acute observations within a broad biological context.

63

JEFFREY "JEFF" SCHWARTZ

Jeffrey "Jeff" Schwartz is a professor emeritus in the Departments of Anthropology and History and Philosophy of Science at the University of Pittsburgh. He studies philosophy in evolutionary theory relating to primate origins and diversification (including humans), the analysis of paleontological and archeological remains, and cranial growth and development in *Homo sapiens* (and other mammals). Jeff's field projects have led him to work in England, Israel, Cyprus, and the major mammal and vertebrate paleontology collections around the globe. One of his most discussed books is *The Red Ape*.[1] He argues that orangutans, not chimpanzees, are humans' closest relatives.

YOUR BEGINNINGS

I think it began in my junior year at Columbia College in New York. I had been pre-med and decided to add anthropology as a second. After my first physical anthropology and archaeology courses, I was hooked, especially after I wrote my first term papers, which was the first time I did any kind of research (albeit literature review) and thinking about research questions. My senior year I studied human osteology and faunal analysis and did a hands-on analysis of bones from a Mesolithic site.

GAME CHANGER

 As part of my biology studies in college, I took courses with bird systematist and functional morphologist Walter Bock (Ernst Mayr's only student) and with Darwinian/Dobzhanskian population geneticist

Frederick Warburton. I was steeped in Darwinian thinking. I spent my 1971–72 academic year at the British Museum (Natural History)—still miss that title—working on my dissertation on prosimian dental development and its systematic relevance, and I was exposed to the "radical" thinking of the cladists there (such as Colin Patterson). While in London I met Ian Tattersall [chap. 66], who invited me to visit at the AMNH. I attended the systematics lectures—old school versus cladism and punctuated equilibria—and I got interested in that literature. All of that changed my thinking forever and allowed me to realize I could question received wisdom.

AMAZING FACT

It's not human evolution but paleoanthropologists and molecular anthropologists who are amazing. I cut my teeth on the systematics of early mammals, then early primates and prosimians, Miocene and extant apes, and for the past twenty-something years the human fossil record. Having immersed myself in the molecular literature, I'm amazed at how removed from all of this human evolutionary studies are in general. It's frustrating. I was trained as a systematist and an evolutionary theorist, and I have studied nonhominid taxa. To me, the human fossil record is so much more exciting and taxically rich than its taken to be. (I use "hominid" instead of "hominin" to refer to humans and their fossil relatives because the latter also constrains seeing the human fossil record as potentially diverse.)

TIME TRAVEL

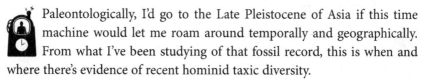

Paleontologically, I'd go to the Late Pleistocene of Asia if this time machine would let me roam around temporally and geographically. From what I've been studying of that fossil record, this is when and where there's evidence of recent hominid taxic diversity.

Intellectually, I'd probably go to the latter half of the nineteenth century so I could meet Darwin and especially T. H. Huxley, St. G. Mivart, and then William Bateson. The latter three thought developmentally and realized that the origin and the survival of species were two different processes. They argued that these processes should be decoupled, not, as Darwin proposed and became cemented with the "modern evolutionary synthesis," as part of the same continuum. These days, developmental biology and genetics is continually providing evidence that they were correct.

DRIVING FACTORS

First, there isn't a road leading to the human lineage. In fact, there isn't a human lineage. That's too unidimensional and gradualistic. There also aren't any driving factors. From the perspective of theoretical and evolutionary developmental biology (evo-devo), altering developmental signaling pathways—the source of morphologically novelty—is serendipitous and not driven by anything. Selection pressures or the like don't change developmental pathways. Evolutionary change is really a matter of "if it doesn't kill you, you have it."

FUTURE EVOLUTION

Humans will be the same, with variation (not diversity) being dictated by the boundaries of the flexibility of biological parameters. As experiments on development and teratological accidents continually demonstrate, any interference with developmental signaling pathways is deleterious, if not lethal.

SPECIAL?

All taxa are special, but in different ways.

RELIGION AND SPIRITUALITY

Nineteenth-century evolutionists were spiritual if not deeply religious. Darwin's "god" breathed life into the first organisms, and then sat back and watched. T. Huxley and Mivart were religious and wrote lots on science and religion. As a strict Catholic, Mivart argued vociferously that religion and evolutionary science could reside in the same individual. The irony is that he was excommunicated on his deathbed. In the twentieth century, folks such as T. Dobzhansky and Teilhard de Chardin conceived of human evolution as continuing into the spiritual.

ADVICE

Train to be a systematist and a theoretical and developmental biologist. Then cut your teeth on any group except hominids. After many years of doing this, and learning about hypothesis testing and the limits of speculation, then maybe take on the human fossil record— but lots of it and in tandem with an extensive comparison with other anthropoids (not just living and Miocene apes).

INSPIRING PEOPLE

Probably Otto Schindewolf. I would also ask him: "From your perspective from invertebrate and vertebrate paleontology and development, what do you think about the way paleoanthropologists have treated the human fossil record?"

64

JOHN SHEA

John Shea is a professor of anthropology at Stony Brook University and a paleoanthropologist and archaeologist studying the relationships between stone tool evidence and major events in human evolution. His books, *Stone Tools in Human Evolution*[1] and *Prehistoric Stone Tools of Eastern Africa*,[2] are essential reading on the subject. John is a proficient stone tool-maker and user of other "ancestral technologies" and an expert in wilderness survival skills. He uses this knowledge to gain insights into prehistoric human behavior. His forthcoming work, *The Unstoppable Species*, examines the evidence for prehistoric human population movements and what they mean for our future.

YOUR BEGINNINGS

When I was twelve (early 1970s), our sixth grade class adopted a curriculum titled, "Man: A Course of Study" (MACOS); it was an experimental educational program teaching principles of evolutionary and behavioral ecology. Through readings, films, and in-class exercises, MACOS introduced us to the lives of herring gulls, baboons, and Netsilik Inuit ("Eskimo") hunter-gatherers. MACOS's focus on behavior changed my life, spurring my interest in science. I grew up in a quiet country town in northeast Massachusetts that is graced with endless woods, fields, streams, and lakes. I spent much of my childhood outdoors, and I became skilled at identifying birds, plants, animal tracks, and the like. MACOS taught me that there was more to natural history than just identifying species. Behavior mattered too. Before MACOS, I looked at animal tracks mainly to figure out which kind of animal left them; afterward I tried to figure out what animals were doing as they made the tracks. MACOS was also the first time I saw ethological and ethnographic films. The Netsilik films, in particular, inspired a lifelong interest in making and using "primitive" (i.e., ancestral) tools. For a time, my friends and I hunted migrating alewife with our hand-crafted reproductions of Netsilik fishing spears. My path from reading animal tracks to becoming an anthropologist had quite a few twists, turns, and dead ends, but MACOS more than anything else spurred me to become an anthropologist.

GAME CHANGER

Calling some paleoanthropological discovery a game changer can simply be a ploy to attract media attention. Real game changers change *how we think* about evolution. Misia Landau's book, *Narratives of Human Evolution*,[3] is one such game changer. Landau [chap. 100] demonstrated how early evolutionists, such as Darwin, Huxley, Haeckel, and others, used frameworks borrowed from prescientific traditions, including the "Hero's Journey" folktale, to organize their scientific accounts of human evolution. She called these works "anthropogenic narratives," and she showed how to detect them in close readings of scientists' written works.

Paleoanthropologists still organize their theories about evolution as anthropogenic narratives. They cast about the contemporary world for inspiration about causes for change in human evolution (e.g., migration, climate change, social networking, divisions of labor) and then construct anthropogenic narratives in which these things play transformative roles. There is nothing wrong with using the present as sources of hypotheses about the past. Doing so is uniformitarianism, the common theoretical touchstone of all natural history. But anthropogenic narratives constrain how we think about evolution. Paleoanthropologists routinely create observations subject to more than one equally plausible interpretation. Narratives are simplified interpretations and require that we choose one among alternative pairs of causes and effects. Inevitably, subjective factors guide our choices and encourage confirmation bias. If one studies stone tools, for example, one naturally seeks out and preferentially accepts evidence that stone tools played crucial roles in human evolution.

How did *Narratives of Human Evolution* influence my career? I first encountered *Narratives* in classes Landau taught at Boston University. Over the course of my career, I have found that awareness of the role anthropogenic narratives play in paleoanthropology inoculates me against them. When I find myself or my colleagues "telling a story" about some major issue in human evolution, I can take active measures to create and retain multiple working hypotheses. This may seem like being a contrarian for its own sake, but science advances when people think differently.

TIME TRAVEL

 I would tour Africa, Europe, and Asia around 400,000 years ago and observe how *Homo heidelbergensis* communicated with each other. Spoken language is our species' most distinctive biologically based behavior. Our species, *Homo sapiens*, evolved out of one or more regional *H. heidelbergensis* populations 300,000–200,000 years ago. Fossil, genetic, and archaeological evidence suggests that our uniquely human capacity for rapid and high fidelity ("quantal") speech evolved around this time as well. Archaeologists, linguists, and others have long wondered about what preceded quantal speech, if only because we find it difficult to envision how any sort of "proto-language" could long persist without incurring strong selective pressure toward quantal speech. Observing social interactions among our immediate evolutionary precursors could provide clues about the selective pressures that encouraged quantal speech's evolution as well as the selective pressures that worked against its evolution among hominins other than *Homo sapiens*.

Alternatively, I would travel one million years into the future to see just how wrong my answer was to the question about what will be shaping human evolution in the future.

DRIVING FACTORS

Hominin canine tooth reduction was an important step from apes to hominins. Apes and monkeys differ from humans in having large self-sharpening canine teeth that overlap and interlock with one another. Apes and monkeys use their prominent canine teeth to pierce and cut things and as agonistic displays in social interactions. Lacking such prominent canines, prehistoric humans used stone tools to pierce and to cut things. (Of all the world's technological primates, we alone are obligatory tool users.) Our social interactions use spoken language and other artifactual symbols rather than bared fangs. Since Darwin's time, evolutionists have seen hominin dental reduction as resulting from ancestral hominin tool use, especially stone cutting tool use. And yet early hominins such as *Ardipithecus* (5.6–4.4 million years ago) and *Australopithecus* (3.5–2.0 million years ago) had canines that overlap to only a minor extent, and *Paranthropus's* canines (2.6–1.1 million years ago) do not do so at all. Clearly, hominin canine reduction was well underway and widespread across multiple hominin genera long before the oldest evidence for stone cutting tool use becomes commonplace (less than 1.8 million years ago). These observations suggest two hypotheses (which are not mutually exclusive): (1) hominin stone cutting tool use may be vastly more ancient than our oldest-dated archaeological sites (about 3.4 million years ago), and (2) something else, perhaps a change in social behavior, may be driving early hominin dental reduction.

FUTURE EVOLUTION

One hundred years from now humans will look more or less the same as they do now, although there will be sharp differences between most humans and the wealthy (taller, better health, longer lives). Rich or poor, all humans will be dealing with runaway global warming and conflicts over water and other natural resources. Urban populations supported by long supply chains will be particularly vulnerable and volatile. These conflicts will divert funds away from space exploration, essentially "grounding" future human evolution on Earth. Disease, famine, and warfare will take their toll, but in 2119 humans will still number in the billions.

One hundred thousand years from now, Earth will be recovering from its Sixth Great Extinction, the one we humans initiated. Global warming will have been reversed, and we will be in the midst of an orbitally forced ice age. As in the past, glaciers will extend south from the Arctic Sea and cover much of northern Canada and northernmost Eurasia. Agriculture may persist in some areas, but petroleum-based agricultural aids will have been exhausted, so farming and herding will be on a local scale and support far fewer people than they do today. All humans will be locavores. Increasing proportions of humanity will rely on wild food (fish, shellfish, insects, birds, and small mammals), and big game hunting will be the stuff of legends. World human population will be less than a billion people. The breakdown of global mass transportation after the twenty-second century's resource wars and the combined effects of natural and sexual selection will cause increasing regional differences in human appearance, but insufficient time will have passed for *Homo sapiens* to split into multiple species.

One million years from now, the Earth will have undergone between seven and ten major glacial cycles, each lasting more than 100,000 years. *Homo sapiens* will be extinct, but divergent patterns of natural and sexual selection will have created "daughter species" that evolved from us. These post-humans will most densely populate sub-Saharan Africa, southern and southeastern Asia, Sahul (Australia and New Guinea conjoined by low sea levels), and the Americas' Mississippi and Amazon river basins. Much as during Pleistocene ice ages, post-Anthropocene Europe, northern Asia, and northernmost North America will be thinly populated and intermittently depopulated by abrupt cold periods. Larger versions of now small game will have evolved, and some of humanity's descendants will practice hunter-gatherer adaptations focused on these larger prey, much as Pleistocene hominins did.

SPECIAL?

 Primate ethology (studies of living nonhuman primates) can be fertile sources for hypotheses, but paleoanthropologists must remember three important things about this information.

1. Living nonhuman primates are not ancestral hominins. They have evolutionary histories too. Just because some apes and monkeys use stones as percussive tools, for example, does not mean their last common ancestors with humans did so as well.

2. Finding similarities between humans and living nonhuman primates was important in the early stages of evolutionary research, but it is no longer news. One expects to find such similarities among closely related species. The important question about these similarities is whether they reflect qualities retained from a common ancestor (homology), independent origins and convergent evolution (homoplasy), or merely perceived formal or functional similarities (analogy).

3. Using the same name for recognizably different things does not advance one's understanding of either of them. Nonhuman primates do not have "cultures." Human culture depends on spoken language, hyperprosociality, and other uniquely human behaviors that are either absent or weakly expressed among nonhuman primates. Ethologists [those who study animal behavior] need a term (or terms) for nonhuman primates' socially learned traditions, but they shouldn't call them cultures because doing so implies a false equivalence.

For me, other species' socially learned traditions are analogous to human culture, not homologous: no spoken language, no culture.

Using the same name for different things never advances understanding. Anthropologists had been using the term *culture* exclusively for human activity for about a century before primatologists started applying it to non-human activities. When primatologists started discovering learned social traditions among apes and other primates in the 1980s, they cast about for a familiar term (more or less as Columbus did in calling Native Americans "Indians"), and they picked culture. The difference is that Columbus called Native Americans Indians because he genuinely thought he was in the East Indies. Recognizing this false equivalence, most anthropologists now use the term Native Americans (United States) or First Nations (Canada) or specific tribal names (Abenaki, Comanche, etc.). Native Americans themselves have mixed views of the term; some use it, others don't. Primatologists, in contrast, chose and continue to use culture strategically to situate their research in mainstream anthropology—more or less as they do with "primate archaeology," which ought to be called "archaeological primatology." Primatologists might have reasoned that calling nonhuman primate activities culture would motivate people to protect them. This was occurring around the same time Peter Singer and other ethicists were arguing for extending human rights to chimpanzees and other apes. In fact, since primatologists started identifying nonhuman primate cultures in the 1980s, primate populations have plummeted.

Archaeologists have already dealt with this culture issue. From the 1930s to the 1960s, archaeologists sought to connect their research more strongly with anthropology, so they borrowed the term *culture*, then in use among ethnographers for discrete ethnic groups, and applied it to groups of prehistoric stone tool assemblages. Treating these archaeological cultures—the Oldowan, Acheulian, Mousterian, and the like—as equivalent to ethnographic cultures (false equivalence), archaeologists constructed "prehistoric culture histories" and traced their comings, goings, and transformations, seeking parallels with the fossil record. As evidence improved, however, archaeologists discovered that these archaeological cultures exceeded the chronological and geographic ranges of any known ethnographic cultures. The Oldowan and Acheulian persisted for hundreds of thousands of years and at intercontinental scales with little or no evidence of significant change over time or variation across space. Some such cultures occurred together with more than one hominin species, which is the opposite of cultures. Stone tool technology is subtractive (tools become smaller with prolonged use), so there is a lot of convergence among the lithic evidence. Since the 1970s archaeologists have generally refrained from calling Pleistocene-age groups of stone tool assemblages cultures. Instead, they use term *industry*, reserving culture for recent (Holocene-age) archaeological assemblages of shorter duration and less geographic extent, and defined by using ceramics and other more culturally sensitive lines of evidence. Archaeological industries are not problem-free, of course, but using a different term for them highlights the difference between archaeological groupings of artifacts and ethnographic groupings of people. Stone tools are not people.

ADVICE

 Always work as hard as you can, and never quit.

INSPIRING PEOPLE

I would like to ask these questions to Omo I, the early *Homo sapiens* individual from the Lower Omo River Valley Kibish Formation in Ethiopia.[4] Dating to around 195,000 years ago, Omo I ranks among the oldest dated humans.[5] It would take some time and effort to explain the historical context and background to these questions, but I think Omo I would provide interesting answers. One wonders what questions Omo I would ask me.

65

TANYA SMITH

T anya Smith is a professor in the Australian Research Centre for Human Evolution and the Griffith Centre for Social and Cultural Research at Griffith University. Tanya is a human evolutionary biologist whose research focuses on tooth growth and internal structure using histology, elemental chemistry, and advanced imaging techniques. She mines records of daily growth, life history, and age at death to resolve taxonomic, phylogenetic, and developmental questions about great apes and humans. Tanya details numerous multidisciplinary studies in this exciting field in her illustrated popular science book, *The Tales Teeth Tell*.[1]

YOUR BEGINNINGS

As a very young child I was fascinated by the natural world, particularly the forest and stream near my home in upstate New York. I would spend all my free time mapping the forest and noting seasonal changes in the stream bank ecology as I caught and studied small fish, frogs, and toads. I knew then that I loved the natural world, but I didn't know any scientists and couldn't understand what a career in research might look like. In my 2018 popular science book, *The Tales Teeth Tell*, I describe my first research breakthrough while in my final year of college, which sparked an empowered excitement that I could discover something than no one had ever seen before or understood. These "ah-ha" moments are few and far between, but when they come, I continue to celebrate how rewarding it is to follow my curiosity as a "day job."

GAME CHANGER

Sequencing the Neanderthal genome and the discovery of interbreeding between humans and Neanderthals has been the biggest evolutionary game changer over the past two decades. During the years immediately preceding publication of the 2010 Green et al. paper[2] [describing the first draft sequence of the Neanderthal genome], I worked down the hall from many of the geneticists involved, and the buzz at the Max Planck Institute for Evolutionary Anthropology was really exciting to witness. The rapid progress of ancient DNA research has revolutionized our understanding of hominin interactions, human genetic diversity, and the spread of modern humans out of Africa. My own studies showing rapid dental development in Neanderthals relative to modern humans have received considerable media attention, due in part to the great public interest in our enigmatic hominin cousins. I look forward to functional studies of the genetic differences between these groups.

AMAZING FACT

The fact that we are the sole hominin remaining is quite amazing when you consider that at least thirty species preceded us over the past 7 million years, and most other primate genera include at least a few species on the planet today. I remain deeply curious about what happened to the other hominins, what role our species might have had in their demise, and what adaptations have helped to make us so evolutionarily successful.

EVOLUTIONARY LESSONS

Human evolutionary studies show us how change is the rule rather than the exception. Hominins came and went for millions of years, leaving their mark for several hundred millennia, ultimately becoming something new or nothing at all. Our species will meet the same fate in time.

RELIGION AND SPIRITUALITY

Science and religion are not in opposition; many people hold both spiritual and scholarly perspectives. Spiritual beliefs and religious faith are approaches to finding meaning in the world that do not rely on gathering evidence. The scientific, or empirical, approach relies on making observations that are subjected to logical verification prior to asserting meaning. Most people have faith in certain things, including scientists, and we are content to hold these beliefs without subjecting them to formal testing. There is no good reason to be skeptical of people who view the world in different ways; our hominin ancestors probably made meaning of the world in ways that scientists struggle to imagine today.

66

IAN TATTERSALL

I an Tattersall is an emeritus curator with the American Museum of Natural History. Since the 1960s, Ian's research has covered three main areas: analysis of the human fossil record and its integration with evolutionary theory, the origin of human cognition, and the study of the ecology and systematics of the lemurs of Madagascar. His current focus is on how modern humans acquired a unique style of thinking and how we managed to dominate the planet while other human species became extinct. In addition to being a prolific public speaker and author in human evolution (his latest book on the topic, coauthored with Robert DeSalle, is *The Accidental* Homo sapiens),[1] Ian is also a wine and beer connoisseur, having written books about the natural history of both.[2]

AMAZING FACT

Within the tenure of *Homo sapiens*, human beings have contrived to acquire a completely unique mode of information processing that is not a simple extrapolation of anything that went before—even though it was obviously based on many millions of years of neural acquisition. The leap to symbolic cognition from the intuitive cognitive mode that preceded it was a qualitative rather than a quantitative one; and as far as the biosphere is concerned, it was very likely the most fateful development in the entire history of life on Earth. What's more, the short time frame in which the new cognitive mode originated makes it clear that it could not possibly have been driven into existence by natural selection. As we know it today, *Homo sapiens* is evidently an entirely accidental creature.

TIME TRAVEL

 It would be tough to choose between Madagascar 3,000 years ago and Europe 43,000 years BCE or thereabouts. If pressed, however, I would have to vote for the latter. How amazing it would be to witness the first encounter between the invading *Homo sapiens* and the resident Neanderthals! The Neanderthals were clearly clever and sophisticated hominids, with a keen understanding of the habitats in which they lived. But for all their advanced intelligence, it seems clear to me that Neanderthals processed information about the world on an essentially intuitive level rather than in the declarative way typical of symbolic *Homo sapiens*. Although that initial encounter was probably most strongly marked by profound mutual incomprehension among the participants, it would provide the most direct mirror we could hope for to observe exactly what it is that separates *Homo sapiens* from every other hominid—indeed, every other organism—that has ever lived.

DRIVING FACTORS

I don't have much faith in driving factors in evolution. It is inconceivable to me that a quadrupedal ape would ever have stood upright on the ground simply to reap any of the many supposed advantages that have been touted for bipedality in the terrestrial context. The only conceivable reason a previously arboreal ape would have stood and moved erect on the ground is because it already habitually moved upright in the trees, using a more strongly suspensory form even than today's orangutans. Quite simply, the ancestral hominid stood upright on terra firma because it found it the most natural thing to do. Having made this choice, it reaped all of the potential advantages—and, equally significant, all of the *dis*advantages of terrestrial bipedality. Had climate change not fragmented the ancestral African forests, there would have been little incentive for any ape to commit itself to the ground. But given those environmental circumstances, postural predisposition was almost certainly the key to the founding hominid adaptation.

FUTURE EVOLUTION

 Demographics. Human beings evolved rapidly in an age of unsettled climates during which an unusual group of secondary predators roamed, in small numbers, across vast tracts of territory. This setting provided the ideal circumstances in which innovations could arise and become fixed in those tiny, scattered, and frequently isolated hominid populations. But since the beginnings of settled life, the demographic dynamic has radically changed. Today's humans live in a continuous and increasingly dense population that has occupied virtually every habitable niche on the planet. Under these circumstances, the prospects for meaningful change in the human population are slim to none, even as gene frequencies slosh back and forth to fool anthropologists into believing that meaningful change is taking place. Short of dramatic demographic change (all too easy, alas, to imagine), we will have to learn to live with ourselves as we are.

EVOLUTIONARY LESSONS

Perhaps oddly for a paleoanthropologist, I am on record as observing that there is little we can learn from our evolutionary past that we cannot learn from observing ourselves today because we have very recently come to think and behave in an entirely unprecedented way. However, one of the most notable products of our unique cognitive style is an intense and analytical curiosity about both ourselves and the world around us. Just how we fit into that world is one of the most important dimensions of that curiosity; and there is no better way of satisfying this fundamental human need to know than by investigating and understanding how we got here. As it turns out, the resulting knowledge carries an important lesson about human nature and human responsibility. The sudden recent origin of the peculiar form of sentience that governs our behaviors makes it evident that our behaviors have not been fine-tuned over the eons, as many like to suppose. If we behave irresponsibly, we cannot blame a vanished, irrelevant, and entirely mythical "environment of evolutionary adaptation." Instead, both as individuals and as a species, we are inescapably accountable for the ways in which we conduct ourselves.

SPECIAL?

"Special" is a loaded word. And, yes, to ourselves we are inevitably special. The fact that we are able to pose the question might even imply this. But it would be more meaningful to ask if we are significantly "different" from other animals, and if so, precisely how. Here the answer is clearly yes, and in more than simply the conventional sense that all species are distinctive in some way from other organisms. Physically we are unusual in many ways, most of them due to our unusual bipedal way of getting around. But more remarkable is the way in which we deal with information about the world around us. Unlike any other organism (as far as we know), we are able to remake the world in our heads and to imagine that things could be other than they are. Observation of our close living relatives, and of a diversity of other clever species out there, can certainly provide benchmarks for our uniqueness. And they may even allow us to create theoretical models for cognitive stages through which our ancestors might have passed; but it is only by close scrutiny of ourselves that we will ever understand just what it is that makes us self-aware.

RELIGION AND SPIRITUALITY

Our unique modern human cognition allows us to form and to develop spiritual and religious ideas, and it is thus compatible with holding them. But these ideas deal with what lies beyond the material world, so they can never be tested in the way that science requires. This makes them inherently unscientific; but science is not the only way we have of comprehending the world. It seems to me most sensible to view science and spirituality as complementary rather than conflicting. Good fences make good neighbors, and it is unwise to allow either to encroach on the other.

INSPIRING PEOPLE

Everyone is going to want to say Charles Darwin. So I am going to choose someone different: Alfred Russel Wallace. Wallace was never afraid (unlike Darwin) to publicly confront questions like this, and it would be really nice to know what he would conclude, more than a century down the line.

67

MATT TOCHERI

Matt Tocheri is Canada research chair in human origins and an associate professor of anthropology at Lakehead University (Thunder Bay, Canada). He is also a research associate in the Human Origins Program at the Smithsonian Institution's National Museum of Natural History. He received his PhD in anthropology from Arizona State University in 2007 and was elected a fellow of the American Association for the Advancement of Science in 2013. His research focuses on the evolutionary history and functional morphology of the human and great ape family, with a special interest in *Homo floresiensis*, the so-called hobbits of human evolution.

YOUR BEGINNINGS

As a child I had a keen interest in science, but it wasn't until I tried some anthropology courses at university as an undergraduate student that I became really hooked. These courses helped me see the world with fresh eyes, and I became more aware of how research and the scientific method enables us to understand the world, past and present. By the time I finished my undergraduate degree, I knew I wanted to build a career doing research, so I started to look at possible graduate schools. Without question, my experiences as a graduate student in anthropology at Arizona State University had the most significant impact on shaping how I think about, approach, and conduct research. There was no single "moment" in my life in which I decided to do research; a long series of experiences ultimately led me down this path.

MATT TOCHERI

GAME CHANGER

 The discovery of *Homo floresiensis* on the Indonesian island of Flores was certainly a game changer for me, and I think it is fair to say that many people who study human evolution would agree.[1] I remember how news of this discovery ripped like a tornado through the scientific meeting I was attending at the time, and there was total excitement and confusion when I returned to graduate school. That semester (Fall 2004) a number of us were taking a seminar course with Dr. Bill Kimbel [chap. 32], and he immediately changed the reading schedule and assigned the new papers about *Homo floresiensis*. Discussing those articles after they had just come out was one of the highlights of my graduate career. Little did I know that two years later I would inadvertently come face-to-face with casts of the wrist bones of *Homo floresiensis* just as I was completing my PhD dissertation on the evolution of the hominin wrist. Analyzing those casts and publishing the results launched my postgraduate career.[2] Since 2010 I have codirected, with Dr. Thomas Sutikna and Wahyu Saptomo, the ongoing excavations at Liang Bua, the site where the original discovery was made. Without question my career would have been completely different had it not been for the discovery of this enigmatic member of the human family tree.

AMAZING FACT

 We have learned pretty much everything we know about human evolution within the past 150 years or so. Just think what we are capable of learning during the next 150 years!

TIME TRAVEL

Having spent more than a decade excavating at Liang Bua, this is an easy question for me. I would go straight there about 75,000 years ago and wait patiently for a glimpse of *Homo floresiensis* or any of the many other interesting animals we find in the sediments there: dwarfed elephants, giant marabou storks, Komodo dragons, and giant rats.

FUTURE EVOLUTION

The forces that have shaped human evolution in the past will drive it in the future: mutation, natural selection, gene flow, and genetic drift. In a hundred years (four or five generations), humans will almost certainly look the same as we do now barring some major cataclysmic disaster on Earth. However, what humans will look like in 100,000 or 1 million years (about 4,000 to 5,000 or 4 to 5 million generations, respectively) is a far more interesting question. If humans survive that long, there will certainly be recognizable change (be it genetic, behavioral, or anatomical) between now and then, and the nature and degree of this change will depend on the surrounding circumstances. Space travel and human colonization of places beyond Earth, in particular, have a strong likelihood of leading to long-term geographical separation of human populations. Such vicariance would have a strong impact on the subsequent evolution of those separated populations. Only time will tell.

EVOLUTIONARY LESSONS

The hominin fossil record reminds us that human diversity (genetic, morphological, and behavioral) was far greater in the past than it is today. There were lots of different kinds of hominin species, and some of them shared this planet at the same time as us (e.g., *Homo floresiensis*, *Homo luzonensis*, *Homo naledi*, Neanderthals, and Denisovans). But all of these hominins have become extinct, and we (modern humans/*Homo sapiens*) are the only ones left. We need to better understand why we are the only survivors so we can make better decisions as a species for how we take care of our planet and each other for the future.

SPECIAL?

No, I do not think humans are special when compared to other animals. However, the rather peculiar evolutionary history of hominins has resulted in modern humans having a number of characteristics that either appear to be unique (e.g., the ability to ponder about our evolutionary history) or are definitely unique (i.e., the ability to actively investigate our evolutionary history as well as that of all living things). Nonetheless, studying great apes and other primates provides the necessary comparative context for understanding human biology, behavior, and evolution, period.

ADVICE

Follow your interests inside and outside of school. Never stop asking questions, but try to learn how to answer some of your questions yourself through your own research and experiments. Read, read, and read some more.

INSPIRING PEOPLE

An expert on human evolution living about 500 to 1,000 years from now. Not so far into the future that nothing is recognizable to me, but far enough that some of the major questions we have today about human evolution have been resolved. If I am lucky, they may have read this book and know who I am :-).

68

MILFORD WOLPOFF

M ilford Wolpoff is a professor of anthropology at the University of Michigan, a fellow of the American Association for the Advancement of Science, and a fellow of the American Anthropological Association. Over his half-century-long career, he studied human evolution from all perspectives (allometry, phylogeny, dental development, tooth wear, sexual dimorphism, and genetics). He is well-known for being the leading supporter of the multiregional evolution hypothesis (he coined the term),[1] which describes Pleistocene human evolution in *Homo sapiens* as a pattern of long-term adaptive changes in central (African) and peripheral populations of *Homo erectus* connected by gene flow. His work earned him the Darwin Lifetime Achievement Award from the American Association of Biological Anthropologists.

YOUR BEGINNINGS

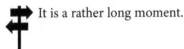 It is a rather long moment.

Philosophy of Science with Joseph Agassi. There was a time in my life when I realized how science is done, and how this could inform my understanding of human evolution. I entered the University of Illinois, Urbana-Champaign, in the fall of 1960 as a physics major, with a vague notion that I wanted to understand evolutionary change by understanding skeletal mechanics. In 1963 when I was a senior, I took a class on the philosophy of science with Joseph Agassi, a student of Karl Popper's. This class was recommended to me by Donald Lathrap, the archaeologist most responsible for bringing me into anthropology.

My Single Species Hypothesis and the way Richard Leakey ended it. Four years later I was a graduate student, and I published the first of several papers on what I called the "single species hypothesis."[2] This was a null-hypothesis testing a Popperian approach to understanding variation in South African Pliocene and early Pleistocene hominids: comparisons with other primates indicated that a single species was the simplest explanation for the South African hominid variation. New information from hominid discoveries in East Africa emerged that showed my hypothesis was wrong, a point convincingly brought to my attention by one of my best friends, Alan Walker. Right after ER 3733 (an early *Homo erectus*) was discovered in Kenya from the same strata that also preserved robust australopithecines,[3] Alan asked Richard Leakey [chap. 54] for permission to show me the cast in Michigan. Richard agreed, and Alan was so excited about ER 3733's implications that he insisted on revealing the ER 3733 cast in the airport bar in Michigan. In 1976 I was invited to present a paper in France on "Natural Selection and Early Hominid Tooth Size Evolution." By this time, I had come to believe that progress in science depended on refutation, so I took this opportunity to change my topic and instead explain why the single species hypothesis was wrong.

What I learned from Vince Sarich. Richard Leakey later called it the best paper I had ever given. For better or for worse, I had thrown my hat into the ring of refutations. I can no longer remember if I felt courageous or was just afraid not to follow through with the approach to science I had accepted and promoted. But I learned one more key thing from that experience. I was at the University of California, Berkeley soon after and was invited to dinner at the home of the most iconoclastic of anthropologists in the Berkeley department, Vince Sarich. After dinner Vince took me aside and scolded me for publicly admitting that my hypothesis was wrong. "Why?" was my surprise answer. "I *had to* because it *was* wrong." Vince responded that "now nobody will ever believe you." Life is complicated.

My Event Horizon. In physics, an event horizon is a boundary that separates observers from events that cannot affect them. I use it here to depict my own boundaries, which prevented a deeper understanding of human evolution, and to describe how some significant boundaries were overcome early in my professional career. I would like to think that I developed some deep strategy, but it was mostly the good luck of having professors, friends, and colleagues who were helpful at just the right moments. These are roles that my wife, Rachel Caspari, and my PhD students (I have graduated twenty-four of them) have come to play. "Science," as my dear friend Jan Jelínek used to say, "is a human activity."

INSPIRING PEOPLE

I would love to go back into the past to 1937 and listen to Billie Holiday when she was first performing with the Count Basie orchestra and sang duets with Lester Young's tenor sax. I wouldn't be there to discuss; I wouldn't dream of talking. I would be there to listen, to set after set after set.

69

CHRISTOPH ZOLLIKOFER

Christoph Zollikofer is a professor emeritus at the Anthropological Institute of the University of Zürich. After obtaining a PhD in neurobiology, he studied music (cello) and then returned to science. Since the early 2000s, he has been one of the world leaders in computer-assisted anthropology ("virtual anthropology"), investigating the patterns of morphological variability and evolutionary diversification in fossil and extant primates, applying computational modeling to evolutionary change, and developing novel image-based analytical tools for anthropology. Christoph has applied his skill set to the study of key fossils, such as the crania of *Sahelanthropus* and early *Homo*.

YOUR BEGINNINGS

 I cannot remember any such decisive moment; research was one among several options, and it gradually became more important.

GAME CHANGER

 Johann Carl Fuhlrott's 1859 communication on the "Neanderthal Man"[1] and the view of an outsider and nonexpert as a germ of scientific innovation was a game changer.

AMAZING FACT

 That we still know so little about how humans evolved. From the perspective of general evolutionary biology, most of the "facts about human evolution" would be deemed guesswork.

TIME TRAVEL

 I would travel to the time when humans and Neanderthals met for the first time. The problem is that we do not really know when this happened, and whether humans were humans and Neanderthals were Neanderthals at that time.

DRIVING FACTORS

 We will never know—mostly because we build those roads from the present to the past.

FUTURE EVOLUTION

 Random events will always be the major force, so any picture we draw about future humans will likely be pointillistic.

EVOLUTIONARY LESSONS

 Like any natural process, human evolution does not tell us what we should or should not do. However, human evolutionary studies are especially inspiring because they use the methods of the natural sciences to explore ourselves.

SPECIAL?

 Great apes provide refreshing perspectives on alternative evolutionary scenarios: how did they and we evolve and diverge over time, each becoming "special" in its own sense?

RELIGION AND SPIRITUALITY

 It's paradoxical. From a materialist's perspective, scientific and religious/spiritual worldviews are compatible because they emerge from the same material basis: the brain. From a nonmaterialist's perspective, they are incompatible because they assume fundamentally different agents governing the world.

CHRISTOPH ZOLLIKOFER

ADVICE

 Be aware of potential cultural bias and naturalistic fallacies. Be critical of categories such as human species, populations, etc. Keep in mind that science is not the same as academia.

INSPIRING PEOPLE

 I would ask these questions to Miguel de Cervantes Saavedra (1547–1616) because he had deep insights into the nature of human diversity and craziness.

PART IV

Now

ANSWERS FROM EXPERTS WHO FOCUS
ON EVIDENCE LARGELY FROM THE
PRESENT (BIOLOGY, ETHNOLOGY,
PRIMATOLOGY, ETC.)

70

SUSANA CARVALHO

S usana Carvalho is a professor of paleoanthropology at Oxford University's School of Anthropology and Museum Ethnography. As a pioneer in primate archeology, Susana has studied stone tool use by wild chimpanzees in Guinea since 2006 and has conducted archeological research in the Plio-Pleistocene of Kenya since 2008. She is the director of the Paleo-Primate Project at Gorongosa National Park (Mozambique), where she leads an international interdisciplinary research team focused on the discovery of novel information about human origins. Research at Gorongosa includes behavioral studies of living primates, their environments, and paleontological fieldwork.

YOUR BEGINNINGS

My awakening moment happened while I was working on a master's in human evolution in Coimbra in 2004. I signed up to join a primatology course in Montepío Forest, Mexico, to study platyrrhine primates—and I returned knowing that the tropical rain forest was the most extraordinary place I had ever seen. It was the first time I felt that we truly belong to nature and to that precise habitat; we are only a tiny, tiny part of the ecosystem. I decided to do my thesis in primatology, and I came up with the idea of doing primate archaeology. For years I traveled between Guinea, to study stone tool use in wild chimpanzees, and the Koobi Fora Field School (Kenya), where I searched for artifacts in Early Pleistocene/Pliocene deposits.

GAME CHANGER

Post Darwin game changers include the Taung child by Raymond Dart,[1] Jane Goodall's discovery of chimpanzees using tools,[2] and all the books by Jonathan Kingdon.[3] The combined influence of these scholars encouraged thinking outside the box, questioning the establishment, thinking of humans evolving as part of an entire ecosystem (and not as some skulls drifting in space!), and the importance of exciting interdisciplinary dialogues.

AMAZING FACT

Every new discovery is more exciting than the previous one and poses more questions than answers. Also, facts about the incredible diversity of hominins and their behaviors, which has been rapidly and steadily emerging since the 1990s.

TIME TRAVEL

Can I have two stops in the machine? First to the Middle Miocene to see the coastal forests of eastern Africa and find out which fauna, particularly which apes, were living there. Then, definitely to around 3.5 million years ago in Africa, when the highest number of hominin species were coexisting. How varied was their locomotion, how did they communicate—and *who was using tools and to do what*?!!

DRIVING FACTORS

The factors that enabled everything else were bipedalism and technology—which I think coevolved, even if we don't yet have evidence for this. Both contributed to our evolution into a generalist ape and increased our ability to disperse and succeed in novel habitats, buffering against climatic variability and periods of scarcity—this ultimately allowed for a significant change of the demographics that would pave the way to the boom we see happening with the advent of *Homo*.

FUTURE EVOLUTION

It will likely be driven more by pandemics and shortage of resources, particularly water, which could drastically change the dynamics of population growth and hypothetically isolate populations located in a few richer and less deprived ecosystems. Given how evolution works, I can only say that as long as we don't become extinct we will keep evolving. Everything else is in the realm of divination.

EVOLUTIONARY LESSONS

We are learning mostly about extinction processes and their causes, but also about behavioral responses and adaptations to rapidly changing environments. This provides perspective for our infamous sense of invincibility as a species. Suddenly you realize that humans represent a few seconds in the whole Earth history, that our species has been around for a mere 300,000 years, and that hugely successful hominin species living for millions of years became extinct. New fields of research such as conservation paleobiology are also taking an impressive step forward, using data from past ecologies to inform future conservation policies and action.

SPECIAL?

After I saw the documentary *My Octopus Teacher*, I find it hard to say that we are special. As a species, our special powers are the ability to imagine and plan long-term, to actively teach, which allows transferring information to all while needing only a few wise innovators, our extended friendship networks, and our dependence on and hyperspecialization of technology.

In the field of human evolution, we are all trying to reconstruct the human paths from bits and pieces. Having an understanding of both extant ecology and animal behavior is as important as knowing anatomy or being able to properly excavate an important site. You cannot propose incredibly overarching theories that require a strong grounding in behavioral ecology and never stop to look at an animal in a floodplain in Africa on the way to dig your fossil or archaeological site. If you spend time observing primates in the wild and learn how they interact and fit in their niches, you are much more likely to see those relationships when thinking about the fossil record. They teach us about adaptation, the limits of behavioral flexibility, responses to selection pressures and, most important, about their relations with other animals in the landscape. This brings ecological validity to our human evolution narratives and brings in aspects of behavior that are not speculations but facts based on empirical observation. I don't believe in doing paleoanthropology without understanding modern ecology and primate behavior.

RELIGION AND SPIRITUALITY

Totally compatible with science if you are an animist! Both human evolution and religion are made of narratives, share an attraction for the unknown, and seek to fulfill a very basic desire to understand our place in the universe and in nature.

ADVICE

Pursue a degree in human sciences, then decide if you want to become a specialist or a generalist—we need both. Don't become blinded by the race for hominin fossils—we need fossil apes and everything else in between. Question the most established assumptions. Search where no one has gone before. Then—and I cannot emphasize this enough—spend some time watching real live nonhuman animals living in their natural habitats. Also, be prepared to update your slides five minutes after you end a lecture.

INSPIRING PEOPLE?

Jonathan Kingdon (but I am very lucky, and I will ask!).

71

FRANS DE WAAL

Frans de Waal is a primatologist and ethologist at Emory University. He is the director of the Living Links Center at Yerkes National Primate Research Center and a U.S. National Academy of Sciences member. Frans studies primate social behavior and intelligence, identifying parallels with humans in critical aspects such as morality, empathy, altruism, and culture. He has published hundreds of scientific papers and more than a dozen books that have been translated into twenty languages, including *Chimpanzee Politics*,[1] *The Age of Empathy*,[2] *Mama's Last Hug*,[3] and his new book, *Different*.[4] Frans's extensive science outreach efforts include two TED talks and his popular Facebook page.

YOUR BEGINNINGS

I have always been attracted to animals and taken them seriously. In fact, I know that children look at animals as their equals. It is only later in life that people begin to differentiate and think perhaps we are not animals or that we are better than them. This change never happened to me, and as a result, I was on my way to becoming a student of animal behavior very early in life. My university education came much later, and it helped me pinpoint what is so interesting about animal cognition and how their emotions compare with ours. When Konrad Lorenz defined the study of behavior as the study of "the liveliest aspect of all that lives," he was right on target. There is never a boring moment when working with animals.

AMAZING FACT

One of the most shocking and impactful studies to come out during my lifetime was by Mary-Claire King and Allan C. Wilson,[5] in which they compared the DNA of humans and chimpanzees. Anthropologists had given humans an entirely separate branch on the evolutionary tree, quite distant from those hairy apes with whom most of them didn't like to compare themselves. We walk on two legs, we have big brains, we have nearly naked skin, so even the fact of placing us in the primate order, as Linnaeus had done in the eighteenth century, was a bit of an embarrassment. This 1975 study all of a sudden positioned us right in the middle of the ape family. Think about it: in terms of DNA, a chimpanzee is closer to us than it is to a gorilla or any other primate. We are sister species! There are even calls to put humans, chimpanzees, and bonobos in a single genus instead of having *Homo* as a genus for us and *Pan* for them. It was an amazing finding, and discussions of human evolution have never been the same since.

DRIVING FACTORS

I really don't think this way. I can't look at an ape as somehow "on its way" to becoming human. Each species, including all of our close relatives, is self-contained. It's right where it needs to be at a given moment in time. This is also why I object to such popular terms as "nonhuman animal" for other animals, because such a term suggests that there is something missing. Poor animal, it is not human! For me it would be the same as calling elephants and tigers "nonhyena animals." Correct perhaps, but unnecessary. Let's not look at other animals as somehow being deficient humans.

After the split from other apes, our earliest ancestors were very similar to them. So similar that for 1 million years they still hybridized with them. They walked on two legs but still had a big toe to be able to climb trees. They had brains of about the same size as ape brains. *Ardipithecus*, who lived more than 4 million years ago, was still like this. Of the extant apes, the one anatomically most similar to *Ardipithecus* is the bonobo, with its long legs and facility to walk bipedally. When *Ardipithecus* was discovered a few years ago, people were struck by its reduced canine teeth and speculated about its possible peacefulness. Bonobos have reduced canine teeth too, and they are remarkably peaceful. I'd suggest, therefore, that we look more at bonobos if we wish to understand human evolution.

SPECIAL?

Like every species, humans have a few traits of their own. In the cognitive domain, I think the most important special human trait is language. Other animals communicate, and I am sure the communication of some species (such as dolphins and orcas) will turn out to be far more complex than we now think, but a symbolic learned language seems to be our unique claim to fame.

The search for what makes us special, however, is often accompanied by a strong desire to set ourselves apart from the natural world. This desire is misguided and dangerous. I consider it the biggest flaw of Western religion and philosophy. The current climate crisis, the virus pandemic, and mass extinctions all stem from the erroneous idea that we are not part of nature and can do with this planet whatever we choose. We are closer to the angels than the beasts, even though the former are merely a product of our imagination. We can eat bats, pollute the oceans, and burn all the fossil fuels, but everything will be fine because we are the masters of the universe. This line of thinking is interlinked with the one that puts humans on an intellectual pedestal. We need a complete overhaul of Western philosophy—with less emphasis on how we are humans and more on how we are animals—to get us out of this mess.

RELIGION AND SPIRITUALITY

There should be no problem except for those who take the ancient books of their religion 100 percent literally. Those people, however, have many other problems as well and probably transgress on a daily basis many of the strict rules set thousands of years ago. However, if these ancient books are taken as an inspiration for a better life, instead of as the infallible truth, there should be no problem understanding how our species evolved and how it compares to other species. This understanding does not keep us from being both human and humane. On the contrary, knowing more about this topic can only deepen our admiration for the complexity of life on Earth and our place in it.

72

ROLANDO GONZÁLEZ-JOSÉ

Rolando González-José is a biological anthropologist at the Argentinian National Scientific and Technical Research Council, director of the National Patagonian Center, and coordinator of the National Program of Genomic Reference (PoblAr). He combines his expertise in quantitative genetics and morphological integration to study past and present human populations. Rolando currently works with a large team of geneticists, medical doctors, bioinformaticians, and other biological anthropologists to consolidate a national biobank of genomic and associated metadata representative of the diversity expressed in Argentina and neighboring countries.

YOUR BEGINNINGS

My family has a farm in northeastern Patagonia (my birthplace) near a marine Tertiary deposit full of fossilized shark teeth, crabs, dolphins, sea urchins, etc. As a child, I remember walking across the hills and the steppe on long wonderful summer days with my brother and friends, picking up my first fossils. Looking back, I realize that these boys weren't walking randomly across the field; they were instinctively looking for patterns in the ground and the sediment layers. Such a natural impulse to systematize the information surrounding us is, perhaps, one of the more atavistic instincts that connect human life with its more ancestral form (such as hunting-gathering). Some kind of scientific systematization of the natural world, in this case the field of paleontology, helps us connect our observations of the natural and social worlds. Incorporating scientific approaches to solve the problems of daily life is the kind of liaison that needs to be cultivated in schools and media to fight against inequality and to promote the advancement of our societies' quality of life.

GAME CHANGER

The idea that humans and Neanderthals, as well as other hominids, admixed in Europe and Asia during its shared history in such hotspots is a key discovery, in my view. This triggered reinterpretation of many of the patterns that we see in the genetic, phenotypic, archaeological, and linguistic record. The evidence indicating that such admixture events were more frequent and complex than previously thought is a striking stimulus to revisit many of our previous models on human micro- and macroevolution. We all were raised in an academic environment that didn't realize that our lineage is as prone to hybridization events as that of any other primate.

AMAZING FACT

The most amazing aspect of human evolution was the onset of culture as a powerful adaptation that, paradoxically, enabled a relative independence from the classical suite of environmental factors that shaped the fate of our sister species [chimpanzees and bonobos] and triggered the emergence of a totally new and amazing environment. Indeed, our approach to environmental factors is permeated by adaptive features that emerged in our lineage. Today, sociocultural traits such as sedentary habits, tobacco consumption, and access to public health services are environmental factors that capture our attention as scientists when approaching, for instance, the evolutionary aspects of human disease.

TIME TRAVEL

I cannot abandon my scientific background, so first I would develop a research protocol to record some field data on food consumption, health status, and anthropometric and metabolic data. Then I would visit a population in the Fertile Crescent during the transition between hunter-gathering and incipient agriculture. This would be a wonderful place and time to explore how early agriculturalists faced such a dramatic transition between two very different lifestyles. Many aspects of modern life, including life in complex societies, nutrition behavior, and disease patterns, emerged as a consequence of the agricultural revolution. However, the details and consequences of this transition are fuzzy when viewed from the archaeological record. A time machine would be one way to recover powerful data in the field!

EVOLUTIONARY LESSONS

Technological shifts affecting the whole population drive important changes at all levels: genetic, phenotypic, behavioral, and symbolic evolution. The main lesson I would extrapolate to modern societies is that we must collaborate for the worldwide advancement of life quality and human development. We should be alert to the importance of population or nationwide shifts that incorporate more and more people in health care systems, improve environmental quality, and guarantee human, cultural, economic, and social rights to everyone. In my view, there are no true solutions at the individual meritocratic level. Helping people understand the importance of shifts at the population level is the most valuable advice that we can provide as scholars of human evolution.

RELIGION AND SPIRITUALITY

They are compatible with evolutionary thought only if we accept that spiritual and religious views of the world are an interesting cultural trait. Perhaps it helped to consolidate early societies around a common belief system and provided rules needed to consolidate networks to guarantee access to benefits or some kind of stability that counteracted external, uncontrolled shifts. In many ways, religions still accomplish this social function. By deeply accepting the evolutionary origin of religions, the study of human evolution is compatible with a religious view of the world. Of course, my reflection only addresses personal viewpoints. I am convinced that states and societies need a strict nonreligious organization to guarantee access to rights to all their inhabitants, no matter their origin or system of thought.

ADVICE

 Via the study of human societies, we can connect the knowledge and understanding of human diversity with an acceptance of it as a point of departure for better, more inclusive, and more fair modern societies. As specialists on human evolution and diversity, I think part of our work is to explain that we do not have to "tolerate" diversity but should accept it as one of the most ancestral and recurrent biological and cultural traits that defines the evolutionary history of our species. Let's use the study of human diversity as a powerful argument for building more inclusive societies.

73

KRISTEN HAWKES

Kristen Hawkes is a distinguished professor at the University of Utah's Anthropology Department and is a U.S. National Academy of Sciences member. Her research combines behavioral ecology in hunter-gatherers with comparisons of our closest living relatives (chimpanzees) and mathematical modeling. Kristen's studies build on the "grandmother hypothesis," which links the evolution of distinctive human longevity to the reliable economic productivity of ancestral grandmothers.[1] This hypothesis may explain key ways that human life histories and social relationships differ from those of great apes, including our postmenopausal life span, slower maturation, shorter birth intervals, pair-bonding habits, bigger brains, and an appetite for cooperation that begins in infancy.

YOUR BEGINNINGS

It was not a moment but a constellation of events. As a cultural anthropologist, I'd done ethnographic research in Highland New Guinea for my dissertation. Like many other cultural anthropologists in the late 1970s, I viewed as mistaken arguments from evolutionary biologists about the power of sociobiology to help explain human affairs. And I thought some of my own field data could show that. A search for help with my planned analysis led me to Eric Charnov, then in the Biology Department at the University of Utah. Charnov, an evolutionary ecologist, was a patient teacher who directed me to readings in a field I had known nothing about. I was gobsmacked by the possibility of actually explaining things like sex differences and aging.

At the same time, Jim O'Connell, a paleolithic archaeologist, joined the Utah Anthropology Department. O'Connell had already recognized that ideas from evolutionary ecology might address some of the big transitions in the human past. He drew my attention to the sequence of shifts that the fossil and archaeological records seemed to document. He had studied Alya-warra foragers in central Australia, aiming to connect patterns uncovered by archaeology to the hunting-gathering behavior that produced them. We began to apply some of Charnov's modeling tools to his Alyawarra data.

I was talking about these ideas in a class in which Kim Hill had enrolled. Hill had just returned from a Peace Corp assignment in eastern Paraguay with Ache foragers. His interest in the ideas and his Ache experience and friendships presented an irresistible opportunity to measure the variables in foraging models Charnov had built for nonhumans to see whether hypotheses from those simple models could explain resource choices made by foraging people. The perspective of evolutionary ecology brought patterns in the wider living world to bear on human behavior, and our Ache work produced results that demonstrated probable payoffs for doing more of it.

GAME CHANGER

When the Ache fieldwork began, I assumed evolution of our human radiation was largely explained by the hunting hypothesis, which dominated the textbooks I was reading. Sherwood Washburn's mid-twentieth-century articulation of that hypothesis seemed persuasively comprehensive, accounting for the stone tools and bones of large animals that comprised the earliest archaeology and the fact that among ethnographically known hunter-gatherers hunting is generally men's work. Pair-bonds, nuclear families, and sexual divisions of labor are absent in our closest living evolutionary cousins, the great apes, but are ubiquitous among modern humans, including hunter-gatherers.

The hunting hypothesis turns on the proposition that our lineage evolved when forests retreated in ancient Africa and hunting was a better way to make a living in the spreading savannas. Because hunting interfered with child care, mothers paired with hunters who brought kills home to provision their mates and offspring. Our Ache observations were, in my view (although not in some of my Ache project colleagues), a direct challenge to the fundamental assumption that hunting is paternal provisioning. Ache prey were cooked, butchered, and distributed widely with no preferential share to the hunter's own partner and offspring. The hunter credited with the kill didn't even have a role in the distribution. This clear finding made the invitation from Nick Blurton Jones to join him in fieldwork with Hadza foragers in northern Tanzania especially welcome. Hadza country may be the best modern analogue of habitats that witnessed the initial human radiations. The big ungulates targeted by Hadza hunters and coresident carnivores are similar to those represented in the earliest archaeology. This was a chance to use O'Connell's ethnoarchaeological eye to document the treatment of large carcasses, including patterns of butchery, transport, assemblage composition, and site structure. Hadza are now far from ancestral hominins, but they face ancient problems as blade-using bipedal foragers in a world of carnivore competitors. The Hadza project was a chance to document foraging and time allocation by sex and age, and to link them with local ecology. (Blurton Jones's 2016 book[2] should be required reading for anyone interested in anthropological applications of evolutionary ecology.)

Our quantitative observations showed meat to be widely distributed by the Hadza, as it was among the Ache. But the much larger Hadza prey drew many to the kill itself—including women and children—to join in eating what Hadza label "the peoples' meat," which approximates what economists call "public goods." Moreover, our accumulating records showed an average daily success rate of 3.4 percent per hunter—about a month of failures for each success. Clearly not a strategy to keep hungry children fed. This posed an obvious question: knowing that most of the meat will go to others and all get shares of any taken, why do men spend time targeting big animals? If household provisioning is their aim, why not focus on smaller more abundant game less subject to attention and appropriation? Men who were never successful at getting a big carcass in our records still got shares. Watching people converge on a carcass, many of them armed, made it obvious that attempts to claim a kill as private property would face costs too high to pay. Remarks at kill sites often included "where's mine?" All know or quickly learn who is credited with the kill. As with Ache, the benefit to the Hadza hunter is not more meat for himself, it goes to all. His own benefit is the credit: his reputation. Others are interested in a man's success as a hunter because that indicates his desirability as an ally, and his danger as a competitor. Of all the ways a man might demonstrate his qualities, hunting reputations are of special interest to others because, in addition to information about him, better hunters bring more shares of meat to claim.

When it comes to eating every day, it's the reliable resources acquired by women that are crucial. The key role of grandmothers emerged from our Hadza data, not because we were looking for it but because we were quantifying foraging return rates and time allocation for both sexes and all ages. The high return rates of postmenopausal women and the greater time they spent at the most energetically expensive task of digging deeply buried tubers was unexpected. We were also surprised at the active foraging of youngsters. Little ones tried, but they were too small to earn enough for their own needs. Accumulating records showed they depended on their mother until she had a new baby; then it was grandmother's productivity that subsidized them.

Hadza dependence on staples that young children are too small to handle effectively themselves, and the role of grandmothers in supporting weaned youngsters, suggested tradeoffs that might explain the evolution of human postmenopausal longevity. Other apes (and mammals generally) wean babies when they can fully feed themselves. Nursing infants begin to pick and eat the same foods their mothers are eating as she carries them along while feeding herself. Human babies also consume more than mother's milk while still nursing, but their supplementary foods are acquired, processed, and supplied by others. Humans are able to bear next babies after very short intervals because weaned dependents are subsidized. Ancestral mothers probably faced the same tradeoffs. They could rely on foods that youngsters can't handle and bear their next offspring before the previous one can feed itself because the weaned children were subsidized. Formal modeling by mathematician colleague Peter Kim continues to investigate the effects of grandmothering subsidies. Simulations of his two-sex agent-based models show that beginning at an ancestral great-ape-like life history equilibrium, with very few females surviving their fertility, grandmothering subsidies drive the evolution of increasing longevity to a new equilibrium with hunter-gatherer fractions of postfertile females.

An unexpected bonus is the explanation that our grandmothering life history provides for the pervasive human habit of pair-bonding. Persistent pairing, which distinguishes us from the other great apes and most other mammals, is attributed to paternal provisioning in the hunting hypothesis. We have repeatedly found that high failure risk and wide appropriation of meat shares make hunting a poor strategy for household provisioning. Ubiquitous pair-bonding in human societies seems better explained by the male-biased sex ratios in the fertile ages that accompany our postmenopausal longevity. As life spans lengthened in both sexes and the end of female fertility changed little, increasing numbers of old males expanded the competition for each paternity. In most mammals, including most primates, sex ratios in the fertile ages are female biased, and males try for multiple mates. When adult sex ratios are male-biased, mate guarding becomes the strategy that gains more paternities. In humans, reputations among men affect whether a man can claim and keep a mate. Attention to this male competition helps explain the recurrent emphasis on hunting reputations among hunter-gatherers, the pervasive importance of male alliances, and status competition among men across societies.

AMAZING FACT

The implications of our grandmothering life history for human features just summarized seem amazing. But I must add to that the connection between longevity and our big human brains. Only recently I began to attend seriously to the work of neuroscientist Barbara Finlay and her colleagues, which continues to confirm that across the mammals—including humans— the final size and composition of brains depends on duration of development. We have big brains because of our slow maturation, which is a consequence of our longevity. Brain components scale for reasons Finlay's work explains: the neocortex is always a bigger fraction of larger brains because stem cell pools are larger after more cycles and the order of events in the process of brain development is astonishingly regular across the mammalian radiation. From this perspective, our neocortex is not larger because of special selection in our lineage for intelligence. It is exactly the proportion expected given our final brain size, which has come with the evolution of our longevity.

This is especially important for explaining distinctive features of human cognition. Our ontogeny combines slower neural maturation with early weaning, notably earlier than expected for a nongrandmothering mammal with our longevity. Sarah Hrdy [chap. 75] has long emphasized the important fact that human mothers rely on allomothers [an individual, other than the actual biological mother, that acts as such] to stack dependents. Consequently, human infants confront a novel cognitive ecology in which mother's attention is distributed. Unlike other ape babies that have full maternal commitment as a birthright, human infants—helpless as they seem—are socially precocious. Very early in neural maturation their actions and responses engage regard from mothers and others, which is crucial for their welfare but not guaranteed. From this perspective, our appetite for shared intentionality and the concern about reputations that make us lifelong cultural learners evolved as survival strategies in ancestral infants facing a grandmothering socioecology.

DRIVING FACTORS

The retreat of forests in the African Pliocene and spread of grass-lands is the presumed initial driver because of the role played by dependence on savanna foods with particular characteristics. In forests, ancestral hominid populations could rely on fruits and leaves as great apes continue to do now. Carried by their mothers, nursing infants could begin picking and eating the same foods that she was eating. As drier, more open habitats spread, and seasonal swings steepened, plants that sequestered nutrients in geophytes flourished: opportunities that ancestral populations could exploit. Ancestral adults could colonize more open habitats to take advantage of savanna resources, pursuing and processing more than enough food for their own consumption. But, as with modern humans today, young-sters were too small to earn high enough return rates to feed themselves. If mothers did not retreat with the forests, they would have to nurse longer and then subsidize offspring past weaning, investing more in each one and lengthening their birth intervals. However, unlike mothers' milk, these foods could be supplied to infants and juveniles by others big and strong enough to acquire and process them.

Of special importance, scramble or interference competition is not an issue with the savanna foods of interest because of their abundance. Amounts available for consumption depend on extraction and processing efforts. Foragers get mutualistic advantages from gregarious acquisition. With deeply buried geophytes, initially increasing rather than diminishing returns for additional effort make it advantageous to accumulate piles for bulk processing rather than extracting, processing, and consuming items one at a time. All such economies of scale multiply with cooking where start-up costs for cooking fires net increasing marginal benefits for mutual processing in piles.

In contrast to hand-to-mouth, eat-as-you-go foraging, production of food in lumps also results in opportunities for others to appropriate shares. This was noted by Richard Wrangham [chap. 95] and colleagues in association with their review of evidence for the importance of cooking in human evolu-tion. They inferred that risks of thievery for batches of resources would be grounds for enlisting a mate as guard. But we've drawn attention instead to the likely appropriators most immediately at hand: dependent juveniles, and (as noted previously) we've found alternative hypotheses about pairing more consistent with the available evidence.

EVOLUTIONARY LESSONS

Understanding the evolutionary roots of our life history and the character of our social cognition is directly relevant to ideas about mating, parenting, education, and politics at neighborhood, regional, and global scales. Issues of male-male competition, status, and sex differences in social and reproductive strategies seem especially relevant in our current political world.

SPECIAL?

The hypotheses here about observed similarities and differences among humans, the other great apes, and our more distant living cousins assume that we have all been shaped by an evolutionary history of natural selection. The hominid family we belong to has the longest life, latest maturation, largest brains, and latest ages at weaning of all primate taxa. Those shared features probably characterized our most recent common ancestors. If so, the even greater longevity and postmenopausal life stage of humans, our even later maturation and larger brains, yet earlier weaning and shorter birth intervals, are distinctly derived. The grandmother hypothesis outlined here can explain that. In this account, the evolution of our grand-mothering life history also resulted in a cognitive ecology for infants that propelled the development of our social appetite for shared intentionality and concern about reputations. Male-biased sex ratios in the fertile ages intensified male status competition and favored mate guarding. It seems unlikely that this can all be correct. But the connections are compelling enough to demand continued pursuit.

74

LESLEA HLUSKO

eslea Hlusko is a research professor at the Spanish National Center for the Study of Human Evolution (CENIEH). Leslea is a multiproxy researcher investigating the genetic causes of mammalian evolution to better contextualize human variation and evolution. Her research combines developmental, genetic, paleontological, and neontological (based on living species) sources of information. Thanks to her research, we can approximate the genetic variation of extinct species and vice versa. In addition, Leslea complements her lab research with active field paleontology research projects in Kenya, Tanzania, and Ethiopia.

YOUR BEGINNINGS

When I was twenty years old and in college, I had the opportunity to participate in the Koobi Fora Field School in Kenya, at the time run by Harry Merrick. He would walk around camp listening to music on his Walkman, and it sunk in that adults are real people with emotions too (and maybe need an escape from us students from time to time). But the most formidable moment of that field school experience for me was following Craig Feibel on the geology walks. The time-depth represented in the rocks, the antiquity of the Earth, was mind-blowing to the point that, for the first time, I was truly distracted from the concerns of everyday life. Jumping from rock to rock—the remains of algae mats from millions of years ago— I realized that I am an incredibly small part of a very large phenomenon. That field experience revealed my need to understand how people fit into this evolutionary history.

GAME CHANGER

My first day of classes as a graduate student at Penn State University included a seminar with Ken Weiss [chap. 94]. He introduced developmental genetics by talking about segmentation in body plans and the then-recent discovery of *Hox* genes. I have never looked at organisms the same way since. After reading Rudy Raff's book, *The Shape of Life*,[1] I decided that there had to be a way to bring this genotype:phenotype approach to the study of human evolution. As my enthusiasm for the new scientific field of evo-devo took off, Alan Walker, my primary PhD advisor, pointed out that fossils are the only evidence of what ancient life forms actually looked like. My entire research career has been unfolding the vision inspired by Weiss, Raff, and Walker—to meaningfully integrate the evolution of form, as documented in the fossil record, into the science of genotype:phenotype mapping.

AMAZING FACT

I am in awe every time I think about how each of us is the result of millions of successful parent-offspring events in an unbroken chain of descent since the origins of life. Every single mother before me was evolutionarily successful, whatever that looked like at the time she lived (be it as a buglike critter in the ocean 550 million years ago, a scurrying little mouselike thing avoiding dinosaurs 100 million years ago, or a woman brave and desperate enough to leave her family and immigrate to the United States a century ago). This fact helped me trust my body during pregnancy and childbirth and have the confidence to listen to my instincts when it comes to mothering my child. Billions of years of evolutionary success stand behind my maternal intuition. How bad could it be?

TIME TRAVEL

Twenty thousand years ago at the height of the last ice age. What was it like to live before agriculture? Before monotheism? Before the patriarchy of our culture, modern medicine, and written language? Before we knew the shape of the continents and the vast expansiveness of our own species? This was the practice of humanity that selection operated on to result in much of the biological variation we see in our species today. I would love to know what it was like to live in that world.

FUTURE EVOLUTION

Plastic, pesticides, and other sources of hormone-mimics are altering our physiologies in ways that we are only just beginning to grasp. These hormone disruptors are everywhere, and their influence on our body composition and fertility are dramatic. Genetic variation in our species very likely makes some people more or less susceptible to these effects; the former will not fare well and the latter will, evolutionarily speaking (i.e., in their ability to have children). The consequences of our passion for temporary plastic bags and flawless strawberries may be the strongest selective pressure on our species for millennia.

EVOLUTIONARY LESSONS

Through the study of evolution, we reveal what the major points of selective pressure have been in the past, and in so doing, our attention is directed to where these points will be in the future. Increasingly, we see that this pressure is on our fertility—from conception through to the next conception. If we really want to make sure we have grandchildren and they have grandchildren, the science of human evolution shows us that we should be paying a lot more attention to factors that challenge our fertility, such as the human-made hormone-mimics in our environment.

SPECIAL?

The forces of evolution have led to a number of extreme biological adaptations, such as the deep sea fish who have many more piezolytes (small organic structures that counteract the crushing pressure of the ocean's depth) than do fish living at shallower depths, or the Antarctic crustacean that can survive long periods of time without food because its body is 70 percent fat (making it the fattest animal on Earth). No other species asks questions about its own evolution and then compiles the answers into book form. Thanks to millions of years of evolution, our biology enables us to have such an extreme culture. Although some people may think this makes us no more special than the deep sea fish or the Antarctic crustacean, I think humanity's ability to bring so many different species to extinction makes us very special indeed—and not necessarily in a good way.

INSPIRING PEOPLE?

 My grandmother 1,000 generations ago, who lived during the last ice age.

75

SARAH HRDY

arah Hrdy is a professor emerita at the University of California, Davis and an elected member of the U.S. National Academy of Sciences. Sarah is an anthropologist and primatologist who has made outstanding contributions to evolutionary psychology and sociobiology, especially regarding human (and primate) female behavior. She has written prize-winning books such as *Mothers and Others*,[1] which won the J. I. Staley and Howells prizes. Among other recognitions, *Discover* magazine named her one of the fifty most important women in science. Currently, Sarah combines working on a new book (on males this time), growing walnuts, and habitat restoration on her family's farm in northern California.

YOUR BEGINNINGS

As an undergraduate at Radcliffe College, then the women's part of Harvard, I worked on summer medical projects in Honduras and Guatemala. In the evenings, I gave "hygiene" lectures to Spanish-speaking adults. After graduation (in 1969) I went to Stanford to learn to make educational films related to public health. While I was there, I audited lectures in Paul Ehrlich's population ecology class. Ehrlich had just published *The Population Bomb*.[2] Listening to him I recalled Irv DeVore (in one of the first undergraduate courses around on primate behavior) mentioning these monkeys in India that, supposedly because they were "crowded," were killing babies. Without any intention of becoming an academic, I applied to graduate school in anthropology at Harvard and Berkeley because these were the only places I could think of where I could study these monkeys. I wanted to go to India and do what I naïvely imagined would be a case study of the pathological effects of high population densities. It was the middle of the year, but because I had done well as an undergrad (graduating summa cum laude), Harvard invited me to begin that January. I dropped all my classes at Stanford and embarked on a career in research without ever consciously thinking it through.

The summer after my first full year of graduate school I went to India with woefully little preparation, and by the end of that brief field season studying langurs at Mount Abu, I realized that my starting hypothesis was entirely wrong.[3] Males were only attacking infants when they entered the breeding system from outside it. This class of infanticide had all the earmarks of Darwinian sexual selection: competition between one sex, typically but not always males, for access to the other, with the result for the loser not necessarily death but less reproductive success, and with the result for a female that her last reproductive choice was canceled. By that time, I was hooked. There was so much I wanted to understand, particularly about females (e.g., why on earth would a female mate with a male who had just killed her infant?). Thereafter, while fulfilling graduate requirements and teaching, I returned to the langurs at Mount Abu and Ranthambhore every chance I got. But as my research became better known and both the National Science Foundation and the Smithsonian started to fund it, our political problems in India began. But that is another story (told elsewhere).[4]

Truth to tell, I don't think I ever "decided" to become a researcher. Questions intrigued me, and I was driven to seek answers. It was years and years before I even thought of myself as a scientist. In 1990, after I was elected to the National Academy of Sciences, I happened to mention my doubts to a revered mentor, Robert Hinde, who told me not to worry: "A scientist is someone who has a question and organizes knowledge to try to answer it." Oh, okay. I can do that.

GAME CHANGER + AMAZING FACT

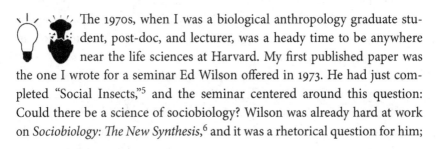 The 1970s, when I was a biological anthropology graduate student, post-doc, and lecturer, was a heady time to be anywhere near the life sciences at Harvard. My first published paper was the one I wrote for a seminar Ed Wilson offered in 1973. He had just completed "Social Insects,"[5] and the seminar centered around this question: Could there be a science of sociobiology? Wilson was already hard at work on *Sociobiology: The New Synthesis*,[6] and it was a rhetorical question for him;

he already knew the answer. Meanwhile, I was hooked, drawn to the power of the comparative method for answering questions about the evolution of allomaternal care and exploitation of infants. The main influences on my evolutionary understanding included the usual suspects. Darwin, of course, especially his *Descent of Man and Selection in Relation to Sex*,[7] as interpreted by Bob Trivers who was teaching at Harvard then, along with evolutionary theorists I met through Bob, particularly George Williams and Bill Hamilton, who became important mentors. (Much has been written about sexist bias at Harvard in those days, all too much of it well founded, which was all the more reason support from Wilson, Williams, and Hamilton meant so much to me.) Back in the seventies, these men were at the intellectual center of the paradigm shift from group selection to focusing on selection at the level of the individual. It was the era of sexual conflict, "selfish genes,"[8] and such vivid aphorisms as Ghiselin's "Scratch an altruist and watch a hypocrite bleed." I am not sure when the intellectual climate started to shift, but by the end of the century there had been a sea change. The new holy grail was figuring out why some creatures (humans in particular) are so eager to help and cooperate with others.

Three children and a broad reading of the literature later convinced me that an ape with the life history attributes of *Homo sapiens* could never have evolved unless mothers had had help, and a lot of it. As I wrote in *Mother Nature*,[9] our ancestors had to have been cooperative breeders. Alloparents as well as parents must have helped care for and provision offspring. What impresses me most about our species is "original goodness," how attentive human babies are to what others are thinking and feeling, especially about them (a terrifying realization to someone charged with caring for them), the extraordinary eagerness of human babies to ingratiate themselves and to please others, and the pleasure people get from feeding or giving things to others. I already knew that babies born in litters or clutches or in species with shared care often have to vie for maternal and allomaternal attention. But I had no theoretical framework for thinking about this until (with help from evolutionary psychiatrist Randy Nesse) I started to understand entomologist and evolutionary theorist Mary Jane West-Eberhard's formulation of social selection. To her, social selection is a subset of Darwinian natural selection based on competition between individuals to be chosen by others as a partner or recipient of help, with sexual selection a specialized subset of social selection. Together with West-Eberhard's view that "the causal chain of adaptive evolution begins with development,"[10] I felt I was beginning to understand why humans evolved to be so much more subjectively interested in and engaged with others, so emotionally different from other apes in these respects. West-Eberhard's views about social selection and the role of development in evolutionary processes are transforming the way I think about my own species, which for an anthropologist is saying something.

TIME TRAVEL

 On scientific grounds, my answer should probably be back to the Miocene to get a better idea about what little-known apes were doing back then, especially just how monandrous or polyandrous females were. How much interaction did males have with infants? Were they anything like siamangs, with males helping out with older infants, or more like gorillas, protective and tolerant of babies but not directly caring for them? Or were they like chimpanzees, aloof from babies but not indifferent to the needs of nearly weaned orphans? But it would be so difficult to follow these brachiators through lush Miocene forests, so hard to actually see what they were doing, and I am too old. So no, I would ignore the big lacuna in primate evolution and choose instead a *Homo erectus* gathering spot in the Early Pleistocene, not right at the beginning, but maybe 1.8 million years ago. I could sit and watch exactly who held newborns? Six-month olds? And especially learn how old babies were when mothers and others (but who?) began to introduce solid foods, and learn how and when infants were weaned. Were they handed food or kiss-fed premasticated treats? I am biased by wonderfully detailed observations of child care among African hunter-gatherers provided by Mel Konner, Nicholas Blurton Jones, Alyssa Crittenden, Barry Hewlett, Courtney Meehan, and others; so I tend to project backward from their descriptions. But this may be ill-advised. Even with what little evidence we have, from teeth for example, I don't know what I need to know about the timing of any of this. Exactly when did little hominins start to be weaned so much earlier than other apes are? And afterward, how did they get fed? What sort of gestures and sounds did they make to solicit food or convince others to feed them? Who responded? What sort of contacts were there between adult males and babies? When did food sharing get started, and how did individuals behave during these transactions? I have no idea when mealtime started to take on the profound social significance shared meals and feasts have today. Definitely by the end of the Pleistocene, but how did this human universal get started? It's frustrating to be so awash in presumptions and guesses but unable to check them. My children remind me that "Mom, the Pleistocene is not the only scene." But I can't help it; that's where I would like to spend some time.

DRIVING FACTORS + SPECIAL?

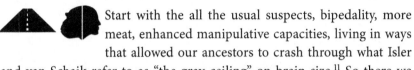 Start with the all the usual suspects, bipedality, more meat, enhanced manipulative capacities, living in ways that allowed our ancestors to crash through what Isler and van Schaik refer to as "the gray ceiling" on brain size.[11] So there we were, bipedal apes with the cognitive and manipulative capacities of the extant great apes confronting the demographic challenge of surviving in the face of the Pleistocene's erratically fluctuating rainfall and temperature shifts. Why are we the only apes around then who made it through, ending up with higher levels of cooperation and sophisticated language? Impressed by how early in life other-regarding, helpful impulses emerged in both sexes, I became increasingly dissatisfied with traditional explanations based on cooperative hunting of big game or intergroup conflict. For me, our unusual path began with our ancestors converging on the same solution as hundreds of other social creatures have when confronted with unpredictable resources or other challenges to successfully rearing dependent young. Other group members, fathers, possible fathers, and matrilineal kin start to help mothers care for and provision their young, what ornithologists call "cooperative breeding." (The French "elévage coopérative" is a better descriptor because in other spheres so-called cooperative breeders can be quite competitive.) Reliance on allomaternal care transforms the terms on which infants develop. With competition for contingently delivered care, youngsters just a bit better than others at appealing to and ingratiating themselves with others fared better. Over generations traits conducive to intersubjective engagement were favored by Darwinian social selection. Although endowed with cognitive capacities similar to those of other apes, we are emotionally very different: not just interested in what others are thinking and intending but actually caring about what they think and feel, and what they are likely to think and feel about us. Apes already pretty good at declarative communication were incentivized to take communication up a notch, laying the groundwork for sophisticated language. Grammar is a handy way of helping others understand what we mean. I think of it as analogous to the whites of our eyes, the white sclera

typical of humans but not other apes. It's a cooperative way of helping others understand not just what we are looking at but how we are looking at it. I agree with Matt Cartmill [chap. 6] that when we look at creatures like other apes "we are less singular than many people like to think." But in our quest for intersubjective engagement with others, we are emotionally very unusual apes with very different needs and preferences.

FUTURE EVOLUTION

Based on current trends, looming possibilities are mostly depressing. For example, if "naked apes" destroy the planet via nuclear war or climate change, the likely mammalian successors will descend from fossorial creatures subsisting in underground bunkers. Given how hardy and resistant to radiation-induced cancers they are, despotically organized eusocial naked mole-rats strike me as good candidates. But let's say bipedal, technologically adept, symbol-generating apes *do* manage to persist, I'm not at all sure that millennia hence the term "human" as we use it today will still apply. Even today, 98 percent of children born in "advanced" societies are likely to survive to breeding age, but their survival is no longer connected to youngsters learning to monitor others or being good at ingratiating themselves with potential caretakers, and energetically costly traits no longer favored by selection tend to fade away. These descendants would likely be bipedal, intelligent, and technically proficient in realms I cannot even dream of, but I am not sure I would like them. This is what I was thinking at the conclusion of *Mothers and Others: The Origins of Mutual Understanding* when I wrote: "If empathy and understanding develop only under particular rearing conditions, and if an ever-increasing proportion of the species fails to encounter those conditions but nevertheless survives to reproduce, it won't matter how valuable the underpinnings for collaboration were in the past. Compassion and the quest for emotional connection will fade away as surely as sight in cave-dwelling fish."

RELIGION AND SPIRITUALITY

No question about it, some religions are totally antithetical to evolutionary views, not to mention extremely harmful to human well-being and even nature generally. For us pagans, however, almost everything about nature is compatible with a spiritual view. Apparently, the same is true for some Buddhists, or at least so I gather from His Holiness the Dalai Lama's interest in science. Spiritual humanism also seems compatible with science and evolutionary viewpoints. I am heartened by this because I believe humans, especially in times of crisis, need the consolations that a religious view of the world can offer: "There are no atheists in foxholes." Right from birth, humans have always depended on support from someone bigger and more powerful than they are, so whether or not "believers" do or don't actually recover from trauma and illness faster, believers clearly derive comfort from the sense that they are not facing adversity alone. I still recall something Ed Wilson said once about humans being "easily indoctrinated. They seek it." For better, and mostly for worse, Wilson was probably right. People, many of them anyway, are desperate to belong, so let's at least make the best of it. A nice aid in that regard is provided by evolutionary psychiatrist Andy Thomson's wonderful little book *Why We Believe in God(s)*.[12] When it came out, I sent copies to each of my goddaughters.

Full disclosure: Some years ago I learned I was to be awarded an honorary degree in humanistics from a university in the Netherlands. Mystified by the news, I called my friend Frans de Waal [chap. 71], my corecipient and also a Dutchman, to ask: "Frans, what is humanistics?" "Something they do in Europe," he replied. I went, and it was a fabulous experience. The more I learned about humanistics and spiritual humanism, the more I found the philosophies on offer tremendously appealing and satisfying, completely compatible with both my understanding of evolution and my reverence for the natural world. Not long after when my cousin's daughter married my son's best friend and I was asked to perform the ceremony, I went online and for $15 was ordained as a minister of Spiritual Humanism. It's a nice fallback.

76

NINA JABLONSKI

ina Jablonski is the Evan Pugh Professor of Anthropology at The Pennsylvania State University, and an elected member of the U.S. National Academy of Sciences. Her research includes the evolutionary history of Old World monkeys and apes, as well as the evolution of modern human skin pigmentation (and its implications for health and social aspects). This last project includes an active field component examining the relationship between skin pigmentation and vitamin D production in the global context of migration and urbanization (she received a Guggenheim Fellowship to conduct this research). Her books on the subject include *Skin*,[1] *Living Color*,[2] and *Skin We Are In* with Sindiwe Magona and Lynn Fellman.[3]

YOUR BEGINNINGS

I cannot remember a time when I wasn't doing research. From earliest childhood, I was looking at nature and wondering. Those early experiences would not qualify as formal research, but they were of seminal importance to me. I realized that I loved looking at things—fossils, leaves, salamanders, paramecia—for long periods of time to try to figure out why they looked the way they did. I was particularly transfixed by fossils because many were "otherworldly" forms of great beauty. The idea that I could see and begin to understand the lives of creatures that lived tens of millions of years ago was enchanting and riveting. I never made a decision to "do research" because I realized I had been doing it all of my life.

GAME CHANGER

D'Arcy Thompson's classic book, *On Growth and Form*,[4] first published in 1917, fundamentally changed how we think about the generation of diversity in biological forms. The book demonstrated through clear narrative and compelling illustrations that different forms of plants and animals could be produced by modification of growth patterns and trajectories. Published long before we knew much about genetics, and even less about evo-devo, Thompson showed that all changes of form—including those occurring in the course of human evolution—are phenomena of growth. As an undergraduate biology major, I was mesmerized, and this book strengthened my interest in studying the evolution of morphology.

AMAZING FACT

Human evolution has been a combined biological and cultural process for at least the last 2 million years. The fact that humans are able to quickly modify their behaviors and the proximate and distant environments they inhabit means that we have extraordinary adaptability. We can modify what we eat, where and how we live, and—to some extent—what we look like. Compared to other mammals, we are truly amazing in these regards!

FUTURE EVOLUTION

Human evolution will continue to be shaped by accelerated transmission of cultural information via electronic and other remote means. The extrasomatic forms of knowledge storage we now use (computers, cell phones, etc.) will be replaced by implanted devices operating in genetically modified human bodies. Information will increasingly be shared involuntarily, resulting in enhanced abilities to think and act collectively, for good or ill. Most humans in a hundred years will be cyborgs. "Normal" humans will be extinct within a millennium as a result of the direct or indirect effects of climate change or the impulsive acts of malevolent despots.

RELIGION AND SPIRITUALITY

 The study of human evolution makes it possible for us to appreciate that shared belief systems have been one of the essential features of human "success." The most widespread of shared belief systems are spiritual, in the broadest sense. As far as we can tell, people have been concerned with spiritual matters for as long as we could communicate flexibly with one another using some form of language. The study of human evolution is sterile and incomplete without study of the origins and manifestations of human spirituality. The fact that this domain of human knowledge has been considered for decades to be unscientific and unworthy of investigation in an evolutionary context is a tragedy for science and the human condition. Holding spiritual beliefs is part of being human and certainly is compatible with studying human evolution!

INSPIRING PEOPLE?

 The Dalai Lama.

77

CLIFFORD "CLIFF" JOLLY

C lifford "Cliff" Jolly is professor emeritus in New York University's Department of Anthropology and a fellow of the American Association for the Advancement of Science. During his fifty-plus-year-long career, Cliff has broadened the study of human evolution by bringing together research on primatology, paleontology, and genetics. His paper, "The Seed-Eaters,"[1] represents a keystone in human evolutionary theory. Cliff's contributions have been recognized by the AAPA's Charles R. Darwin Lifetime Achievement Award (2011), and the International Primate Society's Lifetime Achievement Award (2018).

YOUR BEGINNINGS

I can't identify any such moment! As far back as I can remember, and certainly long before I knew of an occupation called "anthropology," from which one could, maybe, make a living, I have been fascinated by natural history and human prehistory, as well as cultures and societies that were distant and, from an ethnocentric viewpoint, exotic. As a teenager, I remember Carleton Coon's "Seven Caves" as an inspiration, and earlier, a story by Louis and Mary Leakey about Acheulians hunting baboons at Olorgesailie. Even further back, in infancy, I think the genius works of Beatrix Potter provided me with a mental template on which no sharp line separates humans from other creatures.

GAME CHANGER

\ | / I would identify Darwin's *On the Origin of Species*[2] and Thomas
( ) Huxley's *Evidence as to Man's Place in Nature*[3] as the works that deci-
▼ sively brought biology, and human biology in particular, in line with
other natural sciences, showing that supernatural intervention need not be
postulated to explain the diversity and adaptations of organisms. Darwin-
ism has, of course, been enormously amplified and somewhat modified over
the decades, but no subsequent development in biology has had a comparable, immediate impact on mainstream Western culture, both scientific and
nonscientific.

In my own life, semiformal (i.e., college undergraduate) exposure to
evolutionary theory was gained from works by the neo-Darwinians Huxley
and G. G. Simpson. With hindsight, however, I can see that the ground was
thoroughly prepared by the natural histories; guides to seashells, insects,
and plants and birds; and junior-level accounts of fossils and geology that
dominated my schoolboy reading. These works didn't need to refer directly
to evolutionary relationship, adaptation, or natural selection; in the mid-
twentieth century—before neo-creationism had gained a foothold—these
evolutionary phenomena were generally accepted as proven parts of general knowledge.

TIME TRAVEL

For a destination in the past, I would choose Pliocene East Africa, a
setting with faunal diversity unmatched by any on Earth today. Imagine experiencing an ecosystem with four coexisting elephant species!
At least two, and probably more, species of hominins seem to have coexisted,
along with a variety of large, ground-living monkeys—how did they parti-
tion the ecosystem and interact as individuals?

If allowed a second jaunt, I would touch down in the future about half a
million years after the anthropogenic ecosystem collapse and the population
crash that now appears inevitable. It would be fascinating to see which organisms survive and begin to diversify in a recovering, largely human-free world.
Would they include any nonhuman primates? The past few million years of
climatic fluctuation has seen alternating periods of ascendancy of hominoid
apes and cercopithecoid monkeys in the Old World, and it is by no means
certain that surviving humans (if any) would out compete their monkey

cousins. Would the time traveler find that, freed from human competition, some monkey populations of the Old or New World tropics had paralleled aspects of early hominin (or even human) behavioral adaptation?

DRIVING FACTORS

In constructing a plausible scenario, we must avoid the error of accounting for evolutionary novelties in terms of ultimate consequences rather than immediate adaptive value. Consider hands, for example. Hominins did not acquire their adept hands in order to play violin concertos, nor, I'd argue, was dexterity initially acquired as an *adaptation* to tool use. To explain its origin, the relevant question is not "could this hand have made a stone tool?" but rather "what new circumstance would favor a long-thumbed, short-fingered hand in a prehominin ape?" For over fifty years, I've argued for feeding behavior as being key to this and other hominin features. The prehominins, somewhat resembling chimpanzees, lived in a seasonal but relatively moist savanna woodland. As apes do, they spent most of their waking hours seeking, picking, preparing, and ingesting a varied diet of fruits, flowers, leaves, stems, berries, seeds, nuts, invertebrates, and occasionally a small vertebrate. Feeding was a time-limited activity in which efficiency was at a selective premium. As a basal hominin population expanded its range into drier savannas, its dietary spectrum remained broad but de-emphasized larger, fleshy fruits, which are rarer in such habitats. The deficit was balanced by eating more of the smaller, drier, food items common in more arid habitats. These included grass and sedge seeds, and the pods and seeds of shrubs like *Acacia* and *Grewia*, whose bushy structure, often combined with formidable spines, makes them unclimbable. Such foods require skillful harvesting and need to be thoroughly chewed to release their nutritious contents. Reliance on such foods, plucked with a thumb-index grip while sitting upright or standing bipedally on the ground, then ingested and chewed by the handful, would favor the distinctive features of basal hominin teeth, jaws, hands, and hind limbs. This was not a "road to humanity," but it established an adaptive platform from which a distinct, human phase of hominin evolution could launch.

FUTURE EVOLUTION

 The general lesson from the broad sweep of biological evolution is that all species ultimately become extinct, with a minority having left behind one or more "offshoot" populations different enough to be called new species. There is no way of telling what fate awaits *Homo sapiens*. I suspect that humans in a hundred years will look very much like us, but as population movements in response to climate change become more insistent, we can predict more admixture and less geographic structuring of genetic and phenotypic variation.

As for the longer view, we should first discard the common sci-fi vision of future humanity as bulbous-headed, mega-brained geniuses. There is no evidence today that large brains or high IQ are favored by natural selection. Even if the fossils suggested such trends in our immediate pre-*sapiens* ancestors, which they do not, we cannot assume that they would be sustained into the future. By 100,000 CE, the Earth will have experienced at least one glacial-interglacial cycle, with unknowable but undoubtedly profound effects on humans and their artificial ecosystems.

All we can say is that physical evolution will be the outcome of a combination of the basic forces—natural selection (differential survival and reproduction), stochastic change (drift), and mutation, the origin of new inherited variation. The direction that human physical evolution will take is impossible to predict because it will depend on the interaction of these forces with each other and with human cultural evolution and demographic structure, each of which includes a strong element of chance.

EVOLUTIONARY LESSONS

I suppose it depends on what one considers "helpful." On balance, knowledge and factual truth are always preferable to ignorance and falsehood. It should be thought-provoking, and I hope a bit chastening, to learn that the same rules of evolutionary process that produced the diversity of living species also apply to ourselves, and that we sit on one twig on a minor branch of the primate tree. Aside from possibly curbing excessive triumphalism on behalf of the human species, will having a fuller, more evidence-based evolutionary history of the human species and its antecedents increase human well-being? I doubt it, but the same can be said of the vast majority of human productions. From a selfish viewpoint, we can hope that enough of the public, and the holders of purse strings, will agree that knowledge of our extinct ancestors and cousins, and the vanished worlds they inhabited, are a worthwhile source of inspiration, imagination, and entertainment.

SPECIAL?

Most scientists today would agree that humans are clearly vertebrate animals, mammals and primates, situated among the great apes on the branching tree of relationship. In that sense, we are not special. But another vision of humanity's place in nature is expressed every time a clinical researcher, for instance, describes drug trials conducted on "humans and animals." The omission of "other" before "animals" expresses a pre-Darwinian (really, pre-Enlightenment) dualism, rooted in some, but by no means all, religious dogmas. This contrasts humanity, endowed with souls, with animals lacking them. Though the terminology changed, the dualism survived the Darwinian revolution. It has been all too easy and convenient to assume that because nonhuman species cannot name and verbally describe their thoughts, fears, pleasure, emotions, and moods, they do not experience them. Arguably it is more ethical, and no less "scientific," to assume that in the absence of evidence to the contrary, there is continuity between the human species and others in this respect, as in other aspects of their biology. Here studies of great apes and other primates are especially relevant. Similarly, we should assume that cognitive processes, like other aspects of biology, have diversified rather than ascending a ladderlike progression "upward" to humanity.

Then there is the darker ecological sense, in which the human species is indeed "special" among animals in the same way that a malignant tumor is special among the tissues of a living body. By a combination of chance events and favorable mutations in technology and social organization, the human species has broken free of the constraints of the ecosystem in which it evolved, exponentially expanding in numbers, relentlessly invading, subverting, and degrading habitats and food webs one after another with, seemingly, no justification other than to prove that it can be done. Special indeed!

RELIGION AND SPIRITUALITY

Well, it all depends. I don't see how fundamentalist biblical literalism could ever be reconciled with an honest study of the paleoanthropological facts. But a majority of nonfundamentalist believers (including the Holy See) evidently manage to reconcile their faith with acceptance of evolution as a fact. As for spirituality, I think that a sense of awe and wonder at the beauty, complexity, immensity, and strangeness of the universe, even of the tiny speck of it on which we live, is not only compatible with science in general, and evolutionary science in particular, but derives inevitably from it.

ADVICE

To develop an evolutionary view of these topics, immerse yourself in nature, even in your own backyard. If you can, spend time in the tropics, especially tropical Africa; visit habitats as close to their natural state as possible, and observe wild primates, especially monkeys and great apes. As practical advice, read, read, read, seeking out works just above your present level of comprehension. Don't worry about understanding the technical and professional literature completely at a single reading; you can always go back and read it again. Don't discount the power of your intuitive insights to suggest new explanatory hypotheses, but at the same time learn the analytical and statistical skills needed to critically test them.

Beware of triumphalism in science. Certainly, question all accepted dogma, but at the same time, do not dismiss or underestimate the contributions of previous academic generations, back through the eighteenth century and beyond. They may not have had the benefit of digital technology and the internet, but they were just as smart and thoughtful as you! Don't assume that the present answers to paleoanthropological questions are final and optimal. Similarly, don't be triumphalist on behalf of *H. sapiens*—don't adopt a secular version of "man" as the "pinnacle of creation."

78

JON KAAS

Jon Kaas is the Distinguished Centennial Professor of Psychology, associate professor of cell and developmental biology, and professor of radiology and radiological sciences at Vanderbilt University. He is also an elected member of the U.S. National Academy of Sciences and an elected fellow of the American Association for the Advancement of Sciences. Jon has made key discoveries about the brain organization in mammals, including describing new areas in the cerebral cortex. His main research interests are the evolution, development, and functional organization of sensory-perceptual, cognitive, and motor systems, especially in primates, and he has an extensive publication record (55,000+ citations!).

YOUR BEGINNINGS

The direction of my future research dramatically changed at two critical times. Fist, when I decided where to go for graduate school. I wanted to work with Harry Harlow at the University of Wisconsin on the development of behavior in monkeys, and it looked like that would happen. But I learned that Harlow would be on leave when I arrived, so I accepted my second choice, which was to join a newly funded program at Duke on the anatomical and physiological bases of behavior. In my first year, I was invited to join the lab of new faculty member Irving Diamond, and I studied functions of the auditory cortex in cats. This, in turn, led to starting a post-doc in the best laboratory recording auditory neurons in monkeys at a research center headed by Clinton Woolsey. Although I intended to join the group studying the auditory cortex, after I joined the lab Woolsey told me that they had enough post-docs studying the auditory system and that

I would have to study the visual system because no one else in the group was doing that. With the help of others doing short-term rotations from neurosurgery in the lab and a visitor from Singapore, we gradually figured out what worked and what did not. There was no pressure to publish, so my early efforts to study the visual system were not published, to my relief. Back at Duke, the Diamond lab had changed its focus from the auditory system in cats to comparative studies of the visual system. Because of this, and my limited work on the visual system in Woolsey's lab, a former classmate of mine at Duke, Bill Hall, joined me in the Woolsey lab to map the visual cortex in hedgehogs. The next summer I went back to Duke to map the visual cortex in squirrels and tree shrews in the Diamond lab. By this time, we had figured out what we were doing and were able to publish papers on our work.

I returned to Wisconsin to start my own lab, and Woolsey suggested that a visiting graduate student, John Allman, should work with me on his interest in the pulvinar. The rest is history. Over the next few years, we were able to provide visuotopic maps of known visual structures and discovered other visual representations in the cortex. We even mapped part of the pulvinar. Not long after that, I moved to Vanderbilt and expanded my research to include anatomical, histological, and single neuron recording procedures, as well as studies on auditory, somatosensory, and motor systems, mostly in monkeys but also in other mammals. Most important, I had the opportunity to learn Diamond's early approach to studies of the auditory system, and Woolsey gave me great freedom to learn about the visual system. This background made it easy to expand into other approaches and research topics.

AMAZING FACT

Perhaps the most important advances in the evolution of modern humans are the use of language and the control of fire. Language seems almost essential for the formation and functions of large social groups. The evolution of language is hard to study because we are the only speaking primate. However, much of the processing needed for word selection and identification now appears to be the product of offshoots at the well-known dorsal and ventral cortical streams of visual processing, which function in action selection and object identification. These streams are well developed in all primates but appear to be even more so in the large-brained macaque monkeys and humans where they have expanded functions. In all primates, these streams involve somatosensory, auditory cortex, motor and premotor cortex, as well as the basal ganglia. Yet only humans, and perhaps some of our ancestors, have expanded these streams to include language.

In our ancestors, fire was important for processing food so that more nutrients could be extracted to support bigger brains. Fire was also important in protecting our ancestors from predators and providing some light so that activities could be extended into the night. It is hard to imagine the evolution of modern humans without fire and language. The importance of fire for humans is well illustrated in the old movie *Quest for Fire*.

TIME TRAVEL

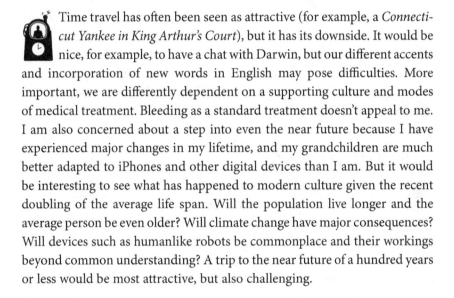 Time travel has often been seen as attractive (for example, a *Connecticut Yankee in King Arthur's Court*), but it has its downside. It would be nice, for example, to have a chat with Darwin, but our different accents and incorporation of new words in English may pose difficulties. More important, we are differently dependent on a supporting culture and modes of medical treatment. Bleeding as a standard treatment doesn't appeal to me. I am also concerned about a step into even the near future because I have experienced major changes in my lifetime, and my grandchildren are much better adapted to iPhones and other digital devices than I am. But it would be interesting to see what has happened to modern culture given the recent doubling of the average life span. Will the population live longer and the average person be even older? Will climate change have major consequences? Will devices such as humanlike robots be commonplace and their workings beyond common understanding? A trip to the near future of a hundred years or less would be most attractive, but also challenging.

DRIVING FACTORS

The line of evolution that resulted in humans simply walking away from the line that produced the large apes. They walked into a newer, drier environment with fewer trees. Even though they walked poorly, walking allowed the forelimbs to be used for carrying and other tasks, and they could see farther when upright and escape up a tree if necessary. This step set our ancestors apart from other apes. It is not surprising that modern-day apes sometimes walk for short distances, but no apes always prefer walking.

SPECIAL?

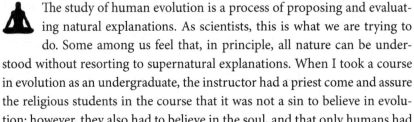 When compared to other mammals, humans are special because we have very large brains that are densely packed with neurons. Our remarkable and unique cognitive abilities result from the way our large brains are organized into systems and subsystems. These systems are largely in the same place but less expansive and subdivided in other primates, so we can study their organization and functions in other primates. We can learn a lot about what is important in the human brain from the simplest and smallest primate brains, and it is often easier to come to an understanding of a complex system by studying less complex versions. We also recognize that some of the complexity and functional capacity of the human brain is present in the brains of the great apes, so it is important to study these apes. As clear differences exist, studies of human brain development and organization, and human abilities, are also rightly studied. We also need further study on how the perceptual and cognitive abilities of apes and humans are similar and different because some of the differences are remarkable.

RELIGION AND SPIRITUALITY

The study of human evolution is a process of proposing and evaluating natural explanations. As scientists, this is what we are trying to do. Some among us feel that, in principle, all nature can be understood without resorting to supernatural explanations. When I took a course in evolution as an undergraduate, the instructor had a priest come and assure the religious students in the course that it was not a sin to believe in evolution; however, they also had to believe in the soul, and that only humans had a soul. Later in life, I became more aware of evidence that humans are very capable of holding two or more incompatible beliefs at once. Thus, students could believe that there are natural explanations for the evolution of humans but also believe in a proposed element, a soul, that is beyond scientific evaluation. Obviously, the need for the supernatural is very strong in humans.

79

LEAH KRUBITZER

L eah Krubitzer is a professor in the Department of Psychology and is head of the Laboratory of Evolutionary Neurobiology at the University of California, Davis. Her research examines the anatomical connections and electrophysiological properties of neurons in the neocortex, the portion of the brain responsible for perception, cognition, learning, and memory. Through comparative studies, her work determines which features are shared by all mammals and which are unique to each species. In this way, she reconstructs how complex brains (e.g., the human brain) evolved from simpler ones. Her work has earned her multiple awards, including the MacArthur Fellowship, commonly referred to as the Genius Grant.

YOUR BEGINNINGS

I have thought about this issue on several occasions and realize that only a few times in life do we intentionally choose a path knowing that our decision will significantly alter the course of our life. For me it was when I chose to do a post-doc in Australia. My path had been meandering, and I sort of stumbled into Jon Kaas's [chap. 78] laboratory at Vanderbilt University through a circuitous route. I was lucky, and I loved what I was doing. I worked hard, published papers, and got my PhD in the lab of a world renown comparative neuroscientist. Following this, I was able to secure external sources of funding to do a post-doc in a well-established laboratory at MIT. I was on my way, topping off my work at Vanderbilt with a pedigree from MIT. What more could an "up and coming" wish for? Well, quite a lot actually.

I realized I was doing what other scientists thought I should do instead of what I wanted to do—go to Australia and work on the duck-billed platypus and echidnas. I believed that understanding the cortical organization of these two extant monotremes would provide insight into the neocortex of the earliest mammals. In Australia I had to catch most of the animals I studied, including the duck-billed platypus. Out in a rowboat in the middle of the night in some remote location on the underside of the world catching platypuses I knew that my life was changed forever. I would spend the rest of my life looking at the brains of as many mammals as I possibly could. Not just for the pleasure of watching them behave in the real world rather than a laboratory but also to understand how complexity emerges in the neocortex: the rules of brain construction and the constraints imposed on evolving brains. To understand these issues, it was critical to look at a variety of brains to appreciate what evolution has produced, common features of organization as well as derivations in brain organization. Catching living animals in their natural habitat also made me realize that I couldn't study the neocortex in isolation if I wanted to understand how it changes over evolutionary and developmental timescales. Rather, it was important to look at the neocortex in the context of the body and the environment in which the animal develops, behaves, and ultimately evolves.

GAME CHANGER

An important game changer for me when I was a graduate student was reading a 1977 essay written by François Jacob titled "Evolution and Tinkering."[1] This essay explores the possible—what could have been had there been a small "tweak" to any aspect of the brain or the body early in the evolution of life on this planet. It was the first time I realized that not "everything goes" in evolution. I was introduced to the concepts of contingencies in and constraints on the evolutionary process, and of being saddled with our evolutionary history. Despite these constraints, evolution has tinkered with existing parts to reconfigure them and co-opt them for different functions to create what appears to be remarkable novelty. I finally understood that evolution is the continuing process of limiting our options. This essay prompted my very first review, subtitled "Are Species Differences Really So Different."[2] Although diversity seems enormous, mammals are still constrained in their body plan and aspects of brain organization. Even though we fly and swim and grasp objects, we are all tetrapods, and a number of homologous brain structures control our limbs. We maintain both of these body and brain features, even in the absence of use, likely due to developmental contingencies in body and brain construction. Jacob articulates this far better than I:

> It is hard to realize that the living world as we know it is just one among many possibilities; that its actual structure results from the history of the earth. Yet living organisms are historical structures: literally creations of history. They represent, not a perfect product of engineering, but a patchwork of odd sets pieced together when and where opportunities arose. For the opportunism of natural selection is not simply a matter of indifference to the structure and operation of its products. It reflects the very nature of a historical process full of contingency.

Even now, as I read his words, I feel that excitement, that first thrill of opening my eyes to the world!

AMAZING FACT

What blows my mind is the infinitesimally small chance that something as complex as the human neocortex could have emerged over the course of evolution. For example, early in the evolution of nervous systems, some 600 million years ago, a small "tweak" in how cells communicate could have fundamentally changed how subsequent life evolved and what brains look like. But the tweak didn't occur, so we are stuck with the weird, seemingly inefficient action potential, which has constrained nervous system evolution and how future nervous systems will evolve. Because of these constraints, evolution can "tinker" with the pre- and postsynaptic elements to modify the conditions under which action potentials occur, but it cannot change the fundamental properties of the action potential (sodium and potassium channels in the cell membrane). It's not exactly humans that I'm impressed with but the thought of all the possibilities that could have been, and the process of evolution itself.

TIME TRAVEL

I would like to travel 60 million years into the future, well after humans have destroyed life on this planet. I think 60 million years is plenty of time for evolution to generate new forms of life, or to reconfigure what may have survived human habitation. I want to know what form life will take—imagine the possibilities!

FUTURE EVOLUTION

It is critical to consider how cultural evolution affects brain development to produce rapid changes to the cortical phenotype that occurs within a lifetime or over several generations. It still amazes me that the industrial revolution occurred less than three hundred years ago but that human brains and behaviors have been altered radically since that time. All behaviors are mediated by the brain, so this means that the brain has changed via nontraditional evolutionary mechanisms (e.g., changes in DNA sequence). Perhaps that is one of the defining features of humans: the ability to alter our brain organization and connectivity rapidly based on the context in which we develop. It is impossible to predict exactly what humans (and their neocortex) will look like in the future. However, given what we know about the factors involved in cortical development and the importance of sensory-driven activity and behavioral affordances in shaping sensory and motor regions of the neocortex, we can make some viable predictions. For example, changes in our manual behavior (e.g., texting, using a keyboard) can alter motor cortex organization. Gaming and other types of virtual reality technologies can alter visual system development and response properties of neurons as well. If we consider social learning, language, and culture as simply complex patterns of sensory stimuli impinging on our developing brain, then changes in our culture will continue to snowball, and our brains will reflect these changes. However, not everything goes, so it is also possible to predict what won't happen to our brains or bodies.

My question to you, Sergio, is this: Do you really think humans will be around in 100,000 years or even 1,000 years? We have disconnected ourselves from tangible resources and have proxies for proxies of resources, and we are destroying some of the irreplaceable resources that actually exist. Furthermore, humans have entwined themselves with their own technologies, making us unique biohybrid creatures. Our future is bound to these technologies, which in an ever-changing physical and cultural context could seriously handicap us.

But wait!

I'm a tweak

You're a tweak

Oh, the possibilities.

SPECIAL?

Not really. Of course, we have specializations, but we follow the same rules of construction and are enslaved by the same constraints as other mammals. Humans are of interest to me as just another example of a mammal—a piece of the evolutionary puzzle. To understand the human condition, in particular human brains, we must study other primates and other mammals. How else could we appreciate what is a human specialization versus a general feature of primates or of all mammals? How could we understand how complexity emerges over the course of evolution by studying one of the most complex brains that has emerged? How could we appreciate the true beauty of the evolutionary process, and the extraordinary variability in brain organization, body morphology, and behavior that have emerged on our planet?

ADVICE

Think beyond the human condition and take a broader picture that includes understanding brain-body-environment interactions of mammals in general. Think about the context, both physical and social, in which the human brain and body develop, including conspecifics and heterospecifics. Think about self-organization and how complexity could emerge, and about the constraints imposed on evolving brains and bodies.

INSPIRING PEOPLE?

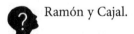 Ramón y Cajal.

80

SUSAN LARSON

S usan Larson is a professor in the Department of Anatomical Sciences at Stony Brook University. Her research centers on the functional mor-phology of humans and other primates, especially our closest living relative, the chimpanzee. Susan uses electromyography (EMG) to under-stand muscle function, often combining it with kinematic motion analysis and force plate studies to analyze locomotor adaptations. Her work and criti-cal perspective continue to be the "gold standard" when interpreting the locomotion of ancient members of the human lineage, such as *Australopithe-cus afarensis* or early *Homo* species.

GAME CHANGER

I believe the discovery of the A.L. 288–1 partial skeleton, dubbed Lucy, and the other associated *Australopithecus afarensis* material was a genuine game changer in how we view our own evolution.[1] Increased brain size and the ability to make tools were always assumed to be the first steps along the human evolution pathway. The *A. afarensis* material clearly showed that upright, bipedal posture preceded any changes in brain size or the habitual use of tools, and it brought new importance to the functional interpretation of postcranial morphology. Certainly people had looked at postcrania prior to that point and tried to infer something about how early hominins got around, but much more attention was paid to skulls and teeth and questions about the relatedness of taxa. The Lucy partial skeleton not only offered the opportunity to say that early hominins were bipedal but also to investigate the nature of their bipedalism. How one goes about inferring function from morphology is the way in which this discovery directly influenced my career. A traditional first step in the functional analysis of any fossil is to compare it to living taxa whose behavior we know, and when similar morphology is identified, infer similar behavior. However, when fossils present as unique collections of features with varying behavior correlates, it becomes critical to understand just how form and function are related. Biomechanical principles allow us to formulate hypotheses about how some morphological feature contributes to the performance of some behavior, but such hypotheses should always be tested before being applied to the interpretation of fossils. The focus of my research is on using laboratory methods and living subjects of comparative taxa to test form/function hypotheses. Muscles and bones often don't work the way we think they do, and time and time again we've had to revise our functional inferences in light of empirical observations on living subjects.

TIME TRAVEL

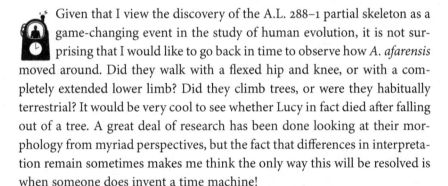 Given that I view the discovery of the A.L. 288–1 partial skeleton as a game-changing event in the study of human evolution, it is not surprising that I would like to go back in time to observe how *A. afarensis* moved around. Did they walk with a flexed hip and knee, or with a completely extended lower limb? Did they climb trees, or were they habitually terrestrial? It would be very cool to see whether Lucy in fact died after falling out of a tree. A great deal of research has been done looking at their morphology from myriad perspectives, but the fact that differences in interpretation remain sometimes makes me think the only way this will be resolved is when someone does invent a time machine!

DRIVING FACTORS

Basically this question comes down to why did the ancestors of humans assume an upright, bipedal posture? For many years it was assumed that orthogrady, which entailed not only truncal erectness but also a broad thorax, invaginated vertebral column, and dorsally positioned scapula, was linked to manual suspension based on the analysis of extant apes. I think one of the most important new insights that has come out of recent Miocene ape research is that orthogrady and manual suspensory postures are not inextricably linked. There was perhaps some hint of this in *Sivapithecus* postcranial material that suggested quadrupedal habits in the ancestors of *Pongo*,[2] but for me it was the discoveries of partial skeletons of *Pierolapithecus*[3] and *Hispanopithecus*[4] that made it clear that orthogrady, probably related to vertical climbing, evolved independently from suspensory behaviors. The adoption of suspensory posture apparently occurred later and possibly multiple times within the hominoid clade.

Therefore, the last common ancestor (LCA) of the African apes and humans was definitely orthograde, but it is unknown whether the origin of the suspensory behaviors of the African apes was prior to or subsequent to this split. The full spectrum of viewpoints exists from those who believe the LCA was chimplike to the degree that it utilized knuckle-walking to those who interpret *Ardipithecus* as having retained a very primitive body plan using palmigrade above-branch locomotion, and presumably climbing, but bipedalism on the ground. There is general consensus that the evolution of habitual bipedalism has something to do with being more terrestrial, but what exactly? If the LCA was a knuckle-walker, why not remain a knuckle-walker? It suits the terrestrial habits of African apes. If the LCA was a palmigrade branch walker, why not continue to be a palmigrade quadruped on the ground? Many monkeys are. There certainly have been many proposals for the origins of habitual bipedalism, including being semiaquatic, needing to see over tall grass, or feeding from bushes, all of which seem like good reasons to stand up but don't really explain why a protohominin would remain standing. Physiological explanations, like reducing the profile of the body to the sun for thermoregulation, or taking advantage of the greater energetic efficiency of bipedalism compared to quadrupedalism, are appealing, but the thermoregulation hypothesis doesn't explain why no other African mammal has abandoned quadrupedal posture for similar reasons. In addition, bipedalism isn't less energetically costly that quadrupedalism. All that is left are proposals related to the need to be able to carry things: tools, weapons, food, water, infants, something. If survival really does depend on being able to bring stuff with you, maybe it is a plausible reason for becoming bipedal.

EVOLUTIONARY LESSONS

 When I was a graduate student, many people supported what was known as the single species hypothesis [Milford Wolpoff (chap. 68)

conceived this hypothesis], which posited that there was only ever one species of hominin that gradually evolved over time from the primitive *Australopithecus* to the more advanced *Homo erectus*, culminating in modern *H. sapiens*. The basic premise was that being humanlike was so unique that it constituted an adaptive niche that only one species could occupy at a time. The niche evolved along with the hominins to include growing dependence on an ever-increasing complex material culture that eventually allowed for the worldwide distribution of humans. Today we know that multiple hominin species existed in the past, some even in the same area at the same time, but most became extinct. After our early evolution in Africa, it is true that hominins began dispersing around the world. This movement was episodic, with populations occupying an area for some time and then disappearing as conditions changed. I believe the takeaway message from this is that (1) there is no inevitability to human evolution, (2) there is no single way of being human, and (3) although our material culture allows us to live in a variety of environments, it can't protect us entirely from changing conditions.

SPECIAL?

Yes, humans are special in that our intellect and the technology it has allowed us to develop have given us the ability to influence, if not control, most of the rest of life on this planet. In the course of our history, we have probably been responsible directly or indirectly for the extinction of many animal species, including other types of hominins. Negative impacts on other forms of life continue today from the local level of habitat destruction to the global scale of climate change. Yet humans are also special in that we ask ourselves questions like this. Because the potential impact of our actions is special, we need to be special in recognizing the great responsibility this places on us. Does human welfare justify anything and everything? In the past, the answer has mostly been yes, but we can consider the effects of past actions and choose to be different in the future. The study of apes and other primates has shown us that some of our best characteristics—love, friendship—as well as some of our worst—lust, warfare—are not unique to humans. This helps us understand some things about ourselves, but more important, it allows us to see ourselves in other animals, and I hope that helps us recognize that their welfare matters too.

INSPIRING PEOPLE?

John Fleagle (if you haven't already). [I did try to have John Fleagle in this book!] John is remarkable in having very broad interests coupled with a real depth of knowledge and understanding in all of them. When the rest of us are getting lost in the details, he's the one we can count on to see the big picture and ask the important questions. I'm guessing he would have really good answers to them too!

81

ZARIN MACHANDA

Zarin Machanda is the Usen Family Career Development assistant professor at Tufts University's Department of Anthropology. Zarin is the director of long-term research at the Kibale Chimpanzee Project. For the last thirty-five years, this organization has conserved and protected the Kanyawara community of chimpanzees living in Kibale National Park, Uganda. She is also on the board of the Kasiisi Project, a community development organization in Uganda that works with more than nine thousand schoolchildren living around Kibale National Park. Her research seeks to understand the factors that shape the quality and development of social relationships among wild chimpanzees.

YOUR BEGINNINGS

I have a distinct memory of watching a documentary when I was five years old about how they used to send chimpanzees into space. In the documentary, they interviewed primatologists and astronauts. I remember not knowing those words, so I looked them up in the encyclopedia and thought that both of those jobs sounded amazing. I couldn't choose between them, so I told everyone I would become an astronaut primatologist and take care of the chimps that lived in space. It turns out that is not a real job, but it's pretty close to what I actually do!

GAME CHANGER

The discovery that chimpanzees and bonobos are more closely related to humans than they are to gorillas has changed the way we think about all the apes. It has certainly changed the way we frame our studies and bring in comparisons with data from humans.[1]

AMAZING FACT

 Parochial altruism—the idea that we exhibit both a high degree of within group cooperation and a high degree of between group aggression—is fascinating to me. In many ways, chimps are very easy to understand. They are selfishly motivated (even in their cooperative behaviors) and are aggressive both within and between groups. Humans are one of the most cooperative species on Earth, and also one of the most aggressive, and the idea that we might be extremely cooperative in coming together to inflict lethal aggression between groups is mind-blowing.

TIME TRAVEL

I was asked this question when I interviewed for graduate school, and my answer remains the same twenty years later. I would go back approximately 7 to 6 million years to see exactly what the last common ancestor (LCA) between humans, chimpanzees, and bonobos looked like and how it behaved. As someone who studies chimps, we are often using chimps as a model for the LCA, but we also acknowledge that chimps have undergone 7–6 million years of their own evolution. It would be really satisfying to me to know exactly how much they have changed.

DRIVING FACTORS

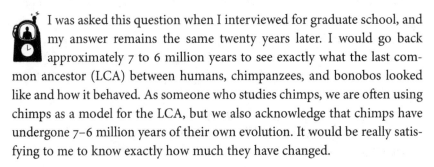 The thing that really sets chimpanzees and humans apart from one another is our capacity for language. Chimpanzees solve their social problems by fighting with each other. Bonobos solve theirs by having sex with one another. But humans have this incredible capacity to solve our social problems and make plans for the future because of our language. Although we were a bipedal ape long before we probably had the capacity for language, I think a lot of the complex social behaviors we see in humans are only possible because of our capacity for language.

EVOLUTIONARY LESSONS

In the last decade or so, there has been a growing interest in evolutionary medicine. Students in my introductory biological anthropology course really find this topic interesting. Evolutionary medicine suggests that we need to consider that our bodies were designed by natural selection to live in very different habitats (at least ecologically and socially different) compared to our modern lifestyles. By understanding how our bodies have evolved, we will have a better understanding of why we might see negative health consequences when our bodies are faced with a very different lifestyle. For example, the increased prevalence of diabetes may be referred to as a mismatch syndrome because we did not evolve to live such a high energy/high sugar lifestyle. We might also examine the evolutionary reasons for difficulties during childbirth and see how we might use this knowledge to improve outcomes for birthing people and their babies.

INSPIRING PEOPLE?

From biological anthropology, I would probably want to ask these questions to either Richard Wrangham [chap. 95] or David Pilbeam [chap. 19]. I have long admired their intellect and ways of thinking about our field. But I would also be particularly interested in how someone far outside of our field might think about these questions. Maybe someone like Carl Sagan, who certainly had a very different perspective on the place of humans in the universe.

82

TOMÀS MARQUÈS-BONET

Tomàs Marquès-Bonet is an ICREA research professor at the University Pompeu Fabra and head of the Comparative Genomics Group at the Institute of Evolutionary Biology (Spain). To truly understand what constitutes unique human features, Tomàs examines the genome variation in living humans and apes and how their population history influences genome diversity. He has contributed relevant work on the evolution of primates and apes, including the genetic evidence for bonobo and chimpanzee past hybridizations and finding a novel and now extinct chimpanzee species from genomic relics in the living species. The pioneering research of his team, always featured in the top journals, is constantly revealing insights into the evolution of humans and our living and extinct relatives.

YOUR BEGINNINGS

I always liked nature, the most zoological application of studying nature, but I never imagined I would do research about it. In my case, it is more about a path (a long path) with a different combination of vital events that finally led me to start a research group. It is true that working in a private company for many years motivated my curiosity toward a more creative style of life.

GAME CHANGER

In my view, genomics as a discipline is a game changer. Whether you work on human genetics or protist evolution, genomics is a tool that has reinvigorated evolution studies and exposed new horizons. Being able to read the sequence information at the population level has indeed transformed our views on population dynamics, conservation, speciation, and many more concepts that are at the root of evolution. If I had to pick one, finding evidence for gene flow among Neanderthals and modern humans has had a big influence in my research.[1]

AMAZING FACT

I am continuously amazed by this question: "How many historical populations of humans were there?" The discovery of Neanderthals, Denisovan, and still today many so-called ghost populations should force us to rethink our own history, which at the end is one of the many fundamental questions of any researcher. I think DNA and the study of ancient molecules (such as proteins) are going to reshape the tree of our species.

TIME TRAVEL

Nonscientifically, I would like to attend any of the premieres of the classical music pieces I still admire today. Can you imagine being among the public for the premiere of the ninth symphony with Beethoven still conducting? But I guess the question was about science, and in that regard, living among dinosaurs would be a fantastic (and dangerous) experience. I would probably pick 60,000 years ago somewhere in Europe to see how Neanderthals and humans did in fact interact.

DRIVING FACTORS

It is, obviously, the big question in my field. Totally open to hypothesis (rigorous or not), the problem is the potential deficit in terms of scientific validation. Curiosity, neoteny, diet changes, genetic changes? Everything is possible and still, today, possibly wrong.

FUTURE EVOLUTION

Evolution is a bit like medicine. It is useful to understand why we are here in the form we have, but it is totally useless in predicting the future. The evolution of the species is highly dependent on the environment, and this is where we lose our capacity of prediction. No one knows how the Earth will be 1,000 years from now (not to say in 1 million years!), hence, it is impossible to know which evolutionary forces will rule over others.

EVOLUTIONARY LESSONS

In general, the words *useful* and *helpful* are not directly linked to basic research. Studying the origins of humans is a scientific endeavor, but it openly overlaps with philosophy, history, and other social sciences and humanities. However, studying our species and understanding processes and characteristics is the base of global education and should (it should) prevent disagreement among populations. Idealistic, I know.

SPECIAL?

We are no more special than *Platypus* is to mammals. Clearly, we have a unique set of characteristics, but in my view, we are just another evolutionary test. We are very social, intelligent, and complex but remain a part of nature after all. Indeed, the only way to understand singularities of the human lineage is by studying the living relatives. The more we know about them, the more precision we will have on our lineage. And this is what I do!

RELIGION AND SPIRITUALITY

That is an easy one! They belong to two different spheres in our culture. I have met incredible scientists who have a strong devotion to religion as well. Fine with me, obviously. I assume the trick is to be able to dissociate the two worlds. If both of them serve your curiosity and existence needs, good for you!

ADVICE

Every three or four years buy general books that summarize the field. Our field is moving extremely fast, and scientific papers need a few years to sediment and consolidate results. There are excellent books out there. Have your mind open, and study humans like you would any other animal in the world. Just with a bit more curiosity.

83

ROBERT "BOB" MARTIN

Robert "Bob" Martin is a curator emeritus at the Field Museum, Chicago. During his fifty-plus-year-long career, Bob has published more than two hundred research pieces on different themes related to humans' biological origins. He has investigated the anatomy of living and fossil primates and their ecology, behavior, reproduction, and molecular evolution. His work has earned multiple prestigious awards, including being elected a fellow of the American Association for the Advancement of Science. In 2013, he published *How We Do It*,[1] discussing the evolution of human reproduction, and he published a regular monthly blog on this topic for eight years in *Psychology Today*.

YOUR BEGINNINGS

I do not recollect making a conscious decision at one particular point in time, but the roots of my research into primate origins were certainly established more than fifty years ago while I was conducting research for my PhD thesis. Because I wanted to learn the German language, as soon as I completed my undergraduate degree in zoology at the University of Oxford, I went to work in the Max Planck Institute (MPI) for Behavioural Research in Seewiesen, Bavaria. Lectures in Oxford by Niko Tinbergen had made me keenly interested in animal behavior (ethology), and when I arrived at the MPI it was directed by another founding father, Konrad Lorenz. [Both were awarded the Nobel Prize in Medicine less than a decade later.] Irenäus Eibl-Eibesfeldt had established a colony of tree shrews at the MPI, and I decided to initiate research into their behavior. At that time, it was widely accepted that tree shrews are the most primitive living representatives of the order Primates, and I hoped that studying their behavior would yield a better understanding of the earliest stages of primate evolution. To begin my research, I had to get the tree shrews to breed reliably, which naturally led to an emphasis on reproduction. I quickly discovered that tree shrews had a highly unusual pattern of maternal behavior; babies were left in a separate nest and suckled only once every forty-eight hours. Primates are characterized by intensive maternal care, so this led me to question why tree shrews had been included in the order Primates. To tackle that issue, I had to review a wide range of evidence on anatomical features of both living and fossil representatives. In particular, I embarked on a detailed examination of the *procedures* used to infer relationships between species in evolutionary trees (phylogenies). Thereafter, reproductive biology and methods for phylogenetic reconstruction remained central pillars in my research. Interests in brain evolution and the need to consider scaling of all biological dimensions to body size (ranging from gestation periods to the size of the neocortex) eventually followed as additional primary interests.

GAME CHANGER

In my own research, one of the most significant breakthroughs—which exerted a pervasive influence on all subsequent developments in the reconstruction of phylogenetic relationships—was recognition of the need to be more precise about similarities shared by species in a tree. Naturally enough, tree-building began with assessment of similarities and differences. Broadly speaking, it seems obvious that closely related species are likely to share more similar features than species that are only distantly related. It was, however, long recognized that similarities may be developed independently in separate lineages through the process of *convergent evolution*. To reconstruct branching relationships within a tree, a concerted effort is needed to eliminate convergent similarities between species and to recognize *homologous* similarities that reflect genetic continuity following descent from a common ancestral condition. However, a decisive breakthrough occurred rather late in the day (not long before I started my PhD research) with the realization that successful discrimination between convergent and homologous similarities is not, in itself, sufficient for reliable reconstruction of phylogenetic trees. The key point is that, for any given target group (e.g., primates), the last common ancestor has an array of characters that can be retained as homologous similarities in any descendants. Such *primitive homologous similarities* (*symplesiomorphies*) provide no information at all about branching patterns in the tree. Only *derived homologous similarities* (*synapomorphies*) that develop later within the tree and are retained in descendants can indicate branching relationships. This crucial distinction was initially clearly stated in a 1950 book published in German by the entomologist Willi Hennig, who first presented his ideas in English in a 1966 paper.[2] Prior to Hennig, shared retention of primitive similarities (often reflecting differential rates of evolutionary change from a common ancestral condition) commonly misled investigators attempting to reconstruct phylogenetic relationships. For instance, earlier evolutionary trees for extant hominoids (great apes and humans) often portrayed the great apes as a single cluster separate from the human lineage. This reflects the fact that all great apes are more similar to one another than to humans because all have retained many primitive features from the ancestral hominoid, whereas multiple novel features have appeared over the course of human evolution subsequent to divergence of our lineage from the great apes. Only an analysis carefully seeking to apply a clear distinction between primitive and derived features can reveal that the African apes (gorillas and chimpanzees) are actually more closely related to the human lineage, and that the orangutan lineage was the first to diverge in the evolutionary tree for great apes and humans.

AMAZING FACT

My primary interests in human evolution in the general context of primate biology have been at the organismal level—morphology, physiology, ecology, and behavior—but my greatest amazement was actually caused by unexpected discoveries arising from investigation of the genome. The most basic surprise was the gradual realization that DNA in the human genome is not entirely, or even predominantly, composed of gene sequences that code for production of proteins. In fact, in a relatively recent development, the Human Genome Project revealed that the number of coding genes is far smaller than originally thought. From initial estimates of about 100,000 genes, the number has now been pared down to 20,000–25,000, and it seems to be roughly similar in other mammals. In fact, only 1 to 2 percent of the human genome consists of genes coding for proteins; noncoding DNA constitutes the remaining 98.5 percent! Some of that noncoding DNA serves well-established functions—notably as *introns* (noncoding sequences included within individual genes), regulatory sequences, and RNA genes—but a very large proportion has no obvious function. Indeed, the term "junk DNA" was initially used dismissively for that seemingly functionless component of the genome. This major component of the genome includes various subtypes, many of which exist as multiple copies of sequences scattered throughout the genome. However, certain inserted sequences known as *retrotransposons* are of special interest because they have viral origins. Viruses typically have a simple RNA genome, and some of them (appropriately known as *retroviruses*) possess genes that permanently insert DNA sequences into the host genome, "hijacking" the genetic machinery of the host cell. In fact, over 8 percent of the human genome consists of *retroposons*—sequences originally inserted by viruses and subsequently inherited along with the rest of the host's genome. Over time, mutations cause functional decay of the inserted viral sequences so they no longer represent a health risk to the host. But here lies a truly amazing twist in the story: one of the few RNA genes possessed by retroviruses codes for an *envelope* protein exposed in multiple copies on the outer surface of the viral capsule. That protein plays a key part in rendering the virus physiologically "invisible" to the host's immune system. Now pregnancy in mammals faces the potential problem that the mother's immune system will identify a developing fetus as a foreign object, leading to its rejection. As an adaptation to overcome that problem, independently in several lineages of placental mammals, the old retroviral envelope genes embedded in the genome have been converted to reduce the mother's immune response to a fetus. As is the case for all higher primates including humans, such "repurposed" viral genes are expressed only in the placenta. In fact, long before their viral origin was recognized, the genes concerned were discovered in the placenta of humans and various other mammals and named syncytins. This example beautifully illustrates how natural selection can act on any available genetic material to permit adaptation to new functions.

TIME TRAVEL

I would dearly love to travel back 80 million years to the Upper Cretaceous period, about 15 million years before an asteroid impact wiped out dinosaurs and many other organisms on our planet. My reason is simple: palaeontologists generally state that primates did not evolve until *after* the asteroid impact, whereas my own research (bolstered by many findings from phylogenetic trees generated with molecular evidence) indicates that primates originated *well before* that impact. This stark difference in interpretation exists because of a general tendency to treat the known fossil record as far more complete than it is in reality. Accordingly, the origin of any group is commonly dated from the earliest known fossil representative. In the case of primates of modern aspect, the oldest known reliable fossil evidence comes from the earliest Eocene epoch, 55 million years ago. However, the earliest well-documented fossils already share numerous distinctive features with modern primates, whereas not a single representative is known from the preceding Paleocene epoch (65–55 million years ago). So a direct interpretation of the fossil record, treating it as fairly complete, is that primates of modern aspect originated at some time between the asteroid impact and the end of the Paleocene. Yet there is no fossil evidence of the earliest stages of their evolution! Together with biological anthropologist Christophe Soligo and mathematicians Simon Tavaré and Richard Wilkinson, I participated in a long-term project to survey the fossil record for primates and develop a way of assigning a date to the last common ancestor, allowing for gaps. Among other things, our research revealed that we have so far sampled, at most, 3 percent of the primate species that ever existed. Allowing for the inevitable underestimation of the time of origin resulting from such a very limited fossil record, we estimated the age of the last common ancestor of primates of modern aspect to be 80–85 million years ago. That is up to 30 million years earlier than the earliest known fossil representative, so skepticism is to be expected. But consider an even more striking alternative that is connected with the observation that *all* placental mammals of modern aspect first appear abruptly in the fossil record at the beginning of the Eocene. A stunning case is provided by bats, which account for a fifth of modern placental mammals, with an estimated 900 to 1,250 extant species, compared with 400+ for primates. The known fossil record for bats is even more limited than that for primates, with a large part documented from the single fossil site of Messel in Germany. Bats first appear in the fossil record in early Eocene deposits around the world, with no convincing evidence of prior relatives. Yet the earliest known bats already possessed all key defining features. Is it conceivable that all of those features, including the origin of flight, evolved within the space of a few million years? If I were able to travel back 80 million years, I would confidently expect to find early relatives of both bats and primates.

DRIVING FACTORS

Over the course of my fifty-plus-year academic career, tremendous progress has been made in tracing the overall trajectory of human evolution from our common ancestry with the African great apes at least 8 million years ago. In tandem, the pace of discovery of highly informative new fossil material has accelerated markedly, including numerous spectacular finds, and the reliability of dating fossil deposits has radically improved. Considerable progress has also been made in the recognition and interpretation of key anatomical adaptations distinguishing modern humans and apes. Moreover, for the investigation of anatomical features of both extant and fossil representatives, major advances have been made in nondestructive 3D imaging techniques—notably CT scanning combined with novel approaches to the reconstruction of fragmentary specimens. In parallel, developments in molecular anthropology transformed our ability to determine evolutionary branching relationships from data that were far more amenable to explicit quantitative analysis. Radical refinement of theoretical foundations for interpreting the ever-increasing array of comparative information permeated all of these new discoveries and developments.

We know far more today than we did half a century ago about the overall configuration of the hominid evolutionary tree (actually more of a bush) and the evolutionary changes that occurred. However, formulation of convincing hypotheses regarding the selective forces that drove those observed evolutionary changes has remained unsatisfactory. Consider, for instance, the three main anatomical distinctions between humans and great apes: (1) radical size reduction in jaws and teeth (notably the canines); (2) evolution of erect, striding bipedal locomotion; and (3) tripling of human brain size within a few million years. One proposal to explain all of these features together was the overarching savanna hypothesis. However, that hypothesis suffers from the fundamental flaw that comparative evidence from nonhuman primates and other mammals has yielded few convincing cases of parallel development. Compared to forest-living relatives, savanna-living nonhuman primates do not have smaller jaws and teeth, extensive adaptation for bipedal locomotion, or significantly larger relative brain sizes. Moreover, it is becoming increasingly evident that distinctive changes in human evolution began among forest-dwelling representatives, notably involving adaptations for climbing. Despite wide acceptance of this, no properly articulated replacement for the savanna hypothesis has emerged. Many one-off hypotheses have been proposed for jaws and teeth, for upright locomotion, or for brain size expansion, but no real consensus has resulted. Hypothetical explanations for a marked and rapid increase in human brain size have invoked a wide and ingenious range of selective factors, such as foraging skills, social complexity, tool production and use, and development of language. One common flaw has been to seek a single selective factor overriding all others. If one thing is certain, the great and unparalleled expansion of the human brain must have been driven by a host of different factors acting in concert and changing over time. To my mind, we still have a long way to go before we have a satisfactory explanation of the factors that drove human evolution.

FUTURE EVOLUTION

Quite often over the course of my career, at the end of presentations on human origins in the framework of primate evolution, I have been asked to comment on likely future developments. I always begin my answer by noting that the study of evolutionary relationships allows us to infer with increasing reliability what happened in the past. And it also allows us to recognize general principles concerning evolutionary change. However, it is exceedingly difficult to predict what might occur in the future. Although the key process of natural selection is not itself random, changes in the environmental conditions under which it occurs are highly unpredictable. Moreover, it has been claimed that evolution through natural selection has largely ceased for humans because of far-reaching environmental modifications and increasingly pervasive medical intervention. But evolutionary change in human biology is certainly continuing to occur at genetic and physiological levels. At the anatomical level, however, future changes will surely be limited. It is often maintained, for example, that human brain size expansion will continue to occur in the future as a positive development influenced by natural selection. Yet that seems highly unlikely. In contrast, it is quite possible that medical intervention is favoring *negative* changes over time. A good example is provided by Caesarean births. Initially a life-saving procedure performed relatively rarely in emergencies, C-sections have become far more frequent worldwide than dictated by medical necessity. World Health Organization guidelines indicate that a C-section rate of about 10 percent might be medically justifiable, but rates in the USA and Europe are now generally around 30 percent, and in Asia they often exceed 50 percent. Indeed, rates above 90 percent have been reported for private clinics in Brazil and South Africa. Human birth is, of course, particularly challenging because of the tight fit between the neonate and the mother's pelvic birth canal, mainly due to the particularly large neonatal brain. In the past, natural selection presumably acted to eliminate neonates with an overly large brain born to mothers with an undersized pelvis. Excessive use of C-sections removes that selection pressure, and it has been inferred that this is already leading to an increased frequency of difficult births because of a mismatch between neonate size and maternal pelvic dimensions. For certain broad-headed dog breeds (e.g., bulldogs), births by C-section are now required in around 90 percent of cases. Is that where the human species is headed? But the birth process is a very complex feature, and many more examples of simpler evolutionary changes with positive benefits can be found at the genetic and physiological levels. Take, for example, the emergence of *lactose tolerance* in certain human populations over the last 5,000 years or so. All mammals are able to digest lactose (milk sugar) as infants, thanks to the specific enzyme *lactase*. However, that gene is typically switched off at weaning, so older individuals can no longer digest lactose. If raw milk is consumed after weaning, this inability to digest lactose gives rise to digestive problems. In at least five different human populations, there has been separate fixation of a mutation reactivating the control mechanism for the lactase gene, permitting lactose digestion in adults. Such relatively minor changes, driven by natural selection, will surely continue to occur in the future.

EVOLUTIONARY LESSONS

Let me first of all make the point that studying the past is worthwhile in its own right, regardless of whether it provides useful guidelines for the future. It is generally accepted that we need to study our own recorded history, extended into the past through archaeological evidence. Yet many people question the utility of inferring human evolutionary history from biological evidence. Even among scientists, there is a tendency to question the value of research into evolutionary relationships. This undertaking is regarded as "soft science" because experimentation is largely ruled out and interpretations are generally based on inference rather than direct tests of hypotheses. Yet reconstructions of our evolutionary past—right back to the earliest stirrings of life on the planet—have much in common with human history, just with a greatly extended timescale. Indeed, bio-logical reconstructions actually have a more secure foundation in that all are based on a universally accepted fundamental paradigm: evolutionary change under natural selection.

With respect to recorded human history, people tend to think of Georg Hegel's famous dictum: "We learn from history that we do not learn from history." That is not entirely true. We can, for example, confidently conclude from numerous historical examples that government by fascist demagogues is highly likely to lead to suppression of democracy and widespread human suffering. Nonetheless, one might understandably conclude that little or nothing can be learned from biological history. However, a crucial difference is that all biology is governed by the paradigm of organic evolution, so cer-tain general predications are possible. In the specific case of human evolu-tion, it has become increasingly accepted that understanding our biological past is a prerequisite for a comprehensive understanding of human medicine. The burgeoning field of *Darwinian medicine* rests on this fundamental tenet. In particular, it is argued that 10,000 years is too little time for modern humans to have become effectively adapted for life in settled communities. Instead, there is often a mismatch between current living conditions and bio-logical features developed during an extensive hunting-and-gathering ances-try. In universities, Darwinian medicine is increasingly becoming established as an important branch of medical education. So we can, indeed, learn useful lessons with tangible practical implications from our biological past!

SPECIAL?

All animals are special because every single lineage has evolved over time to arrive at the modern condition. All are equidistant in time from the common ancestor that gave rise to the entire array of animals present today. However, rates of evolutionary change have dif-fered radically between lineages, and some have accumulated many more novel morphological features over the course of their evolutionary diver-gence from that common ancestor. Note, however, that differences in rates of change over time at the genomic level are far less apparent, so all spe-cies alive today have genomes that have diverged markedly from the com-mon ancestral condition. Although the human species might be regarded as "special" in the sense that striking new adaptations emerged remarkably rapidly since we diverged from African apes over 8 million years ago, this degree of divergence is not evident at the genomic level, so our genome is by no means as distinctive as one might expect from anatomical distinctions. As already noted, at the anatomical/behavioral level, modern humans differ from African great apes (gorillas and chimpanzees) in three particularly striking features: (1) our jaws and teeth have been radically reduced in size; (2) we are characterized by upright, striding bipedal locomotion, a pattern that is unique among mammals; and (3) the human brain is about three times larger, relative to body size, than the brains in apes and monkeys (which differ far less among them).

Ever since my initial research on tree shrews, I have been convinced that a broad-based understanding of primate evolution—and, indeed, of other mammals and even certain aspects of vertebrate evolution in general—is essential for reliable interpretation of the special case of human evolution. Take, for instance, brain evolution. The mammalian fossil record reveals that relative brain size has generally increased over time in all lineages. Even the smallest-brained animals alive today have relative brain sizes larger than that of the common ancestor of mammals. So evolutionary expansion of human brain size is not unique among mammals. What *is* unique among mammals is the extraordinarily high rate of brain size expansion. In searching for a convincing explanation for this very rapid evolution of the human brain, a broad understanding of factors driving brain size expansion in mammals generally is essential. Special case arguments based on very limited comparisons (e.g., restricted to great apes and humans) are virtually untestable and have no real value. Even comparisons across primates are too limited to permit reliable inference of the reasons underlying rapid evolution of that especially large relative brain size in humans.

RELIGION AND SPIRITUALITY

The theory of evolution is a testable interpretation of the diversity of life on Earth. It has survived countless tests of its validity and has led to many major discoveries. Religion is entirely different, being a belief system that is not amenable to scientific testing. Many admirable investigators have combined acceptance of evolutionary theory with profound religious belief, an eminent example being the French Jesuit priest Teilhard de Chardin. Although I do not share them myself, I generally respect deeply held religious beliefs in others. In the past, I have generally expressed my view that evolutionary theory and religion are two very distinct realms and that I firmly support the principle of religious freedom. However, in view of countless transgressions that have occurred in the name of religion, I now feel obliged to add the rider that I respect the religious beliefs of others only to the extent they are not used as justification for any kind of evil behavior, such as slavery, racism, "holy" wars, subjugation of women, or interference with human reproductive freedom. Because evolutionary theory and religion are two very distinct realms, one based on science and the other based on belief, they are mutually independent. In particular, evolutionary theory cannot "disprove" religion, and religion cannot "refute" evolutionary theory. Accordingly, I entirely reject any form of creationism, which I define as "unthinking the thinkable."

ADVICE

My first piece of advice for exploring human evolution and identifying possible future changes is to question everything, take nothing for granted. Although tremendous progress has been made over the past fifty years in collecting and analyzing factual evidence for human evolution, even now airily presented "interpretations" are often speculative and insufficiently supported to be accepted at face value. Hypotheses advanced to explain the evolution of special human characteristics commonly emphasize *plausibility* rather than testability. Many seemingly plausible hypotheses have been advocated to "explain" features of human evolution, but too little attention has been devoted to identifying and applying tests to distinguish between alternative proposals. Once, early in my career, I remarked to an anthropological colleague that the literature on human evolution is rife with plausible explanations of unique features that are readily accepted (particularly by the media) without adequate testing. My colleague responded that, when examining the evolution of features unique to humans, what we really need are *implausible* explanations to explain such extremely rare developments! Especially in the case of striding bipedal locomotion, which has no real counterpart anywhere else in the animal kingdom, there have been major advances in our knowledge and understanding of changes over time and of the biomechanical relationships involved. Yet remarkably little progress has been made in developing a satisfactory explanation for the emergence of this highly unusual pattern of locomotion. Likewise, many different hypotheses have been suggested to account for the uniquely rapid expansion of the human brain, but most proposals have emphasized a single factor at the expense of all others, often simply ignoring the evidence cited in support of alternative hypotheses.

Comparative studies are particularly vital for assessing the relative merits of competing hypotheses. Recent wide-ranging comparisons using large data sets and sophisticated analytical methods have revealed two seemingly plausible hypotheses for human brain evolution that have gained wide acceptance—special requirements of complex social groups and a tradeoff between brain size and gut size because of energetic requirements—but have not been confirmed by testing. However, this has done nothing to stem the flood of novel, single-factor, plausible "explanations" for rapid brain expansion during human evolution.

Last but not least, I would emphasize that theoretical developments are just as important as basic factual evidence in refining our interpretations. And my own experience has confirmed my early conviction that a broad-based approach to primate evolution is essential for reliable interpretation of human evolution. Combining facts with theory, this extended framework permits recognition of general principles that can be applied to the special case of human evolution. The fragility of one-off, untestable, "plausible" explanations developed from narrow comparisons (e.g., confined to great apes and humans) can be avoided.

INSPIRING PEOPLE?

Teilhard de Chardin, the French Jesuit priest and pioneering investigator of human evolution. He struggled throughout his adult life with the interface between evolutionary theory and religious belief (in his case Roman Catholicism). Because of his deeply held convictions and obvious sincerity, I would dearly like to discuss evolution with him. In particular, I would seek his opinion on modern manifestations of creationism and repeated attempts in the USA to have this taught as an "equal time" subject comparable to the science of evolutionary biology.

84

PRIYA MOORJANI

Priya Moorjani is an assistant professor in the Department of Molecular and Cell Biology at the University of California, Berkeley. Priya seeks to understand the impact of evolutionary history on genetic variation to reconstruct demographic history and study its impact on human disease and evolution. Her lab is also developing new methods to analyze large-scale genomic data from present-day and ancient individuals. Her research has profound implications for developing accurate molecular clocks estimating the divergence time among humans and other primates. In 2016, a key study she led makes the case that the chimpanzee and human lineages diverged from each other at some point between 9.3 and 6.5 million years ago.[1]

YOUR BEGINNINGS

My background is in computer engineering and computational biology. I was always fascinated by human history, recent human civilizations as well as the study of human fossils. For me, a eureka moment occurred when I attended a talk two years before applying for graduate school on how we can compare genomic sequences to study population relationships and human adaptation. The fact that our genome preserves a record of our ancestors that we can leverage to learn about human history and adaptation is what drove me to do a PhD in genetics, and this idea excites and amazes me every day.

GAME CHANGER

 I consider the research by Allan Wilson to be a game changer. Allan revolutionized the field of human evolutionary genetics, and key ideas in this field that many of us rely on today were first introduced in his lab. These include molecular clocks,[2] ancient DNA sequencing,[3] and the African origin of modern humans (or Out of Africa migration).[4] Although these ideas were controversial when first published, many have stood the test of time. In addition, technologies such as sequencing have provided new ways of investigating our past that complement archaeological and paleon-tological evidence. This has led to whole new areas of research, such as pale-ogenomics, that have shaped my research program and direction.

AMAZING FACT

The fact that the genetic difference between humans and chimpanzees is very small—only about 1 percent differences in genomic sequence—yet there are profound morphological differences between the two species truly amazes me. It makes me realize how little we understand about evolution, and how much more there is to discover.

TIME TRAVEL

If I had a time machine, I would go 50,000 years back in time to learn about the interactions between modern humans, Neanderthals, and Denisovans. In 2017, I went to the Denisova cave, which is famous for the discovery of a pinky bone of a new hominin species, Denisovans.[5] This cave is truly amazing; it has been occupied by multiple hominin groups for at least the past 200,000 years, and remains of modern humans, Neanderthals, Denisovans, and a first generation mixed Neanderthal-Denisovan individual were discovered here.[6] I would like to go back to the same place to see how these different groups lived and interacted with each other. This could provide some hints about why these groups became extinct and why only *Homo sapiens* survives today.

FUTURE EVOLUTION

Well, why speculate? We can dial back the clock and let the past inform us about our future (e.g., by using ancient DNA and studying human fossils). A hundred years is a very short time for evolution to make major changes, and there are almost no major differences between people from the past and today. We had many cousins (Neanderthals and Deniso-vans) 100,000 years ago—some we interbred with and some we left behind. Most human phenotypes, such as speech and a large brain, are shared with ancient hominins, so these changes probably occurred more than 100,000 years ago. Most phenotypes we associate with anatomically modern humans only date back 300,000 to 200,000 years, which suggests that major changes can occur within 30,000 generations or 1 million years. However, we cannot predict new traits that we may gain in 1 million years because we still lack an understanding of all the changes that have occurred in modern humans.

SPECIAL?

All species are special in their own ways. As an evolutionary biologist, I think of humans as just one outcome of many simulations—starting with the same ancestral variation under certain parameters, we have a human, and under others, we have a chimpanzee. The parameters are currently unknown. Our best chance to uncover these parameters is by studying nonhuman species, which can provide clues about the various shared and unique changes that have occurred in the human lineage. Surprisingly, however, we find few differences (about 1 percent) between humans and our closest relatives, chimpanzees, suggesting that some of these changes may be more complex (duplications, insertions, and deletions, etc.), or occur regularly in nature (operating at expression level), and involve networks of genes rather than single changes. More research is needed to elucidate this question.

Regarding the causes driving the differences between humans and chimpanzees, to understand the *why*, we need to first understand the *how*. For instance, if we understand the genetic basis of some of the unique human phenotypes (bipedalism, advanced cognition, etc.) and determine when these changes occurred, we can start to uncover the evolutionary pressures that could have led to these changes. Did these mutations have a particular selective advantage for our survival? Selection is conditional on the environmental context and the evolutionary pressures the organism faces, so it's entirely possible that the same mutation also arose in chimpanzees but (like most mutations) was lost due to drift. Also, even though we are similar to chimpanzees at a sequence level, there are certainly many complexities of the genome—for instance, duplications, structural variants, and regulatory variants—that are not well characterized yet and could hold the key to answering these questions.

ADVICE

 Always be curious and question every well-established finding. Only by understanding all of the details and reexamining the data can we truly appreciate the nuances and uncover something new.

85

MARK PAGEL

M ark Pagel is the head of the Evolution Laboratory in the Biology Department at the University of Reading (UK). Mark builds statistical models to examine the evolutionary processes imprinted in human behavior, from genomics to culture. His work raises questions in the philosophy of biology, mind, and language. His most recent efforts center around the parallels between linguistic and biological evolution. In addition to his extensive research publication record in comparative studies, he also authored *Wired for Culture*,[1] voted one of the best science books of 2012 by *The Guardian*. Among his numerous honors, Mark was elected a fellow of the Royal Society in 2011.

AMAZING FACT

The word "amazed" is perhaps used too much in modern discourse. But here is a fact of human evolution that should profoundly influence our perspective on human adaptation: owing to their ability to adapt at the cultural level by acquiring skills and technologies, modern humans have been able to occupy nearly every habitat on Earth. Modern genetic studies are revealing that in occupying these varying habitats humans have acquired a remarkable number of different and precise genetic adaptations. Thus, both Tibetan populations and Andean populations acquired genetic adaptations to living at high altitude, but the genetic mechanisms for each differ. Similarly, the ability to digest milk sugars (lactose) as adults has arisen independently at least twice in human populations, and at least two different genetic adaptations have arisen in populations of humans chronically exposed to the malaria parasite. People such as the Inuit of Alaska who occupied cold climates tend to be short and stocky to conserve heat, whereas desert dwelling people such as the Dinka of northern Africa are tall and thin. Genes that influence skin color vary with sun exposure; there are genes for consuming diets rich in marine organisms, genes for shallow-water diving, and the list goes on. No one human has all of these adaptations; rather, they exist in specific groups, which tells us that they have arisen relatively recently and relatively quickly (because humans only began to occupy the world in the last few tens of thousands of years). We don't see this kind of variation within other species because only humans have spread so widely. The lesson we learn from humans is that evolution by natural selection is a highly creative force that can rather quickly match organisms to their environments. We (and other species) are surprisingly well adapted.

DRIVING FACTORS

I haven't much truck with driving factors, but at some point in the *Homo* lineage (the lineage that would eventually lead to our own species, *Homo sapiens*), individuals began to acquire the ability to copy others' actions without the need for specific training or rewards. Many animals can crudely mimic the actions of others, but no animal even comes close to humans' abilities to imitate, and to do so especially with respect to precise motor behaviors such as might be involved in making tools or creating baskets or fishing lures. Added to this is that at some point in time this copying seemed to become "directed," in the sense that we preferentially copy others' actions that led to desirable payoffs. If a chimpanzee watched two people, one of whom was better than the other, chipping ("flaking") hand axes out of a larger stone core, it is unlikely the chimpanzee would copy either of them, much less the one who made a better axe. But humans are different. This ability, probably acquired gradually throughout our *Homo erectus* and even later ancestors, finally came to fruition in us. I am convinced that it was not even present in our evolutionary cousins the Neanderthals who inhabited Europe when we arrived perhaps 45,000 years ago, and that is almost certainly why we (modern humans) outcompeted them. Whatever this ability is—in principle it will be some neural configuration probably unique to our brains and acquired gradually—it set us on a path to acquiring technologies, and we have moved from a slow to what is now a breathtaking pace of technological development. This combination of precise copying and of copying things that led to better outcomes is a form of evolution by natural selection, but it is a selection of ideas rather than genes—good ideas are preferentially retained, bad ideas are discarded. Because ideas can jump from mind to mind, this tournament of idea selection can happen rapidly. No other species can perform this kind of copying; it has created an unbridgeable gap in evolutionary potentials between us and all other animals.

FUTURE EVOLUTION

At least in the near future (next one hundred years) the greatest factor shaping human evolution will be globalization, or more specifically the mass movements of people that globalization is likely to bring about. Our history from around the time our species migrated out of Africa, perhaps 60,000 years ago or so, has been one of relatively small human populations occupying discrete regions and not having anything like the contact with one another that we are witnessing today. We know this because at least 7,000 different languages are spoken on Earth, and these languages would not exist if humans had routinely mixed with one another throughout our past. But globalization is changing all that. Not only are languages being lost at alarming rates but cultures are being homogenized around the world. Just look at photographs of major metropolitan areas such as London, Paris, Tokyo, Los Angeles, Mumbai, Beijing, New York, Moscow, Rome, Nairobi, or Lagos. Apart from a few familiar landmarks (Red Square or the Eiffel Tower), these cities all look the same. This mixing of the world's languages, cultures, and people will eventually genetically homogenize human populations. Far more than any other force, this will shape the immediate future of humanity. So-called lifestyle diseases such as obesity, heart disease, diabetes, and even some forms of cancer will profoundly affect health care systems but will have little influence on our genetics because most of these occur after the age of reproduction.

EVOLUTIONARY LESSONS

Humans, uniquely among animals, evolved to live in small tribal societies with high levels of cooperation even among non-family members. Their high degree of cooperation meant that they could share knowledge, skills, and resources and have a division of labor within the group, and this promoted adaptation to the various habitats these tribal groups encountered. To protect their knowledge and to prevent "free riders" or cheats from taking advantage of others' cooperation, human tribal groups evolved to become tightly bound by systems of morality within the group and to be wary of those outside the group. This evolved social and moral structure made the human tribal group a fearsome and formidable competitor to other tribal groups seeking to occupy the same territories. Elsewhere I have called these tightly knit cooperative groups "cultural survival vehicles." This term is meant to capture the sense in which this particular social grouping—the human cooperative tribe—acts in a manner like a body does to its cells. Our bodies protect our cells and ensure that we can pass our genes on to the next generation; our tribal societies act like social bodies that help to ensure our individual survival in a hostile world full of other tribal societies such as our own. We should not underestimate the psychological power of allegiance to our "tribe"—just think of the joy you feel when your country triumphs in a sporting event or the loss you feel when one of your country's soldiers is killed in battle.

Human history has been the story of these tribal groups, first leaving Africa, and then spreading out around the world. In the current environment, our species lives in a world awash with higher degrees of mixing than at any other time in our history. This mixing is brought about by immigration in response to inequalities in wealth and opportunity that is often at odds with our ancient tribal psychology of favoring "our own," and the result is the rise of xenophobia or other sorts of bigotry. What can be done? It would be glib to offer simple solutions, but one idea that we can draw on is that humans seem remarkably capable of scaling their "tribal" psychology up from the perhaps 50 to 150 people to the thousands, millions, or perhaps billions of people that now live under tribal banners such as the French, the English, the Indians, or the Chinese. It is not beyond our psychological grasp as a species to live harmoniously in large social groupings while at the same time exhibiting a degree of parochialism toward those from outside our group. Key questions for evolutionary and social scientists are how rapidly can such large groupings form, and how much heterogeneity of ethnicity or culture can they accommodate?

RELIGION AND SPIRITUALITY

 It is difficult, maybe even impossible, to reconcile many religious beliefs with the scientific study of human evolution. Religion encourages and even esteems belief without evidence, whereas science garners evidence to support or refute beliefs. But this misses the far more interesting question: Why is there no evidence for the truth of religion? Religious belief has been and still is ubiquitous among human populations, but it is not present in any other species. At least two intriguing strands of thought emerge in answering this question, both of them unique to humans as far as we know.

One is the burden of existential plight: humans can, perhaps uniquely, ponder their place in the universe and their own fate. We can wonder why we are here, what our purpose is, and what will happen after we die. Religion provides reassuring answers to these questions, so holding a false (unproven) belief is nonetheless helpful. This should not be surprising. Most of us hold many false beliefs about how clever we are, how attractive we are, how witty we are, and so on, and these false beliefs smooth our passage through life, making us more resilient.

The second strand concerns group action. Religion can provide an effective mechanism for systems of morality, for encouraging cohesiveness within the group, and for promoting wariness toward those outside the group—parochialism again. Thus, within many religions, nonbelievers or believers of different religions can be considered to be apostates and by implication can be treated harshly without moral opprobrium. Consider the many places around the world where groups of people are in conflict for seemingly no other reason than that they have different religions. This capacity for religion to motivate group action and provide sanctions for prejudicial and even violent behavior is very probably linked to our history as a species. We occupied the world in small cooperative groups that would constantly have been in conflict with other groups. Some anthropologists have even described our past as one of "constant battles." Religion arose as a system of co-adapted beliefs that have specifically evolved to take advantage of our social structure and the way our minds work. Religion is capable of motivating important human actions that promote our survival and well-being.

HERMAN PONTZER

erman Pontzer is an associate professor of evolutionary anthropology at Duke University and an associate research professor of global
health at the Duke Global Health Institute. Through laboratory and
field research, Herman studies the physiology of humans and apes to understand how ecology, lifestyle, diet, and evolutionary history affect metabolism and health. He also studies how ecology and evolution influence
musculoskeletal design and physical activity. His field projects study small-
scale societies, including hunter-gatherers and subsistence farmers, in
Africa and South America. His recent book *Burn*[1] is a sensation among
those interested in understanding how metabolism shapes our bodies, our
health, and our longevity.

YOUR BEGINNINGS

In late August 1995 I was a wide-eyed freshman at Penn State University taking an introductory seminar in human evolution cotaught
by two amazing professors, Jeff Kurland and Warren Morrill. I grew
up at the end of a dirt road in a very rural part of Pennsylvania, and my
first semester at college felt like I'd entered an enormous new world—like
Dorothy waking up in Oz. That seminar was even more mind-expanding.
Exploring the fossil record, evolutionary ecology, and the living primates
was like pulling the curtain aside to get a good look at the Wizard behind it
all. If there was a way to keep thinking about human evolution for a living, I
knew that was what I wanted to do.

GAME CHANGER

 The biggest game changer of them all has to be *On the Origin of Species*,[2] of course. Without natural selection and a common origin to unite the life sciences, biologists (including paleoanthropologists) would just be stamp collecting. But to pick something closer to home for me, I'll go with "The Expensive-Tissue Hypothesis" by Leslie Aiello [chap. 20] and Peter Wheeler.[3] That landmark paper, published in 1995, proposed that the advent of hunting in the genus *Homo* permitted a smaller digestive tract and led to selection for larger brains because hunting is cognitively challenging. As a result, Pleistocene hominins traded guts for brains, one "expensive tissue" for another. This idea has come up now and again in human evolution—Arthur Keith tried to demonstrate something similar with orangutans and leaf monkeys in the 1890s, but allometry hadn't been invented yet and he couldn't pull off the analysis. Aiello and Wheeler's take on it was a tour de force, uniting the fossil record, brains, digestion, dietary ecology, and the physiology of living humans and apes into a coherent and testable framework for understanding one of the true sea changes in human evolution, the evolution of big brains.

TIME TRAVEL

Going into the past would be fun and would allow me to settle some bets I've had with colleagues for years (e.g., origins of hominin bipedalism and the amount of meat in the diet of Pleistocene hominins). But we know the broad outlines of the past, and the details are coming increasingly to light. We have no inkling of the future. I'd love to know what the Earth will look like 1,000 years from now, post–fossil fuels. Is it Mad Max? Is it a full blown climate disaster? Have we blown ourselves up with nuclear weapons or something even worse? Are there 20 billion people on the planet or 20,000? How about 10,000 years from now? 100,000? 1 million? (Obviously, my notebook on this journey would be the Captain's log$_{10}$.) Are we even still here? With the benefit of real foresight, perhaps we could get ourselves on a more sustainable track.

DRIVING FACTORS

 Revenge. I had this scene in my head when answering your questions: https://www.youtube.com/watch?v=uM-BE3p9TUI.

FUTURE EVOLUTION

Our species' relationship with energy, both internal and external, will determine our future. Right now the food environment we've built for ourselves—the ultra-processed foods packing our supermarket shelves—leads too many of us to overconsume, and that excess internal energy makes us obese and sick. We rely on a steady supply of external energy as well and have done so since Pleistocene hominins tamed fire. Today we meet roughly 80 percent of our enormous external energy demands (something like 200,000 kcal/d for the average American) with fossil fuels, which are wrecking the planet through climate change and pollution. If we can figure out how to manage our energy more responsibly, in a hundred years we'll be a vibrant species on a healthy planet, and in 100,000 years who knows? Perhaps we'll be responsibly exploring and colonizing the solar system and beyond. If we can't get a better handle on our energy environment, we'll be in a desperate situation in a hundred years (out of oil and fighting wars over clean water) and likely extinct in 100,000.

EVOLUTIONARY LESSONS

Evolution reveals a couple of key insights that are essential for understanding ourselves and caring for our planet. First, evolution shows how tiny changes can have an enormous impact over long timescales. A fractional change in fertility or survivorship can lead a species to total extinction or worldwide expansion. The carbon emitted from a coal-fired power plant seems insignificant in the vastness of our atmosphere, but a few thousand power plants over just a few decades can change the climate. Second, evolution teaches us about the forces that shaped our bodies, which in turn helps us understand how our bodies work and how to keep them healthy. An evolutionary perspective provides the clearest vision of what is "normal" for our species—the diet, daily activity, and social environments in which our bodies evolved to thrive.

SPECIAL?

Our bodies aren't special, but our environments are. Studying apes and other primates is essential if we're to understand how our own bodies have changed, and they provide a useful model for the ways our social and physical environments can affect our mental and physical health. The big difference, of course, is culture and the degree to which we construct our own environments. We've created fantastic Disneyworld landscapes, both real and imagined, and we're so deeply embedded in them that our relationship with even basic essentials such as food feels untethered from their biological roles. We eat unicorn cupcakes because it's our daughter's schoolmate's birthday, not because we're optimally foraging to survive in a limited resource environment. But if you look closely, you can still see our evolutionary biology at work—those cupcakes are sugary and colorful because our primate brains are drawn to those traits, and we're sharing food because that's what hominins have been doing since the dawn of hunting and gathering.

RELIGION AND SPIRITUALITY

It depends on one's cosmology I suppose. If your religion requires you to believe that the universe was created a few thousand years ago, that species can't change over time, or that humans are special and separate from the natural world, that worldview is not compatible with the science of human evolution. You'll have trouble with the chemistry underlying geological dating, the physics underlying modern astronomy, and the biology of modern genetics as well, all of which point to an ancient and interconnected world. However, if your religion doesn't require you to reject modern science, there needn't be any conflict. The Catholic Church, for example, has written very clearly that there's no conflict between its spiritual teachings and the science of human evolution.

87

HOLGER PREUSCHOFT

olger Preuschoft is a professor emeritus at the Institute of Anatomy (Ruhr-University; Bochum, Germany). With six decades of research, Holger has published more than two hundred scientific papers and edited nine coauthored books around a central question that is the basis of his recent book, *Understanding Body Shapes of Animals*.[1] Holger seeks to understand why and how animals evolved into the specific shapes that we see today or in the fossil record. He tries to explain morphology through the lens of biomechanics, and he pays particular attention to primates to shine a light on human evolution.

YOUR BEGINNINGS

When I started studying biology in 1955, I had just made a big change in my life and decided to only do things I really like to do. So I drifted through the fields of science, lectures, and libraries and became familiar with physical anthropology and later with comparative anatomy. Both carried me slowly, but continuously, into research work.

Before 1955 I had finished school (gymnasium), where I learned English (very useful!), Latin (you will be surprised, also very useful for my future life!), mathematics (weak), physics (from which I did *not* forget simple but essential approaches), and history (which I still like very much), and I passed through a stage in which earning money seemed to be my major and only task. This stage lasted three years and included the formal degree of a "Kaufmännischer Angestellter" (which I cannot translate). It did not lead to a really rewarding income, but it seemed to determine my future life. I was in contact with friends from school who studied medicine or arts or (less attractive) economics, and one day in early spring 1955 I decided to quit my appointment and do something I found really interesting: getting close-up views into biology, history, Germanistics, and later medicine. All this took place in Frankfurt, where I pretty well understood the opportunities, but this step did not seem reasonable in view of the complete absence of money or other resources (remember, this was ten years after World War II), but it worked. At about the same time, I met my future wife, Lotte, who also had a background lacking in resources. In her case, both parents and the rest of the family objected strenuously against her intention to go to university, but she did. Years later we married and had a family. This was after my promotion to Dr. Phil. Nat. (at Frankfurt University) in Tübingen, where I had my first fully paid position in physical anthropology, but with a doctoral thesis in descriptive anatomy and a "Habilitationsschrift" in functional morphology.

GAME CHANGER

 New fossil material unearthed in East Africa since 1959, and at about the same time, my contacts with "functional anatomy" in a strict sense, which allowed a novel approach to fossils.

AMAZING FACT

 That this species developed such a huge brain (with all its consequences) and changed from its former affinity to tree-living to a specialization toward long-distance walking on flat ground. The question is *why*?

TIME TRAVEL

 Pliocene-Early Pleistocene, where the essential evolutionary changes took place. Similarly of interest would be the period in which Neanderthals and modern *H. sapiens* met in the Near East and in Europe.

DRIVING FACTORS

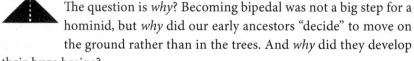 The question is *why*? Becoming bipedal was not a big step for a hominid, but *why* did our early ancestors "decide" to move on the ground rather than in the trees. And *why* did they develop their huge brains?

FUTURE EVOLUTION

 No idea in view of my deficient talents in the field of prophecy. It should not be lost, however, that widely spread conventions (among living societies) about how a human should look and behave are difficult obstacles to overcome when thinking about evolutionary changes!

EVOLUTIONARY LESSONS

 Humans, as a species, behave unreasonable, as do all other animal species. We are sawing on the branch on which we sit (exhausting resources, overcrowding).

SPECIAL?

Yes, we are special in terms of bipedality and poor climbing abilities in combination with our enormously efficient brain and speech! The study of our living relatives reveals how much we have in common with other primates—but also how much we diverge from them.

RELIGION AND SPIRITUALITY

This question has two aspects. Science simply does not need a religious or spiritual background, and it does not attempt to "explain" facts by supranatural agents. But the study of human evolution does not exclude them.

ADVICE

Most important: do not follow authorities, regardless of their reputation, nor accept the currently predominant opinions. Make your own inquiries, ask questions, and search for your own answers in line with other fields of science: statistics, mechanics, physics, chemistry, etc. Never discard a question as "not being scientific." Nonscientific questions usually turn out to be questions that are particularly difficult to answer.

 INSPIRING PEOPLE?

Ernst Haeckel.

88

JOAN RICHTSMEIER

Joan Richtsmeier is a distinguished professor of anthropology and a member of the Center for Neural Engineering at Pennsylvania State University and is an elected fellow of the American Association of Anatomists and of the American Association for the Advancement of Science. She won the 2019 Henry Gray Scientific Achievement Award of the American Association for Anatomy and was awarded the 2021 Rohlf Medal for Excellence in Morphometric Methods and Applications. Joan studies the molecular basis of craniofacial variation in growth and evolution, quantitative morphology, and ontogenetic mechanisms related to phylogenetic change. Her current research focuses on the role of the embryonic cranial endoskeleton on dermatocranial formation and variation.

YOUR BEGINNINGS

Two experiences prompted me to do research. My original interest in research occurred when I was a teenager sitting in a movie theater watching a very late showing of something (*Rocky Horror Picture Show*?) and the double feature was *Chariot of the Gods*. For those of you lucky enough *not* to have seen this film that strives to prove mankind's ancient civilizations had been visited by extraterrestrials—about which film critic Phil Hall said, "They don't make films like this anymore, and we should be glad for that"—it is a joke of a film. But I didn't know anything about extant primitive peoples and had never thought about how or why humans could have built something as consequential as the pyramids or the great Moai of Easter Island. I had no fascination with extraterrestrial visitations, but I wanted to know more about primitive cultures.

The second time was while taking a lithics technology summer course in Kampsville, Illinois. We learned to knap tools from chert and obsidian. As a class, we made a toolkit, analyzed the shapes and sizes of the tools, butchered a goat (and had a great cookout), and then analyzed the changes in shapes and sizes of the tools and their cutting edges. It was the first time I understood how one might tease human behavior out of artifacts left behind.

GAME CHANGER

 I read Gavin De Beer, John Bonner, and C. H. Waddington as a graduate student, and I was interested in development when teaching anatomy from an embryologic perspective, but my experiences were not deep enough at that time to appreciate the insights offered by these scientists. It was not until much later, when reading Ken Weiss's [chap. 94] essay "A Tooth, a Toe and a Vertebra: The Genetic Dimensions of Complex Morphological Traits,"[1] that I seriously considered that DNA evolves. I was trying to learn genetics simultaneously, and this essay convinced me that identification of specific genes was far less important than understanding how complex traits were constructed by cells using the instructions provided by genes. This really moved me toward wanting to understand evolution from a developmental perspective. Ken Weiss's essay reinforced my growing awareness that genetic programs drive developmental processes and that the capacity for morphology to change can only come about through changes in development mechanisms, which are under genetic control.

AMAZING FACT

 The breadth and depth of variation across time and place and the consistency of the vertebrate form.

TIME TRAVEL

I would travel to the proto-contact period in North America, specifically to the Middle Missouri River valley. I spent summers excavating villages along the Knife River and doing surveys in the Dakotas. There are historic writings available for the same time period, and Europeans witnessed how certain implements were used, but most of this was from a male perspective (e.g., Lewis and Clarke). The women played important roles in these villages and communities too, and I would love to understand their perception of the world, their way of life, and how visits from European trappers affected their worldview.

RELIGION AND SPIRITUALITY

 I was raised Catholic and brought up in a Christian community and home, and I was taught to live by the Golden Rule. Until I read an essay (I don't remember the author) by someone who felt evolution and religion were incompatible, I had never even considered this a problem. Faith is an aspect of a being's spiritual essence. You cannot test hypotheses or prove that God does or does not exist. If God exists for you, that is enough. If God does not exist for another person, that is enough. Spirituality is personal. Evolution has nothing to do with faith. I don't "believe" in evolution. I accept it as the only viable, testable hypothesis (now with tons of data to support its validity) about how life began, is maintained, and changes. Whether or not God exists, and my opinion on that matter, has no impact on whether or not evolution is real.

ADVICE

Follow your interests and be true to them. Do not do what is in vogue or fundable. Do what you love and enjoy, and do it well.

89

ROBERT SAPOLSKY

R obert Sapolsky is a professor at Stanford University, holding joint appointments in biological sciences, neurology and neurological sciences, and neurosurgery. He studies issues related to stress, neuronal degeneration, and gene therapy. Robert also studies wild baboons in Kenya to identify the sources of stress in their environment and the relationship between personality and stress-related disease. He has been featured in the *National Geographic* documentary "Stress: Portrait of a Killer," in the *New York Times*, and in the podcast *The Joe Rogan Experience*. He has authored several books, including the popular *Why Zebras Don't Get Ulcers*[1] and *Behave*.[2] Robert's research has received numerous awards, including the prestigious MacArthur Fellowship "Genius" Grant.

GAME CHANGER

A huge game changer is the chimp literature, starting with Goodall, showing tool use, multigenerational transmission of culture, and organized intergroup violence that becomes genocidal.[3] Those findings did in a zillion different "humans are the only species that . . ." bits of dogma and transformed our sense of continuity with other species. They sure transformed my thinking. I'd say another was Sarah Hrdy's [chap. 75] discovery of competitive infanticide (initially in langur monkeys).[4] This was the final nail in the coffin of classical group-selectionist thinking in evolution.

TIME TRAVEL

 Oh, why not—I'd travel back 60,000 years or so to the Indonesian island of Flores to hang out for a while with *Homo floresiensis*: tool making hominins less than four feet tall with brains about the size of chimpanzee brains, but with parts of the frontal cortex approximating the size found in modern humans. I assume their behavior would be beyond fascinating.

EVOLUTIONARY LESSONS

To appreciate the extent to which humans evolved to be freer of genetic influences than any species on Earth. In the future, we must constantly keep in mind our potential for behavioral and societal flexibility. Suppose you have a gene that codes for some neurotransmitter that influences behavior. On a mechanistic level, in a species "tied" tightly to its genes, this might take the form of the gene expressing constitutively at roughly the same level all the time. What would things look like in a species with more behavioral flexibility? The gene wouldn't be constitutively expressed. Instead, it is under the control of context-dependent transcription factors—lots of different ones activated in different circumstances, and with only partially overlapping networks of genes under their control. For example, Transcription Factor A induces expression from this gene, plus genes 2, 4, and 6; Transcription Factor B induces this gene, plus genes 6, 7, 8. . . . There are more ways for the environment to transduce flexibility, unlike in the first inflexible species, and there are lots of opportunities for epigenetic regulation of our gene. Furthermore, there is environmental regulation of the levels of receptor for that gene's neurotransmitter. Thus, in the latter species, the same gene will have very different affects on different circumstances, freeing that species from its more deterministic effects.

SPECIAL?

Primatology is helping in this regard—of course. The thing about humans that most strikes me is the balance between ways in which we are on a clear continuum with other primates and ways in which we are utterly unique. The continuity comes in the form of similarities in capacities for empathy, sense of justice, cooperation, patterns of aggression, and so on. Where the astonishing uniqueness comes in is how we make these traits abstract over space and time—like no other primate, we can press a button and kill another human whose face we never see, or press a button and contribute money to a charity, thus saving the life of another human whose face we never see. This is why both our best and worst moments are so much more extreme than what we see in other primates.

RELIGION AND SPIRITUALITY

Vast numbers of people would disagree with me about this, and there's no way to prove my deeply felt view, but I don't think they are remotely compatible. Human evolution, like all of evolution, and like all of biology, has led me to a purely materialist view of life that leaves no room for spirituality/religiosity.

90

CHET SHERWOOD

C het Sherwood is a professor of anthropology at The George Washington University's Center for the Advanced Study of Human Paleobiology, a member of GW's Mind-Brain Institute, and a member of the U.S. National Academy of Sciences. He also serves as a director of the U.S. National Chimpanzee Brain Resource. Chet compares the anatomy and molecular function of the human brain to that of the great apes and other primates. He studies how brains differ in size, internal structure, cellular composition, and gene expression. Chet's lab also investigates how brain variation is correlated with behavior and how brains change during the life span of different primate species.

YOUR BEGINNINGS

I watched fleas jumping up and down on a filthy sofa in a West Hollywood bungalow and thought to myself, "I can't do this much longer." I was playing guitar in a lousy punk band, traveling around in a van, and crashing on sofas in random places at the end of the night. Sure, it was a lot of fun; but my back was aching, and I had a constant ringing in my ears. With the extra time I had on my hands between band practice and working as a shipping clerk, I was devouring nonfiction books about evolution and brain science. Although I had no specific plans after finishing college a couple of years earlier, I suppose the classes I took in biological anthropology had gotten to me. I decided to walk up to the volunteer office at the American Museum of Natural History in New York City. I offered to do whatever they wanted, just to get a foot in the door. They had me moving dinosaur fossils from the collection into a new shelving unit in the basement of the museum—and that was it. I was awestruck to be in such close contact with the evidence of evolution on Earth.

GAME CHANGER

 René Descartes's *Meditations on First Philosophy*[1] made a huge impact on me personally, especially because it was the first time I had encountered any writings on the philosophy of mind. Modern science still grapples with these metaphysical themes today in our pursuit of understanding the human experience: How can we trust our perceptions of the world around us? What evidence is sufficient to dispel skepticism? What is the relationship between mind and body? One doesn't need to agree with Descartes's conclusions to recognize their influence. His aim was to prove the existence of an omnipotent supernatural god and an immaterial soul, which many would reject today. Nevertheless, Descartes's writings were a significant catalyst for later explorations of consciousness and cognition in humans and other animals.

TIME TRAVEL

I would like to travel back approximately 50 million years to a shallow oceanside location around India or Asia. I would encounter the ancestors of whales and dolphins in their initial transition to an aquatic life. I would sit quietly and observe the various species of Archaeoceti in their natural habitat, some wolflike and others crocodile-like in appearance. Knowing that their descendants will ultimately become so radically different, it would be fascinating to see hints of the wonderful adaptations to come. Evolution is a process. In a lifetime, we can only glimpse snapshots as it unfolds. It would be magnificent to observe the combination of familiar and strange in the archaeocetes.

EVOLUTIONARY LESSONS

Past evolutionary pressures caused modern human brains to be exceptionally plastic and adaptable to change throughout the life span. A lot of scenarios have been proposed to explain this, but they all generally boil down to the advantages of promoting cultural learning for the purpose of group cooperation, technological innovation, language acquisition, or some other benefit. The evolutionary roots of humanity have endowed us with brains that have enormous capacity for plasticity early in life. The ugly downside is that this makes human individuals extremely vulnerable to early adversity due to societal inequities such as systemic racism and sexism. These harms become embodied in the function of the immune system and the brain and have lifelong consequences for health, well-being, and prosperity. The bitter irony is that what distinguishes us as a species—our slow neurodevelopment and brain plasticity—is also at the core of the insidious harm we can do to each other through acts of discrimination.

SPECIAL?

To borrow a term from the great Lee "Scratch" Perry, humans are the "Super Ape." We are special in obvious ways—we're the only species writing books on the topic after all. But what is most humbling and amazing to me is how humans, for all our oddities, are really not inexplicably peculiar. In many ways, we are exactly what one would expect from an ape with an enlarged brain, free hands, and a slower pace of development and aging. As we learn more about the cognitive abilities of other species, we see increasing similarities to ourselves. Studies of wild great ape populations have demonstrated their capacities to plan for the future, use and make tools, interact with complex communication systems, and transmit local cultural traditions in behaviors. Human brains take the basic blueprint of the ape brain wiring diagram and extend it by adding neuronal computational units, amplifying the later developing components that are shaped by social and environmental experience, and strengthening the connectivity that makes us feel rewarded when we participate in coordinated social actions. The beauty is to recognize how what appears to be special about humans is actually deeply rooted in our biological continuity with patterns and processes that join us with other animals.

INSPIRING PEOPLE?

Tilly Edinger was a brilliant scientist who was instrumental in establishing the field of paleoneurology, an area of research that melds the approaches of neuroscience and paleontology. She also had tremendous personal strength. She faced persecution as a Jew and was expelled from her professional position at the Senckenberg Museum by the rising

Nazi regime in Germany in the 1930s. She emigrated to the United States, but she continued to face challenges as a woman in science. Nevertheless, she made enduring contributions to our understanding of brain evolution through her studies of the correspondence between endocranial morphology and the external structure of brains in vertebrates, brain size change over time in the equid and sirenian lineages, and sensory evolution in whales. It would be fascinating to hear her perspective on these questions about evolution and the human condition.

91

CRAIG STANFORD

C raig Stanford is a professor of biological sciences and anthropology at the University of Southern California. Craig's work focuses on the ecological relationships among the primate species sharing a tropical forest ecosystem, and he has conducted extensive field research on wild great apes, monkeys, and other animals in East Africa, South Asia, Southeast Asia, and Central and South America. He is best known for his field studies of the predator-prey ecology of chimpanzees in Gombe National Park (Tanzania) and his study of the chimpanzees and mountain gorillas in Bwindi Impenetrable National Park (Uganda).

YOUR BEGINNINGS

As a natural history-obsessed kid, I had been doing field "research" since I was a preteen, writing down observations of the natural world around me in suburban New Jersey. My adult research interests were inspired when I wandered into a secondhand bookstore during my first week of college and bought a tattered copy of Robert Ardrey's *African Genesis*.[1] Ardrey's idea that human nature was formed by hunting in the crucible of the African savanna may not have entirely stood the test of time, but it kicked off my intense interest in understanding the lives of the creatures that were our earliest direct ancestors. Those variants on the ape lineages that evolved into early human ancestors intrigued me from that day forward.

GAME CHANGER

I began graduate school a decade after the discovery of Lucy[2] and a few years after the original interpretations of the *Australopithecus afarensis* skeleton (by Johanson, White [chap. 45], and Lovejoy)[3] were being challenged (by Stern [chap. 92], Susman [chap. 42], and Jungers).[4] Was Lucy a fully terrestrial biped that walked very much as we do today, or did she have more apelike adaptations to tree climbing and a more general arboreal-terrestrial life? I found that debate absolutely fascinating and also central to understanding our own ancestry. I was a graduate student at the University of California, Berkeley, so imbued with the terrestrial view, but once I was exposed to the alternatives, I came to see Lucy as an obvious climber and walker. My subsequent work on wild great apes has only reinforced that view.

AMAZING FACT

The very thing that evolution deniers are unable to see—how myriad random changes to the genome can over time lead to evolutionary change in a lineage—is exactly the beauty of the evolutionary process. In human evolution, it means an ape ancestor gives rise to a human ancestor in a process that was in no way preordained or predictable. It occurred when contingency upon contingency accumulated, resulting in a new set of adaptations, and eventually a new life form. This degree of complexity is beautiful, powerful, and explanatory in a way that specious, simplistic narratives are not.

TIME TRAVEL

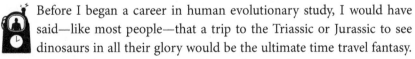

Before I began a career in human evolutionary study, I would have said—like most people—that a trip to the Triassic or Jurassic to see dinosaurs in all their glory would be the ultimate time travel fantasy. But traveling back in time to East Africa at the very dawn of humanity, 6 to 4 million years ago, would be both awe-inspiring and provide profound insights into who we are today. In all likelihood, we would not be able to distinguish an ancient ape whose descendants would one day be considered human from a very similar ancient ape whose evolutionary trail would ultimately die without leaving modern descendants. The complexity of the human fossil record increasingly appears to involve multiple lineages, only one of which survived to give rise to the genus *Homo*.

DRIVING FACTORS

This is the million-dollar question in the study of human origins, and after a long career studying living great apes in hopes of answering it, I am convinced that the answer is far more complex than we ever believed. There was almost certainly no one driving factor or unitary explanation: meat-eating, bipedalism, and social evolution all played roles. These behaviors led to anatomical evolution, but cognitive evolution related to the rise of big-brained hominins came much, much later.

RELIGION AND SPIRITUALITY

I don't find a spiritual worldview to be compatible at all with the study of human evolution. Some would say that believers and nonbelievers can share ideas about human origins with mutual respect and find some common intellectual ground. I disagree entirely. I don't believe in being shrill or defensive about religion versus science, but I don't think a spiritual worldview has anything of substance to offer that could advance our understanding of the natural world. Some scholars try to shoehorn religion and science together; I don't find value in that approach. I was raised religiously and accept that spirituality provides a moral compass and emotional support for billions of people. But I don't see anything good coming of attempts to push religion and science together. Religion is about values; science is about truth.

ADVICE

I advise my students to work on issues that are central to our understanding of human origins. Don't nibble around the edges; tackle the big questions head-on. Of course, a doctoral thesis has by definition a narrowly defined topic, but that topic should bear directly, not indirectly, on reconstructing the lives of our distant ancestors.

INSPIRING PEOPLE?

Darwin is the obvious answer. But Darwin labored in an era in which information about genes, a decent human fossil record for which dates were known, and the behavior of living apes did not exist. This makes his contributions all the more impressive, but he's not really someone I would go to for answers about the past. The ideal person with whom to discuss these issues would be a prominent human evolutionary scholar of the twenty-second century. I would ask this scholar-of-the-future a few basic questions: What does the human family tree look like in the year 2120? It may well look very different than it does today. What have we learned about the origins of upright walking based on these new fossils? What about their diet, their level of sexual dimorphism, the reconstruction of their ancient habitats? Sadly, this conversation with the top human evolutionary scholar of the twenty-second century won't happen with me because she hasn't been born yet.

92

JACK STERN

Jack Stern is a distinguished teaching professor emeritus at Stony Brook University (New York). He is interested in the origin of bipedalism in the human lineage. During more than forty years, Jack has studied the functions of specific muscles in primates (including humans), partly to explain how bones have evolved to promote such functions and how to use this knowledge to interpret the fossil record. The most recent research of the Stony Brook team focuses on using muscle recruitment patterns in chimpanzees to improve mathematical models of bipedalism in *Australopithecus*.

YOUR BEGINNINGS

For most of my early life, I had expected to be a physician. Today applicants to medical school are wise to spend time in activities that allow interaction with the sick or needy. In my day, having research experience added more to one's application package. I spent the summer after my second year of college working in a gastrointestinal physiology research lab at a local Chicago hospital. My role was menial and the experience unsatisfying, but it helped me get into the research-oriented University of Chicago Medical School after my third year of college. I spent the first summer of my medical school career working on a ventricular assist device in the lab of a cardiac surgeon. Nonetheless, I saw my future as being a clinician; that is, until I started my clerkships and discovered I very much disliked dealing with sick people. I quit medical school a few months into my third year. (Lest you think I flunked out, I was first in my class and scheduled to receive an award for this distinction.) It wasn't long after I left medical school that I realized I needed a plan for my future. The subject I had most enjoyed in medical school was gross anatomy, so I applied to, and was accepted in, the doctoral program of the Department of Anatomy at the University of Chicago. I had absolutely no concept of what research in anatomy would be and would have struggled mightily to find a dissertation topic, but the department fortuitously hired Charles Oxnard. His brilliance and charisma awakened in me a love of comparative functional morphology: "The only clear conclusion I have come to is that what have appeared to be my own freely taken decisions provide very little evidence of more wisdom than the blind dictates of destiny."[1]

GAME CHANGER

There are two commonly accepted facts that I still find difficult to accept. One is the closer systematic relationship of chimpanzees to humans than to gorillas. I keep hoping this will eventually lead to the discovery of a flaw in assigning greater systematic weight to similarity in DNA than to morphology, but I am not holding my breath. My second problem is with the ancient coexistence of multiple bipedal hominids. I want there to have been only one way to achieve bipedalism, for that to have occurred in one lineage, and for the nature of this bipedalism to have steadily improved under natural selection. All evidence argues against this, so I am left with the likelihood that evolution, like so many other facets of existence, does not feel obliged to accommodate itself to my preconceptions.

TIME TRAVEL

People familiar with my work might conclude that I would like to go back in time to view *A. afarensis* to see if any of the things I and my Stony Brook colleagues said about its locomotor behavior were true. But this issue has already been definitively resolved by a different time traveler (22 min, 41 sec into the video at https://jackstern.org/Lectures/LeakeySymp.mp4).[2] My choice of a time to visit is in the future. It is based on the fact that I suffer from the condition known as "physics envy." I have spent many hours in retirement trying to learn about particle physics and cosmology. I would like to go far enough into the future to learn if a successful Theory of Everything (uniting quantum theory with general relativity, understanding the nature of dark matter and dark energy, having the ability to deduce the twenty parameters of particle masses and elementary forces that must now be measured, and discovering definitive information on the existence or not of multiverses) was achieved. How far will I have to go to do this? I will settle on five hundred years. If scientific progress continues at its current pace, that should be long enough. If, in the intervening years, humanity has entered a new dark age, or ceased to exist, then that too will be interesting, although in a depressingly different way.

DRIVING FACTORS

I am a believer that the greatest driving force of evolution is food acquisition. This is a topic on which I have done no original thinking and pretty much rely on a gut feeling (no pun intended). The firmness of my belief is reflected in a lecture I gave to medical students, in which I sarcastically dismiss theories that have bipedalism arising in response to energetic, social, or thermoregulatory factors. I have favored Kevin Hunt's [chap. 31] notion that bipedalism arose in an ape forced to stand and feed on low-hanging branches because diminishing forests placed limits on the amount of food that could be found in the canopy.

EVOLUTIONARY LESSONS

I watch lots of news on television and hear much discussion of the present state and possible future of humanity. I cannot recall anyone bringing forward a fact about human evolution specifically, as opposed to evolution in general, when contributing to such discussions. Does it make any difference to how we now live, or will do so in the future, if human evolution was a bush or a tree? Does it make any difference if *A. ramidus* was bipedal, or if Lucy climbed trees? Does it make any difference if *Homo* originated in Africa or Asia? Maybe the diversity of genetic variation in Africans is meaningful for solving some current or future problem, but I don't know why. One might claim it makes a difference if some modern humans have Neanderthal genes that predispose them to certain diseases, but it is the existence of such genes, not their source, that is relevant. I will allow for one broad exception to my conclusion. It does make a difference that we descended from animals that were not bipeds and that we still suffer from our less than perfect adaptation to erect positional behavior.

The acceptance that evolution exists as a process does play a role in our present and future. It demonstrates the possibility of extinction unless we behave ourselves. It helps to understand how variation in resistance to diseases has arisen and how microbes can overcome this resistance. It leads to a better appreciation of chance as the determining factor in many fates. I suspect there are several more examples of how the mere existence of the process of evolution guides thinking about the present and the future, but I am tired.

I believe that the details of human evolutionary history serve the purpose of satisfying our desire to hear a good story. This desire characterizes all people. It is the driving force behind much research in the sciences and humanities. A good story makes life more interesting, and there are few better stories than how we arose. I previously mentioned my current interest in particle physics and cosmology. I also have been learning about the American Civil War. In both cases, these topics are no more significant to me than the good story they tell.

RELIGION AND SPIRITUALITY

 If it were demonstrated that there is no scientific way to account for human origins and characteristics, that would be a big incentive to become religious. However, scientific explanations do exist, but they have little bearing on religious views because the nature of religion allows it to be made compatible with any set of facts. The only firm conclusion I have come to is that the existence of Donald Trump as president is incompatible with the existence of a benevolent God.

INSPIRING PEOPLE?

George Gaylord Simpson.

93

ANDREA B. TAYLOR

A ndrea B. Taylor is a professor in the College of Osteopathic Medicine at Touro University and a research associate in the Department of Anthropology, Institute for Biodiversity Science and Sustainability, California Academy of Sciences. Currently, she is also the coeditor-in-chief of the *Journal of Human Evolution*. Andrea studies comparative morphology and biomechanics, focusing on primate feeding systems. She is the cofounder of the American Association of Biological Anthropologists' Committee on Diversity Women's Initiative. Her initiatives to support mentoring of women and other underrepresented minorities in STEM were recognized in 2014 by a grant from the Elsevier Foundation New Scholars program.

YOUR BEGINNINGS

My path to research, like my research career, did not unfold in a linear fashion. I was fascinated with the natural world from the time I was old enough to be let loose outside on my own with a jar and a butterfly net—which is to say, around the age of five. Growing up in Evanston, I had easy access to the Field Museum, the Museum of Science and Industry, and the Shedd Aquarium. I hounded my parents to take me to these amazing museums as often as possible. You couldn't spend time in these places and not develop a profound sense of intellectual curiosity about nature. My first foray into research occurred when my middle school science project placed in the Greater San Diego Science and Engineering Fair. I'd had an interest in science, but like many girls of my generation that interest hadn't been encouraged until my middle school biology teacher persuaded (read "coerced") me to participate. The winners were invited to spend three days touring the University of California, San Diego and the Scripps Institution of Oceanography, meeting and talking with researchers and seeing firsthand what they did in their research labs. The entire experience was exciting and inspiring. Math, however, was always my albatross. My father, a civil engineer, tried to help me through the mysteries of decimals, fractions, and percentages, using chalk and a chalkboard, but I seemed to have a "Do Not Enter" sign planted firmly across my forehead when it came to math that no amount of fatherly dedication could penetrate.

I questioned my ability to do research, and by the time I entered the University of California at Berkeley in 1979, I'd decided on a career in investigative journalism, which I thought would be a good way to marry my interests in writing and research—without the math. When I took an introductory course in biological anthropology to fulfill a prerequisite outside my major (this was only five years after the discovery of "Lucy"),[1] the fossil evidence for reconstructing our evolutionary history was like working on one long and spatially complex puzzle, and I was hooked. This course led to another in historical archaeology with the late James Deetz and another in primate behavior with Phyllis Dolhinow, and before I even realized what was happening I wasn't a journalism major any longer.

When Katie Milton took me aside one day at the end of a class and asked if I had considered going to graduate school, I decided to pursue a career in research. When I was navigating my career path, I always felt as if I was the only person in the field who hadn't followed the conventional trajectory from undergraduate to graduate school, to post-doc (and maybe more than one), to assistant professor, to promotion and tenure. What I learned is that there were many others like me, and many trajectories that can lead to a productive and fulfilling research career. In my case, these "deviations" from a linear path profoundly enriched my personal and professional life in ways I could not have anticipated.

GAME CHANGER

A game changer for me was Stephen J. Gould's *Ontogeny and Phylogeny*.[2] Publication of this seminal book in 1977 (two years before I went to college and nearly fifty years after de Beer's "Embryology and Evolution")[3] reinvigorated the concept of developmental transformations in evolution and led to decades of research and debate about the role of heterochrony (and neoteny) in human evolution. Much of my undergraduate education in biological anthropology had been delivered from a paleonto-logical perspective. When I went to graduate school in 1985 at Northwestern University, I worked with Brian Shea, whose own research in ontogeny, allometry, and heterochrony greatly broadened my perspective about the processes that influence evolution and the mechanisms underlying those processes. (Long after I finished graduate school I kept a picture of Gould's clock models taped to the wall above my desk to keep these concepts at the forefront of my own work.) Even if heterochrony has since come to be viewed as less important in evolutionary developmental biology, the influence of developmental evolution in shaping the adult phenotype, and the potential to increase morphological evolvability and produce new features across species, profoundly changed the way I approach the study of evolutionary morphology.

FUTURE EVOLUTION

We already know that technology is now shaping the human experience and influencing human evolution much more than natural selection. We only have to look at the influence of "smart devices" on our children's brains to know that technology is influencing our evolution at a much faster rate than natural selection. Computer science, robotics, biomedical engineering, the use of electrical brain activity to directly control neuroprosthetic devices—these have made us ever-more dependent on technology. In addition, the science of gene editing through tools such as CRISPR and "prime editing" are shifting control of our DNA from natural to artificial selection. There are good reasons to think this may be different in the near future with climate change, global pandemics, and so forth. Unfortunately, although humans have become extremely skilled in developing and adopting new technologies, we seem to be no better at predicting how technology will affect our species in the long term than we are at predicting the long-term impact of natural selection.

SPECIAL?

The question of whether humans are "special" depends on how one defines the term. A number of characteristics we once thought separated humans from nonhuman primates (as well as other animals) have not stood the test of time; culture, tool use, behavioral complexity, and innovation are some of the characteristics that come to mind. As biological anthropologists seeking to understand what sets humans apart from other animals, we're always looking for differences—it's what we're trained to do— and we're really good at it. When we don't find them in the ways we expect, or when they don't hold up over time, we seem determined to find them by applying "newer" methods, more powerful technology, and looking for more nuanced distinctions. I do think we differ in biologically meaningful ways from other animals, including our closest living relatives. In terms of numbers alone, humans comfortably inhabit six continents whereas gorillas and chimps have become progressively restricted to parts of west and east Africa (largely at the hands of humans). If we want to be able to continue to study the African apes to better understand our differences, we may need to follow more closely in the footsteps of Jane Goodall and focus on our similarities, and this may be our best chance to keep these species alive.

RELIGION AND SPIRITUALITY

I see absolutely nothing incompatible about the study of evolution— human or otherwise—while at the same time holding spiritual values.

There's plenty of evidence that spirituality has played a fundamental role in our evolution, and it seems logical that we study the origins and evolution of spirituality alongside the origins and evolution of increased brain size or bipedalism or cooking. Spiritual values may even be one of the reasons we have yet to bring our species to the point of extinction, despite our having the means to do just that. I'm not sure we've realized the full potential of spirituality as an important part of our survival as a species, but we can always hope.

ADVICE

There is a tendency to specialize in science, and evolutionary anthropology is no exception. This specialization comes with trade-offs, one of these being deeper insights at the potential expense of more limited cross-boundary connections. Many of our greatest scientific discoveries—the use of stem cells for "growing" human organs, the development of neuroprosthetic devices that can be directly controlled by human electrical brain activity to restore movement to paralyzed patients, the discovery of the role of iron in preserving proteins in fossil tissues—resulted from cross-boundary connections among diverse fields such as molecular biology, paleontology, anatomy, biomechanics, neuroscience, material science, and engineering, to name just a few. There is strength in collaboration—and as I'm constantly reminded, you never know where you'll gain your next insights. I came to collaboration late in my career, and it has made all the difference. My advice to anyone interested in pursuing research related to our past or our future (or both) is to pursue your passion, and don't limit yourself. There are plenty of researchers interested in collaborating, and together you might make those cross-boundary connections that advance the field in ways you could never have imagined.

94

KENNETH "KEN" WEISS

Kenneth "Ken" Weiss is the Evan Pugh Professor Emeritus of Anthropology and Genetics in the Department of Anthropology at Pennsylvania State University. He studies the evolutionary processes and genetic basis leading to complex traits (e.g., teeth, skull) and how these vary within and among populations and species. His research includes how specific gene families are involved in different aspects of human disease susceptibility, such as cardiovascular problems or diabetes. Ken is also interested in the intersection of evolutionary biology history, bioethics, and society. He has coauthored two popular books: *Genetics and the Logic of Evolution*[1] and *The Mermaid's Tale*,[2] the latter accompanied by a blog of the same name.

YOUR BEGINNINGS

I became interested in anthropology in the 1960s while in the Air Force and doing some light reading. I was a math major and a meteorologist, with no undergraduate anthropology background. But by chance I became interested in evolution and how it works—indeed, in the fact of evolution itself, about which I'd previously thought very little.

GAME CHANGER

W. W. Howells's then current textbook, *Mankind in the Making*,[3] introduced me to a new topic. The subject of human evolution caught my interest, and I pursued it in graduate school at Michigan. Decades later I am still exploring how evolution works—and arguing against the seriously mistaken, if widespread, notions about it, even among biological scientists.

AMAZING FACT

Nothing specific really amazes me, given the general amazement at the universe and life itself; but the evolution of self-awareness and objective curiosity about the world are, perhaps due to the abstract nature of language in our species, of course, fascinating.

TIME TRAVEL

Maybe during the time of Lyell and Darwin, when worldwide facts were becoming available, and it was—suddenly—clear that biblical origin stories were not literally true. Life and geology are local *processes*. The age of the Earth, and of life, are mind-boggling to think about. I don't know that some particular time in deep history would be as interesting—but maybe it would be more frightening!

DRIVING FACTORS

We can, I think, only guess about the circumstances. Driving factors probably include abstract thought connected to language, the usefulness of explicit mental symbols, and interpersonal communication—all enabled our realization of the fact of, nature of, and amazing truth of evolution as our origin story.

FUTURE EVOLUTION

Anyone answering this question is indulging in pure science fiction. Mutations cannot be predicted, nor can we know the Earth's future ecology and biosphere, so there is no way to make such predictions. Of course, it is not unimaginable that we as a species will destroy ourselves . . . and, finally, leave the rest of life in peace!

EVOLUTIONARY LESSONS

The fact of evolution and the local nature of its work and hence the inherent local diversity of our biological nature is key to many things. In particular, we want universals (e.g., in medicine), but we know we evolve locally. We want genomic risk scores to be universal, for example, and we promise "precision" in such respects (even ignoring the central causal effects of unmeasured and unpredictable environments); but this flies in the face of the central fact of evolution, unlike physics and chemistry: *locality*. There may be "laws" of evolution in terms of its processes, but unlike physics, they apply ad hoc and locally. We need a different kind of science to understand life and, in particular, genetic effects.

Also, human evolution has often had a warlike history among territorial groups. If this can't be contained, with new weapons civilization is doomed. Likewise, our inability to conserve resources is ultimately self-dooming due to our massive population size.

SPECIAL?

Our language and aspects of our "intelligence" are unique to us. Of course, every species has unique aspects—or it wouldn't be its own species. Our symbol-based intelligence leads to culture, and that in turn leads to *accumulating* knowledge and legacy beyond the immediate local surroundings. No other animal or plant has anything close to that.

RELIGION AND SPIRITUALITY

A deist view, although having no actual empirical support one way or another, that "God" started things knowing how they'd unfold allows there to be a true God, without any actual material evidence. But love, care, and interactions lead to spiritual guides about how to live. Our attribution of some of this to "God" rationalizes it, and whether true or false, at least this helps keep selfishness under some restraint.

ADVICE

Be objective about what you see or choose to look at. Don't cook your evidence by choosing only what will give the answer(s) that you want. This is one of the hardest things to do in science, but one of the most important, especially in this context. View our past as it really was, not as you would like it to have been, and acknowledge the essential unpredictability of the future.

INSPIRING PEOPLE?

Darwin or Wallace, just to see their particular answers and how they reached them, especially if they could see the twenty-first-century world in the context of the world at their time. But their answers would, of course, reflect their particular perspective (even with another century's worth of evidence).

95

RICHARD WRANGHAM

R ichard Wrangham is the Moore Research Professor of Biological Anthropology at Harvard University, founder of the Kibale Chimpanzee Project (in 1987, Uganda), and cofounder of the Kasiisi Project (in 1997), which promotes conservation among young Ugandans. His research on chimpanzee ecology and social behavior translates into inferences for human evolution. He summarized his work on the evolution of warfare in *Demonic Males*,[1] coauthored with Dale Peterson; described the vital role of cooking in human evolution in *Catching Fire*;[2] and presented a theory of the evolution of morality and *Homo sapiens* in *The Goodness Paradox*.[3] He is also a MacArthur fellow.

GAME CHANGER

I was galvanized by Sibley and Ahlquist's[4] research claiming that chimpanzees are more closely related to humans than they are to gorillas. Their claim was so challenging to conventional thinking that several years went by, and many follow-up studies, before it was widely accepted. I absorbed the lesson in 1989 thanks to David Pilbeam [chap. 19]. The discovery's shock value came from the fact that chimpanzees and gorillas are very similar to each other. To a large extent, a gorilla is like a bigger version of a chimpanzee. Some scientists even used to place gorillas and chimpanzees in the same genus, *Pan*. So it was challenging to accept that these two kinds of ape do not form a natural evolutionary grouping separate from humans.

The meaning of Sibley and Ahlquist's discovery depends on why chimpanzees and gorillas are so similar. There are two possibilities. One is that chimpanzees and gorillas were originally rather different from each other and have become more similar over evolutionary time. That makes little sense, however. Among other problems, the two species are broadly sympatric (meaning they share the same habitats) and have very similar diets. In animals generally, closely related species that are sympatric tend to become more different, not more similar—a logical response to between-species competition called the principle of character divergence.

So the other explanation is overwhelmingly more likely. It says that after chimpanzees and gorillas differentiated from each other (around 11–9 million years ago), they changed very little. Biologically, that makes sense, and from the perspective of human evolution, it is fascinating. It means that the last common ancestor of chimpanzees and humans, the one that gave rise to the australopithecines, was a member of a conservative group of species including gorillas and chimpanzees. What was that ancestor like? If it was relatively big, it would have been more gorilla-like. If it was small, it was more chimpanzee-like. The australopithecines were more chimpanzee-sized (or even a little smaller than chimpanzees), so the species that gave rise to the australopithecine lineage can be inferred to have been very like a chimpanzee. Fossils will eventually test this hypothesis.

Sibley and Ahlquist told us that our last rain forest ancestor was almost certainly very similar to a chimpanzee. With that strong hypothesis in mind, I began to think much more clearly about the selective pressures and evolutionary processes that could explain how humans became so different from an ancestor that made its living similar to that of a living, breathing chimpanzee.

AMAZING FACT

Considered from afar, it seems remarkable that the most intelligent species on Earth should also be exceptional in its violence. Admittedly, we understand much about why this happened. We evolved as a territorial species; individuals maximize fitness by taking advantage of power; in the past, we used our intelligence to find power differentials and take advantage of them; and our evolutionary psychology does not change quickly. Nevertheless, the fact that we find it so hard to use our intelligence to control the devastation of violence seems objectively as amazing as it is sad and troubling.

TIME TRAVEL

 I would go to the end of the Miocene to see what happened as a population of chimpanzee-like forest apes were changing to walking upright. I would like to go in the early or middle dry season, which I would expect to be the time when the new adaptation of bipedalism was most important. And I would hope to see that the apes were spending a significant amount of time in the shallow, barely moving waters of an estuary or inland delta, wading up to their chests, tugging at buried rhizomes, getting their food staples by spending many hours a day bipedal in water.

DRIVING FACTORS

At the end of the Pliocene, between 3 and 2 million years ago, an australopithecine ape transitioned into the closest we have to a missing link, the mysterious *Homo* (or *Australopithecus*) *habilis*, a little-known species with apelike bodies and jaws and a brain that had taken a leap toward a larger size. Around 2 million years ago, *habilis* evolved into the earliest "full" *Homo*, *Homo erectus*, with bodies the shape of ours today. Two broad kinds of logic make me think this transition came about through an ape learning to control fire and cook, and eventually (with *erectus*) becoming biologically adapted to using fire. First, no other explanation accounts so well for various critical features of *erectus*, such as the small chewing teeth, small guts, and commitment to sleeping on the ground. Second, whenever our ancestors first learned to control fire, the effects must have been huge; but there is no later period in human evolution when such a large change in anatomy occurred. After *erectus* evolved, there is no evidence of a big evolutionary jump of the type that should have occurred when our ancestors became adapted to the control of fire. For these reasons, I think human evolution began when an ape learned to cook. In the years since 1999, when my colleagues and I started publishing this idea, the evidence for early humans using fire has been pushed back from about 300,000 years ago to 780,000 years ago at Gesher Benot Ya'aqov in Israel, one million years ago at Wonderwerk [South Africa], and possibly earlier in Koobi Fora [Kenya]. But the archaeological verdict is still out. It will be fascinating to see how future evidence unfolds.

SPECIAL?

Yes, we are special. Sophisticated language, I believe, makes us *sapiens*. Among its many consequences, one seems particularly important. Language enables us to coordinate in making plans to kill each other. Chimpanzees, gorillas, baboons, and almost all species living in multifemale, multimale groups have to put up with alpha-male despots because they have no way to get rid of them. Humans are different. When our ancestors became sufficiently skilled with language to conspire against resented members of their own group, they were able for the first time to control the power of despots. Our species uses leveling mechanisms that in prestate societies basically came down to subjecting outcasts, rivals, enemies, and nonconformists to capital punishment. The result, I believe, was a reduction in reactive aggression, an intensified fear of being viewed as "different," the development of a new group-focused style of morality, the evolution of the genes underlying these new psychological traits, and the origin of *Homo sapiens*.

ADVICE

To understand our species, it is vital to know general principles of evolutionary theory. So postpone your focus on humans until you have done a lot of evolutionary biology. If you are interested in evolutionary ecology, study another species in nature before spending time with humans. In some cases humans might represent exceptions to general biological rules, but knowing how and why the rules work will always be helpful.

INSPIRING PEOPLE?

With every generation we have new and better answers. So, if I can time travel but have to ask someone living or dead, I would like to ask whoever was born this year and grows up to be, by the end of the twenty-first century, the most well-informed expert on human evolution. By then many of the problems we are still uncertain about will be much more clearly resolved—such as what selective pressures favored increasing brain size, when fire was first controlled, how similar the last common ancestor of chimpanzees and humans was to a chimpanzee, the role of boats in peopling the Earth, how human energy budgets differ from other species, or what genetic changes tend to occur in domesticated animals that are shared with humans. It would be exciting to hear the answers to these and many other questions.

PART V

Outro

VIEWS FROM THOSE WITH EXPERIENCE
IN OTHER FIELDS (E.G., MEDICINE,
PSYCHOLOGY, MATHEMATICS,
AND PHILOSOPHY)

96

DANIEL GILBERT

D
aniel Gilbert is the Edgar Pierce Professor of Psychology at Harvard University. He studies how humans understand and interact with others, predict their own happiness, and make decisions. In addition to his scholarly work, Daniel is a bestselling author whose 2006 book, *Stumbling on Happiness*,[1] has sold more than a million copies worldwide. He is also the host and cowriter of the award-winning NOVA television series, *This Emotional Life*. His three TED talks have been viewed more than 30 million times.

YOUR BEGINNINGS

My father was a scientist, my mother was an artist, and I was torn between those two worlds. In college, I accidentally stumbled into a social psychology course and discovered to my amazement that there was actually a discipline that asked the questions artists ask about the nature of our lives together here on Earth, but that used the techniques of science to produce precise, elegant, and objective answers to them. I was instantly hooked—and have remained hooked for forty-five years.

GAME CHANGER

Nisbett and Ross's 1980 classic book, *Human Inference: Strategies and Shortcomings of Social Judgment*,[2] was a game changer for me. Prior to this, social psychologists had explained inferential mistakes in motivational terms: "People stereotype because they are lazy." Or "People overestimate their abilities because it makes them feel good." Nisbett and Ross took a completely different approach, arguing that inferential errors are the predictable by-products of the cognitive systems that produce them. The mistakes we make about ourselves and others are just the price we pay for using inferential strategies that are simple, fast, and right most of the time. I was in graduate school when that book appeared, and it was a total revelation. It changed the way I thought about everything, and every bit of work I've done in the thirty-five years since then can be traced to its themes.

AMAZING FACT

We are the only animal that can see far into the future. This ability is either absent or limited in every other species on Earth. Despite the fact that our minds can predict the future with great certainty, our behavior remains "stuck in the present." For instance, we know precisely what kinds of catastrophes climate change will produce and precisely when it will produce them, yet we can't summon the will to take collective action to prevent them. What kind of animal can see a train in the distance but still can't convince itself to get off the tracks?

TIME TRAVEL

The future, of course! We already know about the past, so why waste a one-time miracle? I'm dying to know how this story goes—what discoveries we will make, what myths we will explode, and how our social structures will change. In a few thousand years, people won't be all that different than they are now, but the way they live certainly will be. Exactly how far into the future would I go? Far enough to be amazed by what I see, but not so far that my species is extinct. Two thousand years separates us from the ancient Greeks, so I guess I'll use that number and set my time machine for 4021 CE.

ADVICE

My advice is: Do it! We need you! Understanding human nature is not just the most interesting problem of our time, it is also the most urgent. Every significant problem our species faces—from war to famine to racism to climate change—is a behavioral problem. Every one of these problems could be solved overnight if people would behave differently. So why don't they? That is the question we must answer if we are to have the luxury of answering all the others.

INSPIRING PEOPLE

The Greek philosopher Epicurus has my vote for Famous Guy Most Likely to Have Been an Alien. How else can we explain his thinking, which came out of nowhere and was literally thousands of years ahead of its time? Epicurus argued that the universe was made of atoms and nothing more, that all things were just combinations and recombinations of these atoms, that humans were merely animals, that there was no life after death, and so on. He foresaw the entire worldview of modern science in a way that was just uncanny. On top of all that, he understood that the only thing that matters for human beings is happiness, and that everything else that we think matters is just another way of talking about happiness. He wrote many books, all of which have been lost. Nothing remains of his work except a few pages of maxims and some letters to his friends. In fact, we know about his ideas now only because the poet Lucretius described his theories in the epic poem "On the Nature of Things," which was also lost for centuries but luckily was rediscovered in a monastery in 1417. I'd very much like to sit down with Epicurus and ask him questions. "What planet are you from?" would be the first, and then I have a very long list of other questions.

97

HENRY T. "HANK" GREELY

Henry T. "Hank" Greely is a professor of law, professor by courtesy of genetics, director of the Center for Law and Biosciences, and chair of the Steering Committee of the Center for Biomedical Ethics at Stanford University. Hank specializes in the ethical, legal, and social implications of new biomedical technologies, particularly those related to genetics, stem cell research, assisted reproduction, and neuroscience. He summarized his ideas on the topic in his books *The End of Sex*[1] and *CRISPR People*.[2] He has also discussed other sensitive issues, such as the "de-extinction" of past species, on the TED stage.

YOUR BEGINNINGS

In 1991 Stanford was planning a three-day symposium on the then new Human Genome Project. The planning committee—Paul Berg (chair), Lucy Shapiro, and David Botstein—wanted to add another member, someone from the law school, and I was the only Stanford law school faculty member who could spell DNA. I had a great time on the planning committee and ended up giving a talk at the symposium, which turned into a book chapter. I met lots of fascinating researchers, and it changed my career forever. The most powerful consequence of meeting many great researchers was getting to know Stanford's Luca Cavalli-Sforza, who introduced my panel at the meeting. He and our colleague Marcus Feldman later called on my advice for the Human Genome Diversity Project, which truly launched me into law and the biosciences. Until then I was still considering a focus on U.S. health policy.

AMAZING FACT

That we're here. I don't think that's inevitable but is instead the contingent result of a lot of chance events over the last 4.55 billion years, most of which we'll never be able to identify. The Cretaceous/Tertiary boundary event, though, is probably one clear example.

TIME TRAVEL

I would travel 10,000 years into the future, just to see what, if anything, we have become. I'm sixty-seven, and what bothers me most about the increasing reality of my finite life span is that I want to see how the movie ends. Ideally, I'd go forward 1,000 years, 10,000, 100,000, 1 million years, and so on. If I get only one shot, my guess is that in 10,000 years we will have changed enormously but will still be connected to 2019. In 100,000 years, probably not.

FUTURE EVOLUTION

We will be shaping our own evolution, both in the ways we have inadvertently used in the last 10,000 years or more and in newer ways. When we change our environment, we change the selective pressures on us; for example, agriculture changes allele frequencies. We'll keep changing our environment, but we will also be able to modify ourselves in more direct ways—biological ways by changing our alleles as well as allele frequencies, and nonbiological ways by increasing the power of the tools that give us power, some of which will become even more incorporated into our "selves" than our smartphones have become. If our civilization survives the next one hundred years—by no means certain—we will be starting to modify ourselves, genetically and through ever-more integrated devices. In 100,000 years, I doubt that *Homo sapiens* will be around, although I suspect we will have one or more successor species. In 1 million years, about 0.02 percent of the time since the Earth was born, I haven't a clue—though I'd love to know.

For the next several decades we will mainly use genetics in reproduction to avoid passing on serious genetic diseases and disease risks. A few nondisease traits that people care about are easy to pick out from the genome, mainly "boy or girl," and I do think parents will increasing do that. Most of what we care about is the result of the interactions of many genes, many aspects of our environment, and chance. It will be a long time before we could be confident that we can safely select for better athletic ability or strong math skills. In fifty to a hundred years, maybe. Even then, though, one has to ask why change a human's genes rather than a human's tools? Why try to give a person great distance vision through genetic modifications instead of giving her powerful and easy-to-use binoculars?

My advice to our children, grandchildren, and great grandchildren is to keep a keen eye on the safety of any interventions involving children. Do not get lured into overly risky interventions based on speculative benefits and underexplored risks. Second, monitor the effects of any direct interventions in the human genome (and, for that matter, in human/device integration), and watch for unanticipated problems. I do not think it makes sense to give specific advice three generations in advance . . . nor would they (or should they) listen to detailed or specific advice from the grave. The circumstances, of both science and culture, inevitably will change in ways we cannot foresee; the people on the spot will need to make their own assessments.

EVOLUTIONARY LESSONS

First, studying the past, like studying *anything*, is its own reward. Interrogating, knowing, understanding anything is, for me at least, deeply satisfying, whether it is human evolution, a historical event, or a novelist's work. By looking at the past—the deep past of the biosphere, the rise of our species, or the short blip that is our historical past—I think we can get a better sense of the interactions and complexities of life and the world in ways that can alert us to problems today and in the future . . . if we are willing to pay attention.

SPECIAL?

We tell jokes. We care about totally imagined people and worlds. We talk, we write, we commit genocide, we make ice cream. We are different and in that sense special in a vast number of ways. Are we inherently more valuable than other species? I think only to ourselves.

A truly objective observer might not find our cities and cultures much different from termite mounds and societies. But we will care more about ourselves than about the "other," and to the extent that we get to make the rules—to the extent that nothing more powerful starts making the rules—we will continue to put on airs and give ourselves privileges.

ADVICE

 Have fun!

98

KEVIN KELLY

Kevin Kelly is the founding executive editor of *Wired* magazine, an avid blogger (*The Technium*), and a YouTuber (*Cool Tools*). He is also the author of several books on the future of the economy, culture, and technology, including the *New York Times* bestseller *The Inevitable*.[1] Kevin has launched multiple projects ranging from cataloging all living species (All Species Foundation) to fostering long-term responsibility in human enterprises (he's a board member of the Long Now Foundation). Interested in the visual arts and traveling, his most recent project, *Vanishing Asia*,[2] compiles almost 9,000 photographs collected over forty years, capturing scenes that are disappearing.

GAME CHANGER

The discovery of DNA was a game changer. Although the gene was long expected, its highly computational nature was not. The core of humans turned out to be digital, which meant we are informational beings and could be engineered. We have only begun to confront both of these disruptions.

AMAZING FACT

The most amazing fact of human evolution is that there is only one sentient being on the planet, us. Our closest cognitive cousins are not very close. Where are all the other sentient and semi-sentient creatures?

TIME TRAVEL

 I'd go 1,000 years into future; 500 years is not enough because our roads and infrastructure would be very familiar. But 10,000 years is too far—it might be disorienting or disquieting, and other "humans" may be unrecognizable. Am I allowed to bring back knowledge from the future? If so, I will bring all I can carry.

DRIVING FACTORS

 Language. Language gave us access to our own minds and is the source of our self-consciousness, which is the source of our imagination and invention, which is the source of civilization. Once the earliest hints of protolanguage emerged in a few individuals, they were highly attracted to anyone else with a hint of language. Language is mutually sexy—those who can communicate crave others to communicate with—and this attraction greatly accelerated the development and dissemination of language.

FUTURE EVOLUTION

Soon humans will speciate into many different varieties. Some will be purely organic, others will be cyborgian. On average there could be one new species of humans every year in the future. The term *human* will not be precise enough to encompass them all; and in the far future it may only be used to describe the unaltered set of Amish farmers growing corn on small plots around the globe.

SPECIAL?

Sort of, but not really. Through better study of our animal cousins and the creation of sentient artificial minds and robots, we'll come to see that the space of minds is vast. And in this vast space of possible minds, the human being will reside in a corner off to the edge, not in the center, like our star at the edge of the galaxy. Humanity will be seen to be a very specific and very peculiar creature that has evolved to survive on this planet. Just like all living creatures, humans are unique. Each one is hacking its own special niche. Our mind and being are unique, but also just one of a million others.

ADVICE

 Pretend you are ET.

99

DAVID KRAKAUER

D avid Krakauer is the president and William H. Miller Professor of Complex Systems at Santa Fe Institute. He has held fellowships and appointments at Oxford, Princeton, the University of Pennsylvania, and the University of Santa Barbara. David's research focuses on the evolutionary history of information-processing mechanisms in biology and culture. David focuses on the evolution of intelligence and stupidity on Earth. David's work is essential to elucidate major evolutionary transitions by simultaneously studying multiple organization levels: genetics, cell biology, microbiology, organismal behavior, and society. In 2012 David was included in the *Wired* magazine "Smart List" as one of fifty people "who will change the world."

YOUR BEGINNINGS

All people trained in evolution learn the Latin phrase *Natura non facit saltus* (nature does not make jumps). Looking back across the twentieth century, this is something that Max Planck would dispute. Discovery of the quantization of radiation places jumps at the foundation of nature. And more recently, interest in the information-theoretical basis of life makes "it from bit" the rule not the exception. Nevertheless, in my own life, it has been gradualism all the way. I very much live on the real number line of development. I do, however, remember when I was seventeen spending an endless summer in Portugal, where my father lived, reading Jean-Pierre Changeux's *Neuronal Man*[1] alongside Friedrich Nietzsche's *Human All Too Human: A Book for Free Spirits*.[2] Somewhere in between physiology and the aphorism, a certain idea began to extend that I now see as a connection spanning information and matter. It is empirically true that since that summer this has been my primary preoccupation. I do not want to credit a singular fusion of nihilism and neuralism with any life-altering decision. But there is little doubt that events like this one were special extended moments.

GAME CHANGER

Probably learning assembly language, in this case 6502. The hands-on discovery that all high-level programs (and by extension thoughts) can be conceived in assembler mnemonics and hexadecimals compiled into binary gave me, through practice, what reading Turing and his sequels would much later give me through theory. And more profoundly, what Darwin had in mind with natural selection—the iteration of a rather simple rule to produce innumerable varieties. Many years later, now out of high school, when I read books like Dawkins's *The Blind Watchmaker*[3] or Dan Dennett's *Darwin's Dangerous Idea*,[4] I was a little surprised that such obvious ideas needed to be published. I like these books now, but back then, having implemented the Game of Life using ADC, BIT, TAX, TAY, and LSR, etc., they struck me as the contributions of an establishment not hip to the bit.

AMAZING FACT

The way we collectively encode schemata of reality in material. There is something about cubism, the Fortepiano, complex numbers, moveable type, Lisp code, etc. that is amazing. All of these have been created through a cultural dynamic that discovers generative abstractions. It is vertiginous to look at an Akkadian cuneiform tablet from the third millennium BCE and realize that those vertical declivities in clay describe the first recorded dream interpretation—when Enkidu explains to Gilgamesh on Tablet IV that a falling mountain foretells the defeat of the guardian of the Cedar forest. Myths recorded in saturated dust.

TIME TRAVEL

About 100,000 years into the future. Let's say that the modern human lineage is on the order of 100,000 years old. All human art and artifacts no more than tens of thousands of years old. Human writing and machine technologies are a few thousand years old. In the last few thousand years, cultural evolution has outpaced organic evolution. The question I am interested in answering is what is the limit on the forward chronological capacity of the human imagination? When does Upper Paleolithic hardware—human brains—no longer support modern technological culture? The idea is to give culture as much time as biology has had to reach some kind of asymptotic state. The reason to time travel is to determine whether we have evolved into self-delivering Soylent Green for robots or reached a new level of hybrid enlightenment.

FUTURE EVOLUTION

Human biological evolution is being shaped by directed genetic engineering. There was a quarter century interval between the purification of restriction enzymes and Dolly the sheep. The Asilomar conference in 1975 called for banning recombinant DNAs derived from pathogenic organisms. In 2020, following the most devastating pandemic in over a century, Johnson and Johnson engineered a new recombinant DNA virus based on DNA from a pathogenic coronavirus fused with a pathogenic adenovirus. The world celebrated. It took less than fifty years to transition from a proscription of genetic dystopia to a prescription for genetic utopia. It is seriously difficult to imagine what we will be doing with recombinant genomes in a thousand years. But if very recent history is anything to go by, we will have rewritten a significant portion of the human genome for increased resistance and to reduce sporadic and heritable disease. Is this a good idea? I am not so sure. But the use of fast cultural evolution as a defense against fast organic evolution (microbial, metastatic, etc.) seems inexorable.

Culturally the obvious change is the emergence of a variety of forms of non-human-like machine intelligences. These will guide human reason into the future just as traditional education has guided human reason in the past. Recent developments are both alarming and reassuring in equal measure. The outsourcing of elements of human reason to competitive cognitive artifacts such as Facebook and Google is alarming. Experimentation with complementary cognitive artifacts, including some games and crowdsourced world creations (Mine Craft, Terraria, etc.), is encouraging.

SPECIAL?

Horizontal gene transfer is special. Echolocation is special. Flying is special. Ultrasonic perception is special. Living thousands of years is special. A song that travels ten thousand miles through the ocean is special. And yes, the twelve-tone technique is special, as is the Rubik's cube. The most profound implication of Darwin's theory is that there is a principled way to speak of special things; namely, in terms of synapomorphies and apomorphies—cladewide traits and species-specific traits. The human apomorphies are fascinating, and sleuthing for their "atomic" homologs in our phylogenetic relatives is often illuminating. After all, who would have thought that the labeled-line code for numerosity representation would be primate-wide? Or that chimpanzees can be better at discovering Nash equilibria in coordination games than human beings. What makes comparative cognition so interesting is the way it shows how very misguided many of our assumptions have been. We need to find a way to study human cognition in a way that does not make it an "upgrade" while recognizing its unique character. It is certainly an interesting fact that the one species for which we have unequivocal evidence of self-conscious knowledge production is uniquely sociopathic and on course to destroy the planet and all life thereupon.

ADVICE

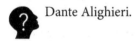 Read George Eliot, Jonathan Swift, and Edward Gibbon. Meditate on the existential implications of Darwin and Wallace. Rehearse justifying your career decision to Alan Turing and William James. Imagine at the end of your life having to explain your contributions to Martin Luther King Jr. and Simone Weil.

INSPIRING PEOPLE

Dante Alighieri.

100

MISIA LANDAU

M isia Landau is an anthropologist, writer, and artist. She taught at Wellesley, Dartmouth, Harvard, and Boston University, where she won the Metcalf Award for Excellence in Teaching. In 1991 she published her classic book, *Narratives of Human Evolution*,[1] which showed that scientific theories of human evolution are, in their basic narrative structure, versions of the hero myth. From 1994 to 2009, she was senior science writer at Harvard Medical School and published more than five hundred articles on a wide array of biological and medical discoveries. In addition to writing, she studies drawing and painting at the Academy of Realist Art, Boston.

YOUR BEGINNINGS

As a child, I was captivated by the idea of time. I loved science fiction and stories of future worlds but also myths and fairy tales, which had a kind of timeless quality. In college, I thought about studying literature, but a friend told me about a class she was taking called Introduction to Physical Anthropology taught by Ted Steegman. I remember finding my seat in the large auditorium and being immediately transported. I was fascinated by the imagery—the exotic fossil skulls and bones—but also by the mystery of the remote human past that in Steegman's gifted hands had a foot in both myth and science. The question that fascinated me above all else was what lay behind the expansion of the brain of early humans like Cro-Magnon. While researching my senior thesis, I learned that a hallmark of the human brain was a great profusion of dendrites, the feathery looking receptive portion of neurons. As a first-year anthropology graduate student, I began working in a neuroscience lab looking at dendritic growth in chickens. I loved the magic of histology—the blue and magenta stains used to illuminate the brains were beautiful—but not the methods used to harvest the brains (which consisted of cracking open fertilized eggs and snipping the heads off the innocent chicks inside). More important, I soon realized it would be years before I could hope to answer the questions that had captivated me in the first place. I left and began looking for a way to explore the big picture questions that attracted me to paleoanthropology and would draw on my strengths and passions. The path I found entailed time travel on a smaller scale; namely, through history. I immersed myself in nineteenth- and early-twentieth-century writings about human evolution—the works of Charles Darwin, T. H. Huxley, Arthur Keith, Elliot Smith, and the Americans William K. Gregory and Henry Fairfield Osborn. I was searching for unifying themes, but these scientists, working in the days before genetics, disagreed not only about which fossil forms gave rise to humans but also how evolution actually happened. I struggled to find a theoretical framework that would help me make sense of the incredible diversity of viewpoints. In the midst of my confusion, I had a conversation with Keith Hart, a social anthropologist in my department who, on an intuition, suggested that I look at the writing of two schools of literary thought: the Russian formalists and the French structuralists. I was immediately excited. How amazing to be able to combine literature with science! I would ultimately discover, in a kind of aha moment, that behind their differences these theories displayed a common narrative pattern—the hero story. It was a controversial thesis, but it was not meant as a criticism of the field. Paleoanthropologists address some of the most difficult questions we can ask. When talking about change over time, we can't help but use narrative. My goal was to help scientists tell better stories.

GAME CHANGER

One of the books I came across during my foray into literary theory was a slim volume by an obscure Russian folklorist, Vladimir Propp, called *Morphology of the Folktale*.[2] Despite their many different details, he found that Russian folktales more or less told the same story, one that consisted of thirty-one invariant elements he termed "functions." I can remember sitting in a green leather chair in an alcove in Yale's Sterling Library reading Propp's description and having a flash of insight. The story he was delineating felt so familiar. I ran upstairs to the stacks where decades and even century-old books on anthropology were kept—and rarely read judging by their often uncut pages—and two titles jumped out: *Mankind So Far*[3] and *The Story of Mankind*.[4] I realized that the body of writing I had been struggling to understand could be usefully approached as a genre of literature. Propp's book was not about human nature per se, but it would change everything for me, and ultimately how I looked at our species. It would spark a lifelong preoccupation with the role of storytelling, not just in science but in the very process by which we became human.

AMAZING FACT

Rather than a single fact, it's the grand sweep that amazes me. If I had to home in on a single aspect, it would be the power of human adaptability. Evolutionary biologists talk about exaptation—how traits that evolved for a particular use in a particular setting have gone on to serve so many others during human evolution. Natural selection, operating on slight genetic variations and the occasional beneficial mutation, produced a large brained, hypersocial, intensely willful, creative, ruminating, storytelling primate capable of expanding its reach beyond the scattered patches of Earth on which it originally evolved to inhabit nearly every corner of the planet and beyond. We have the ability not just to visit the moon and Mars but to see what's going on light years away in the far corners of the universe. Many of these adventures began with an idea. Years ago, my friend Shep Doeleman dreamed of turning Earth's observatories into one vast telescope and using it to image a black hole. In 2019, he and his colleagues revealed the first ever photograph of a black hole. Now they are planning to make movies. That's the really incredible journey—from deep inside our brains to the vast stretches of space. Of course, the flip side is that we have the power to destroy the very planet that gave rise to us—an ending, I might add, that occurs over and over in myth.

TIME TRAVEL

So many possibilities, but to satisfy my anthropological curiosity I would land at the mouth of Lascaux, the cave complex sometimes referred to the Sistine chapel of the Paleolithic, and accompany the artists as they make their way, presumably by torchlight, deep into the interior.

Was it one or a few lone artists, or did they work in pairs or groups? Women or men? How did they work—with what kinds of implements and supports? How did they plan their work? What was their intention as they painted? I'd want to be there as they made their way back to their camp. What stories did they tell—possibly around a campfire—on their return? On my way home, I would stop, if possible, in 1860 and make my way to Oxford for the famous debate between Thomas Henry Huxley and Bishop Samuel Wilberforce, which is often cast as a battle between evolution and creationism. Wilberforce is said to have asked Huxley, on which side, maternal or paternal, he was descended from an ape. Huxley answered that he would prefer an ape ancestor to a person who used his intellectual gifts to deceive. No notes were taken, so this account is hearsay. I would want to set the record straight, but most of all I'd want to meet and interview Huxley, who is a personal hero. I might then make a brief trip to London to speak with some of my favorite artists—the pre-Raphaelites Dante Gabriel Rossetti and Edward Burne Jones—and maybe a quick trip across the channel to France to meet Jean-Francois Millet, William Bouguereau, and the young Claude Monet. I might tell them about their ancient artistic ancestors—if only they'd believe me.

DRIVING FACTORS

I'm wary of speaking about driving forces because multiple interacting factors—bipedalism, a shift in diet, possibly rudimentary tool use—were probably at play, but another factor that may have played a role relatively early on is the ability to pattern time. Tool making depends on the ability to learn and teach a series of actions, which is essentially a temporal pattern. Humans are not the only tool-using, pattern-seeking primate. Many animals' existence depends on finding patterns in the outside world. But our ancestors' ability to connect events through time, to find beginnings, middles, and endings that made sense and helped them survive, would be elevated during millions of years of evolution to a fine art, literally resulting in the great works of literature. At first, these causal chains played out in the theaters of our ancestors' minds. They encountered a lion in a stand of trees and replayed the event over and over in their minds and possibly their dreams, reliving not just the encounter but the feelings of fear they experienced in the moment. At some point, and language may have evolved in part to enable this, they shared those patterns, essentially stories, conjuring up in their listeners the same emotional response that they had in the moment. This sharing of vicarious emotional experience in the form of stories—it all comes back to narrative!—is an essential aspect of who we are. Humans love stories for a reason. They are, from an evolutionary perspective, good for us. They help us survive. Later the ability to arouse emotional states in others would gain a dark side as humans used stories to control one another. But that is a longer story.

EVOLUTIONARY LESSONS

Although the fossil record may have little to say, the archeological record and the comparative study of living humans and primates speak volumes about species' tendency to draw distinctions between groups—to see the world in terms of who belongs and who doesn't, in terms of us vs. them. This deeply rooted tendency, reflected and reinforced in ritual and stories, had biological value when humans were living in small face-to-face groups competing for scarce resources on a finite landscape (although even then there was significant cooperation and probably interbreeding between groups). As they grew in size, and with the rise of the nation-state, human societies cultivated that sense of group identity through rituals and the telling of grand narratives, myths, and origin stories, which depicted members of the group as heroic while demonizing outsiders. Donald Trump is not a biologist, but he did have insight into this deeply rooted aspect of human nature, our tendency to see the world in binary opposites: good vs. evil, them vs. us. In the lead-up to the 2016 election, he played on this ancient vulnerability with his fabricated and fear-inducing stories of Mexican rapists and Muslim terrorists and also his narrative of a return to a golden (white) age in American history, which is what "Make America Great Again" was all about. It's how he got his foot inside the door to the White House. Many see stories as entertainment or ways to express, understand, or communicate about the world. But they also are powerful tools for influencing, shaping, and controlling human behavior. Trump's Big Lie that he won the 2020 election, along with the outlandish conspiracy theories perpetuated by QAnon and other groups (including outside forces like Russia), was designed not only to rouse fear and anger and other intense emotions in his followers but also to rouse them to action, as we saw on January 6, 2021. We need to pay attention to the stories we tell, not just in science but in the wider world.

SPECIAL?

Yes, for the reasons I've already described and for so many others. But the term *special* could be applied to a vast array of species. Just watch one of David Attenborough's incredible explorations of the natural world.

RELIGION AND SPIRITUALITY

Just contemplating the process by which every species evolves fills me with awe, and awe is a fundamental aspect of spirituality. But there are other reasons to see a connection. The questions that drive paleoanthropology are fundamental: Who are we? Where do we come from? They are at the heart of all religions, as well as literature and art, although I prefer to regard human evolution as my cosmology. It gives me a deep and satisfying feeling to know, in broad outlines, how we got here. I also believe that spirituality is compatible with the process of human evolution. Whether the human brain evolved to be spiritual or not, it is clearly a widespread exaptation. Most human groups have a well-defined set of religious or spiritual beliefs and practices that bind—and often control the emotions and behaviors of—their members. Indeed, the feelings of transcendence—of being part of something greater than oneself—at the heart of spirituality could be seen as our social sense writ large. Of course, those feelings of empathy and connection may be encouraged only toward those who share our beliefs, although some might argue that those are not truly spiritual feelings. In his famous essay "Evolution and Ethics,"[5] Huxley described the eternal battle between nature and civilization: how behaviors shaped by the natural process of evolution are often at odds with and undermine our moral progress as a species. We must continually be on guard against our natural impulses, even though Huxley was not entirely optimistic. Given the events of recent years, it appears to be as true today as it was when Huxley wrote his melancholy treatise.

ADVICE

I would caution you not to project yourself onto your subject matter, but I encourage you to plunge *into* your subject. Find a question that captivates you and that draws on your natural talents and curiosity, and follow it where it leads. The world will open up to you.

Apparently remote and unrelated areas of knowledge will gain coherence and meaning. The subject of human nature is vast. To understand it in a way no one else has done, be open to the possibility that apparently unrelated areas of knowledge are deeply connected.

INSPIRING PEOPLE

 If the time machine doesn't work, I would ask the gods to grant me a day—better yet a week—with Thomas Henry Huxley.

101

GEOFFREY MILLER

Geoffrey Miller is an evolutionary psychologist at the University of New Mexico (USA) and CUHK-Shenzhen (China). He works on sexual selection, mate choice, alternative relationships, virtue signaling, intelligence, consumer behavior, effective altruism, existential risks, and long-termism. His books include *The Mating Mind*,[1] *Spent*,[2] *Mate*,[3] and *Virtue Signaling*.[4] Geoffrey is active in popular science outreach, especially on Twitter as @primalpoly.

YOUR BEGINNINGS

In 1989 I was a graduate student in the Stanford psychology department. My mentor, cognitive psychologist Roger Shepard, welcomed two new post-docs to our lab: Leda Cosmides and John Tooby. I read their visionary early papers on this new field of evolutionary psychology they were developing, which aimed to integrate the evolutionary functional analysis of adaptations with information-processing models of human cognition, emotion, and motivation. I thought this was the most exciting intellectual project I'd ever encountered, and I felt an irresistible calling to make evolutionary psychology my life's work.

That year we were also fortunate to have several brilliant visitors to Stanford, including David Buss, Gerd Gigerenzer, Martin Daly, Margo Wilson, and Steven Pinker—who all went on to be leaders in the field. My best friend and grad student collaborator, Peter Todd, also got interested in this nascent field, and we all had intensive weekly discussions about how to make evolutionary psychology awesome and successful. Ever since then, it's been my passion to understand human minds as sets of psychological adaptations shaped by ancestral selection pressures.

I've been "red-pilled" by a series of books that led me further and further down the rabbit hole of understanding human nature in this evolutionary "deep time" perspective. I grew up in the 1970s and 1980s reading a lot of science fiction, so I always viewed the twentieth century as a very thin slice in a vast river of time stretching from prehistory to the distant future.

GAME CHANGER

Later, in the 1990s, I was especially inspired by the "Culture" novels of Iain M. Banks, which present one of the few science fiction futures worth aspiring to.[5] Reading *The Descent of Man and Selection in Relation to Sex* by Charles Darwin[6] convinced me that sexual selection through mutual mate choice was a neglected tool for explaining human psychology, and it directly inspired my Stanford PhD dissertation (1994) and my first book, *The Mating Mind*. *The Extended Phenotype* by Richard Dawkins[7] and *The Handicap Principle* by Amotz Zahavi and Avshag Zahavi[8] showed me how evolution can shape behaviors, signals, and cultures, which influenced my book *Spent* about the evolutionary psychology behind consumerism. *The Blank Slate* by Steven Pinker,[9] *The g Factor* by Arthur Jensen,[10] and the *Behavioral Genetics* textbook by Robert Plomin and colleagues[11] convinced me that evolutionary psychology needs to become better integrated with behavioral genetics, intelligence research, and the study of individual differences, which led to a lot of my empirical research as an assistant professor (2001–2008).

More recently, I've been influenced by books such as *Doing Good Better* by William MacAskill[12] (on effective altruism), *The Precipice* by Toby Ord[13] (on existential risks to humanity), and *The Infinite Machine* by Camila Russo[14] (on crypto/blockchain)—all of which made me think about how we can stretch our limited human minds to understand evolutionarily novel opportunities, threats, and technologies that will dominate the twenty-first century.

AMAZING FACT

 I'm astounded that our human minds, which were shaped almost entirely to live in small-scale tribal clans with a few dozen people, invented ways to achieve large-scale cooperation on a civilizational scale of millions of people. How did human nature become so incredibly scalable? If I was an extraterrestrial looking at our planet 50,000 years ago, I might not have predicted that these newly encephalized bipedal hominids would quickly develop agriculture, cities, governments, markets, science, and technology.

I feel like I have a pretty good grasp of how we got from one-pound brains to three-pound brains, but I'm still amazed that we got from three-pound brains to SpaceX, Bitcoin, AlphaGo Zero, YouTube, and OKCupid.

TIME TRAVEL

 If you go backward in time, you're putting yourself in the role of participant/observer scientist—you'd be the best-informed sentient being around. If you go forward in time, you'd be putting yourself in the role of research subject for much smarter future scientists to study. Assuming technological progress and cognitive enhancement continues, a person from our era who travels a few hundred years into the future would probably be the dumbest sentient hominid anybody had ever seen. But we might still be welcomed, interviewed, and cared for. We might become minor celebrities. Future life would probably be more interesting, blissful, and weird than we can possibly imagine. I hope there would be some way I could be cognitively upgraded so I had some idea about what was going on.

DRIVING FACTORS

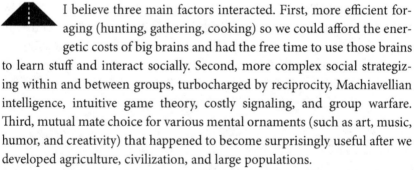 I believe three main factors interacted. First, more efficient foraging (hunting, gathering, cooking) so we could afford the energetic costs of big brains and had the free time to use those brains to learn stuff and interact socially. Second, more complex social strategizing within and between groups, turbocharged by reciprocity, Machiavellian intelligence, intuitive game theory, costly signaling, and group warfare. Third, mutual mate choice for various mental ornaments (such as art, music, humor, and creativity) that happened to become surprisingly useful after we developed agriculture, civilization, and large populations.

As I argued in *The Mating Mind*, the interplay between sexual selection and group selection was probably crucial in the emergence of creative individual intelligence that can organize itself into formidably capable tribes and awesomely inventive civilizations.

FUTURE EVOLUTION

 We've been semiconsciously shaping our own evolution for millions of years through mutual mate choice. That trend of self-directed evolution will continue, just with a bit more help from new genetic technologies that allow direct selection and manipulation of traits. Humans will push to become smarter through cognitive enhancement and more virtuous through moral enhancement. Meanwhile, artificial intelligence systems like AlphaZero and brain-machine interfaces like Neuralink will continue to make progress—especially given the likely AI arms race between geopolitical superpowers.

It's hard to predict whether the most common smart entities a century from now will be genetically enhanced humans with some AI augmentations or whole-brain emulations (like those in Robin Hanson's book *The Age of Em*[15]) or AIs flying around in physical drone bodies or disembodied AI tribes living and working in a Matrix that's only loosely connected to what we perceive as "base reality." If I had to bet, I'd predict that there will still be billions of humans running around—but also trillions of sentient AIs of various sorts. And the AIs will be hard to count because they won't have the same discrete, bounded phenotypes that individual humans have.

The likelihood that most of our descendants in 100,000 years will resemble current humans is basically zero. Evolve or become extinct. Those are the choices.

RELIGION AND SPIRITUALITY

 Until a few years ago, I was a hard-core Darwinian "New Atheist" in the style of Richard Dawkins, Sam Harris, Dan Dennett, and Christopher Hitchens. I had nothing but contempt for organized religion. I thought it was a memetic plague upon humanity, with huge costs and few benefits. I also had a pretty skeptical view of spirituality, which seemed to involve mostly Eastern mysticism repackaged into New Age consumer goods and services: overpriced yoga centers, crystal healing, and spa treatments in places like Sedona, Arizona, Boulder, Colorado, and Ubud, Bali.

However, ever since college I'd been interested in psychedelic research. As a psychology major, I read a lot of Timothy Leary's academic papers from the early 1960s on LSD experiences and LSD therapy. Long before he became a hippie guru, Leary was a very smart Harvard psychology professor, and he took psychedelics seriously as windows into human nature and the cosmos. That whole line of research taught me a lot of epistemic humility—that what we perceive as "ordinary life" might be a very superficial cross-section through a much deeper, much more complicated reality. I've stayed interested in psychedelic research, and I saw Timothy Leary talk at Stanford when I was in grad school (he was in his seventies, but still vivid, funny, and charismatic). I've been to Burning Man. I have lots of friends who have taken LSD, DMT, psilocybin, or ayahuasca, and I've paid attention to the revival of psychedelic research over the last few years. It's hard to dismiss spirituality completely if you have any familiarity with powerfully psychoactive chemicals.

Also, ever since I read about "the Matrix" in the 1984 novel *Neuromancer* by William Gibson[16] (long before the 1999 movie *The Matrix*), I have been interested in the simulation hypothesis—the idea that we might be living inside some type of virtual reality computer simulation. As a grad student and post-doc I published a lot of computer simulations myself, using genetic algorithms to simulate how evolution could shape autonomous agents to solve various adaptive problems. It didn't seem that far of a stretch to imagine much smarter beings with much more powerful computers doing the same for us.

If you put together psychedelic experiences and the simulation hypothesis, it seems intellectually arrogant to dismiss all spiritual views of reality. However, it's unlikely that schizotypal cult leaders in ancient civilizations had any better insights into spiritual reality than modern people can discover for themselves.

ADVICE

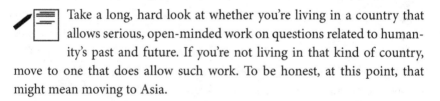 Take a long, hard look at whether you're living in a country that allows serious, open-minded work on questions related to humanity's past and future. If you're not living in that kind of country, move to one that does allow such work. To be honest, at this point, that might mean moving to Asia.

INSPIRING PEOPLE

 Elon Musk, because he'll probably be the one human from our era who's still remembered and admired a thousand years from now.

102

EUGENIE "GENIE" SCOTT

E ugenie "Genie" Scott is the former executive director of the U.S.
National Center for Science Education, a nonprofit organization aiming to improve the teaching of evolution, climate change, and science
in general. Genie is a world expert on the evolution controversy, science
denialism, and explaining science to the public. She authored *Evolution vs.
Creationism*[1] and coauthored with Glenn Branch *Not in Our Classrooms:
Why Intelligent Design Is Wrong for Our Schools*.[2] Genie has received numerous awards for her work from science organizations and educator organizations and has received ten honorary degrees. Asteroid "249540 Eugeniescott"
was named for her in 2014.

YOUR BEGINNINGS

At the age of ten or eleven, idly flipping through my older sister's
college textbook, I stopped, stunned at the reconstructions of Java
Man, Neanderthal Man, and Cro-Magnon Man. This was in the mid-
1950s, so these must have been the classic American Museum of Natural
History busts by McGregor. *These* were our ancestors! We once looked like
this! One of them looked a little like her boyfriend! How and why did we
change? The book cover said "Anthropology." I decided then that I wanted
to become an anthropologist and study fossil humans. It didn't quite work
out that way: yes, I became an anthropologist, but my atypical career took
me away from academic work and toward explaining science—especially
evolution—to the public. But human evolution is still my favorite "kind"
of evolution.

GAME CHANGER

You'll be surprised: Ernst Mayr's *Animal Species and Evolution*.[3] Yes, we've learned more about biology since 1963, but this book undergirds the two cores of biological anthropology: human evolution and human variation. It's about genes and how they move around within a species. A population is a group of individuals that exchange genes. Species are aggregates of populations. Populations within a species that have reduced genetic contact with other populations often die out, and some may change so much that they can no longer exchange genes with other populations. A new species arises: speciation occurs. This "population thinking" explains why anatomically modern humans, Denisovans, and Neanderthals (and maybe others not yet discovered) could exchange genes: these large-bodied, bipedal, tool-using, highly social hominines were only incompletely speciated. It's why we have tiglons and ligers too. The mess that is early hominin taxonomy might be a function of the same thing; only creationists expect nice, crisp "kinds" of organisms. Evolutionists understand that "good" species that can no longer exchange genes sometimes take a long time to develop, and in the meantime, genes are swapped among populations.

Human variation and "race" similarly are enlightened by population thinking. There was an explosion of scholarship in biological anthropology in the 1960s, rethinking the concept of race. Montague, Livingstone, Brace, and Brues—many of our forebears fought hard against the public concept of race. Even today the general public concept of race is that races are discrete (bounded units), stable or permanent over time, are "pure," and that individuals can reliably be placed in a race.

Biologically, races of song sparrows or lizards or other vertebrates are *groups* of populations living in the same geographic area that, because of geographical propinquity, are probabilistically more likely to exchange genes with one another. Yes, humans—especially today—hop around geographically; this Wisconsinite exchanged genes with a male from California. Still, it's statistically more likely that individuals—even humans—will choose mates from the local population. Sometimes (not always) these groups of populations share some (not all) phenotypic or genotypic characteristics because of this tendency for breeding within the local geographic region. Race is at best a squishy concept, generally not necessary, but marginally useful for communication about variation within a species, which has implications for conservation and some other research. Is the Florida panther an endangered race of cougars? Are giraffes one species with multiple subspecies or several species? And so on.

Mayr laid out basic population dynamics: as members of the same species, populations are open genetic systems. It is thus impossible for such groups of populations to be boundable units, nor is there "racial purity." Because of evolution, as well as genetic exchange among populations, races (groups of populations in a geographic area) will not be stable through time: a race in the biological sense is a temporary phenomenon. Much less can race be viewed on an individual level: if race is biology, it's a statistical concept. White supremacists will have to find something other than biology upon which to base their hatred.

Like most biological anthropologists, I avoid the word *race* (much as I avoid the word *theory*) because it takes too long to explain why what people think the word means bears no relationship to how biologists use the term. Several times when I've been on radio call-in programs on stations with predominantly African American audiences, I hear that Darwin was a racist or that evolution is a racist idea. Of course, the callers aren't really interested in biological variation in song sparrows or giraffes: the concern is racism and discrimination. With severely limited time (when do they next cut to commercial?), I'm not about to try to explain how intragroup variance rates disprove their understanding of race, a la Lewontin. My sound bite is: "If you understand the modern *genetically based* concept of evolution, it's your best weapon against racism." With luck, a door may be cracked and curiosity may lead a listener through it to find out more. Population thinking, based on intraspecific gene flow, is foundational to understanding both evolution and human variation, and dispelling the misconceptions of race. And Mayr illuminated that path long ago.

AMAZING FACT

In comparing humans with our closest primate kin, chimps and gorillas, no matter what behavior we examine, if we look at it carefully enough and study it long enough, we see differences in degree rather than differences in kind. When I was a baby anthropologist in the 1960s, primate field and laboratory studies were dismissing claims of human exceptionalism such as tool use, symbolic communication, cooperative hunting, problem solving, etc. Harder to fall from human exceptionalism have been behaviors such as altruism and deceit and the emotions of empathy, a sense of justice, compassion, and grief—much less love and friendship—but that's the way the science is headed, and the sooner the better, in my opinion.

TIME TRAVEL

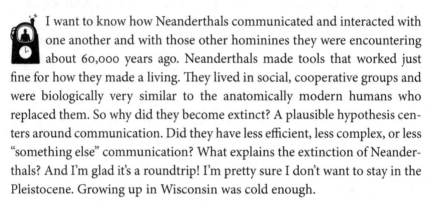 I want to know how Neanderthals communicated and interacted with one another and with those other hominines they were encountering about 60,000 years ago. Neanderthals made tools that worked just fine for how they made a living. They lived in social, cooperative groups and were biologically very similar to the anatomically modern humans who replaced them. So why did they become extinct? A plausible hypothesis centers around communication. Did they have less efficient, less complex, or less "something else" communication? What explains the extinction of Neanderthals? And I'm glad it's a roundtrip! I'm pretty sure I don't want to stay in the Pleistocene. Growing up in Wisconsin was cold enough.

DRIVING FACTORS

To answer this question, you have to look at the traits that distinguish us from our closest kin, chimps: obligate bipedalism, reliance on tool making, and reliance on symbolic, phoneme-based communication. Here's what we know: we were bipedal before we had stone tools, and we had stone tools before we got big brains. We have evidence of symbol use with burial of the dead, but that isn't until very late in human evolution. It may be impossible to know when language became phonemic, and that would have occurred later than our split from the chimp lineage anyway. But we know a lot more now about both chimps and humans—the end points of two evolutionary lineages that once shared a common ancestor—than we did fifty years ago, and we'll learn more in the future.

FUTURE EVOLUTION

A hundred years from now we will look (and largely behave, alas) the same as we do today. In 100,000 years into the future, it is impossible to say, but we'll still be featherless bipeds with a very efficient communication system and will be dependent on our technological ability for survival. In 1 million years, as a species humankind probably will be extinct.

SPECIAL?

Of course we're special! We are very good at being human. Barnacles are also special, and we'd be very poor at being barnacles. My former professor Jim Gavan discouraged students from using the phrase "higher primates." ("Can you hang by your feet?!") All organisms today are the result of the same processes of evolutionary adaptation and probability. Human exceptionalism is a toxic excuse to justify actions that benefit us but may be detrimental to other organisms. My brother-in-law liked to fish, but he didn't like to eat fish, so he practiced catch-and-release. I recall him casually saying, as he pulled a hook out of a thrashing perch and threw it back into the lake, "Good thing they can't feel pain." Um, it's a vertebrate. Its nervous system can't do calculus, but it provides very well for feeling pain. It's special.

RELIGION AND SPIRITUALITY

Some religious views indeed are incompatible with an evolutionary understanding of the origin and development of humans and other living things on Earth: the literal interpretation of the Bible, Quran, or Torah, in which God specially creates everything in its present form all at one time is incompatible with the understanding that the universe unfolded over time and continues to change. But even within the Abrahamic faiths, this is not the only view nor, in Christianity, is it the predominant one. The Catholic theologian John Haught, channeling the geneticist Theodosius Dobzhansky, wrote, "Nothing in theology makes sense except in the light of evolution." The Protestant theologian Robert John Russell similarly has proposed that religious views must be compatible with science to be coherent and defendable.

To creationists, religion trumps science. Given the revealed truth of the Bible, scientific data and theory supporting evolution cannot be correct; an arrow of causation runs from religion to science. For scientists who accept evolution, the arrow points in the other direction: science informs religion. That science informs a religious perspective does not mean science determines religious views. Religion first and foremost is about the relationship of humans to the divine, or the supernatural. It's about meaning and purpose, and it usually incorporates the ineffable—none of which characterizes science. Religion is not primarily about explaining the natural world: that's what science does, and does extremely well—especially compared to revelation. Religious individuals who accept evolution tend to accept the methodological naturalism of science, where only natural forces are used to explain natural phenomena. But religious scientists and other individuals reject that science compels philosophical naturalism: the view that matter and energy are not only sufficient to explain the natural world but that the universe is comprised *only* of matter and energy leaves no room for God.

Conflict between evolution (or any other science) and religion arises when one side or the other attempts to have it all: to both explain the natural world and provide answers for bigger questions of purpose and meaning. When atheists claim that science "proves" that there is no God, religious people rightly bristle; when creationists bend the data and distort science to prop up biblical literalism, scientists—whether believers or nonbelievers—rightly are outraged.

That said, there are philosophical positions such as humanism that are highly informed by science and draw meaning from this contemplation. I have friends who are evangelical Christians who look at the geological column or the human fossil record exactly as I do, a humanist. We accept the same data to explain the history of life, but the *meaning* we draw from these data vary greatly and are reflections of our respective theistic and nontheistic worldviews. Get the arrow pointed in the right direction and, yes, evolution is compatible with religion.

103

ANIL SETH

A nil Seth is a professor of cognitive and computational neuroscience and the codirector of the Sackler Centre for Consciousness Science at the University of Sussex. He is also the codirector of the Canadian Institute for Advanced Research (CIFAR) program on Brain, Mind, and Consciousness; codirector of the Leverhulme Doctoral Scholarship Programme: From Sensation and Perception to Awareness; and a Wellcome Trust Engagement Fellow. Anil studies the biological basis of consciousness, bringing together neuroscience, mathematics, computer science, psychology, philosophy, and psychiatry. Anil is also a best-selling author (e.g., *Being You*[1]), a popular blogger (NeuroBanter.com), and a regular contributor to various media (his TED talk has more than 12 million views).

GAME CHANGER

The philosopher Daniel Dennett published *Consciousness Explained* in 1991[2]—an injection of first-principles rationality and neuro-philosophical clarity that hit me like a freight train as I was beginning my undergraduate studies in natural sciences at Cambridge. Like many kids, I'd been fascinated by the big questions surrounding human consciousness: How does it happen? What is it for? What happens when we die? and What is free will? Dennett's book, although not actually providing the answers its title promised, showed me—and many others—a way to think about the problem and provided a license to think about it from many directions at once. It influenced me probably more than I realize.

AMAZING FACT

The paradox that we are both less special and more special than we tend to think we are. Less special because we are not in fact at the center of the universe, part animal and part angel. The history of science and of rational thought has traced a progressive dethronement, a dissipation of human exceptionalism, that has allowed us to see ourselves increasingly as part of, rather than apart from, the rest of nature. And yet also more special. Every species is unique in its own way, but we are unique in our uniqueness. No other species has been able to hold up a mirror to the universe as we have. There seems to be no reason for the cosmos to deliver its secrets to us, and to us alone, but so far this seems to be what's happening.

TIME TRAVEL

It would be a hell of a gamble, given the reasonable chance of materializing amid smoldering ruins, but I guess I'll gird my loins and jump 1,000 years into the future. I'd like to know whether we make it through this current bottleneck in our survival prospects, and if we do, what kind of world—and what kind of people—lies in wait. I'd also like to know the answers to my questions about consciousness, but who knows whether the answers would be comprehensible for a mind like mine. (On my glorious return, I'd take the credit anyway.)

FUTURE EVOLUTION

Human evolution is becoming increasingly decoupled from the selective pressures that shaped our ancestors, and that shaped this history of all life on this planet. Cultural and technological influences determine our prospects more profoundly every year, and the rapid uptake of gene editing technologies promises to shortcut the evolutionary cycle entirely. That said, we're always one nasty virus (one *more* nasty virus) away from being cut back down to size by good old biology. Of the many ways future humans will be different from the current variety, one fascinating dimension will be the extent to which we merge with our machines. This process is already well underway (think cochlear implants), and absent some cataclysmic event there will be a lot more cyborg in our collective future.

ADVICE

 Keep your mind open to insights from as many directions and disciplines as you can. Something as complex as human nature will never reveal itself through one path alone.

INSPIRING PEOPLE

Can I cheat and ask someone who hasn't been born yet? If so, I'd ask an evolutionary biologist or neuroscientist (if such categories still exist) a hundred years from now. If not, I'll ask Douglas Adams to answer these questions.

ACKNOWLEDGMENTS

First and foremost, I am eternally grateful to the top-notch world experts featured in this book. Thank you for your generosity and your trust in the project. Thank you all for taking the time to put your thoughts into words and share your personal experiences, insights, advice, and wisdom.

In addition, many of the book participants helped me in different ways during the process. Dan Lieberman and Ian Tattersall provided exceptional guidance to help me find the right place for this book. Ashley Hammond brainstormed various ideas about the book during the whole process, and as my wife, she patiently listened for hours to the various emerging challenges. John Shea, Sonia Harmand, David Alba, and Salvador Moyà-Solà gave helpful feedback while designing the figures.

The beautiful figures in this book and the icons summarizing the questions are the great work of Christopher Smith, who is both an experienced artist and a colleague in the field. Thank you for your laborious work and your patience over the many iterations of each figure. You are a rock star.

The entire team at Columbia University Press, especially Miranda Martin, Brian Smith, and Michael Haskell; and at KGL, especially Ben Kolstad and Kay Mikel, who helped me nearly seamlessly navigate the creation of this humongous work. In addition, the anonymous reviewers of the book gave constructive criticism and ideas to improve it. Thank you all for your phenomenal praise and your help.

My research institution, the American Museum of Natural History, facilitated some of the resources and time to develop this project as a part of their efforts to increase science popularization and outreach. Thank you.

My colleague and good friend Nathan Thompson helped me brainstorm some of the questions in the book during our trip to central Borneo to study orangutans at Tuanan in 2019. Thanks, dude! (And also to Erin Vogel for arranging logistics for getting us to Borneo!)

Tim Ferris, Seth Godin, Kevin Kelly, and Tim Armstrong are some of my life mentors in the distance (Kevin is also a book participant). Your work over the years inspired me to go ahead and embark on the journey leading this book. Thank you for doing what you do.

I am thankful to my friends and family, especially my parents, Pili and José Luis. Thank you for believing in me, always. I love you all so much.

Finally, I'm grateful to little Max for adding that spark of pure joy to everything that makes every day just right no matter what.

RECOMMENDED READINGS

This list includes the "game-changing" works recommended by the experts in the book.

Abbey, Edward. *The Monkey Wrench Gang*. New York: Perennial, 1975.

Aiello, Leslie, and Christopher Dean. *An Introduction to Human Evolutionary Anatomy*. London: Academic, 1990.

Aiello, Leslie C., and Peter Wheeler. "The Expensive-Tissue Hypothesis: The Brain and the Digestive System in Human and Primate Evolution." *Current Anthropology* 36, no. 2 (1995): 199–221.

Ardrey, Robert. *African Genesis*. London: Collins, 1961.

Banks, Iain M. *Consider Phlebas*. New York: Macmillan, 1987.

Berger, Lee R., Darryl J. de Ruiter, Steven E. Churchill, Peter Schmid, Kristian J. Carlson, Paul H. G. M. Dirks, and Job M. Kibii. "*Australopithecus sediba*: A New Species of *Homo*-Like Australopith from South Africa." *Science* 328 (2010): 195–204.

Blurton Jones, Nicholas. *Demography and Evolutionary Ecology of Hadza Hunter-Gatherers*. Cambridge: Cambridge University Press, 2016.

Broom, Robert. "The South African Fossil Ape-Men. The Australopithecinae, Part I: The Occurrence and General Structure of the South African Ape-Men." *Transvaal Museum Memoirs*, no. 2 (1946): 7–144.

Brown, Peter, Thomas Sutikna, Mike J. Morwood, Raden P. Soejono, Jatmiko, E. Wayhu Saptomo, and Rokus Awe Due. "A New Small-Bodied Hominin from the Late Pleistocene of Flores, Indonesia." *Nature* 431 (2004): 1055–61.

Cann, Rebecca L., Mark Stoneking, and Allan C. Wilson. "Mitochondrial DNA and Human Evolution." *Nature* 325 (1987): 31–36.

Cartmill, Matt. "Arboreal Adaptations and the Origin of the Order Primates." In *The Functional and Evolutionary Biology of Primates*, ed. Russell H. Tuttle, 97–122. Chicago: Aldine-Atherton, 1972.

Conroy, Glenn C., Martin Pickford, Brigitte Senut, John Van Couvering, and Pierre Mein. "*Otavipithecus namibiensis*, First Miocene Hominoid from Southern Africa." *Nature* 356, no. 6365 (1992): 144–48.

Conroy, Glenn C., Martin Pickford, Brigitte Senut, and Pierre Mein. "Diamonds in the Desert: The Discovery of *Otavipithecus namibiensis*." *Evolutionary Anthropology* 2, no. 2 (1993): 46–52.

Coppens, Yves, F. Clark Howell, Glynn Ll. Isaac, and Richard E. F. Leakey, eds. *Earliest Man and Environments in the Lake Rudolf Basin: Stratigraphy, Paleoecology, and Evolution*. Chicago: University of Chicago Press, 1976.

Dart, Raymond Arthur. *Adventures With the Missing Link*. London: H. Hamilton, 1959.

——. "*Australopithecus africanus*: The Man-Ape of South Africa." *Nature* 115 (1925): 195–99.

Darwin, Charles. *Descent of Man and Selection in Relation to Sex*. London: John Murray, 1971.

——. *On the Origin of Species. Or the Preservation of Favoured Races in the Struggle for Life*. London: John Murray, 1859.

Darwin, Charles, and Alfred Wallace. "On the Tendency of Species to Form Varieties; and on the Perpetuation of Varieties and Species by Natural Means of Selection." *Zoological Journal of the Linnean Society* 3, no. 9 (1858): 45–62.

Dawkins, Richard. *The Extended Phenotype: The Long Reach of the Gene*. Oxford: Oxford University Press, 1982.

——. *The Selfish Gene*. Oxford: Oxford University Press, 1976.

Day, Michael H., and John R. Napier. "Hominid Fossils from Bed I, Olduvai Gorge, Tanganyika: Fossil Foot Bones." *Nature* 201, no. 4923 (1964): 969–70.

De Beer, Gavin Rylands. "Embryology and Evolution." *Journal of Philosophical Studies* 5, no. 19 (1930): 482–84.

Dennett, Daniel C. *Consciousness Explained*. New York: Back Bay, 1991.

Descartes, René. *Meditations on First Philosophy: In Which the Existence of God and the Immortality of the Soul Are Demonstrated*, trans. Elizabeth S. Haldane. Cambridge: Cambridge University Press, 1911.

Dobzhansky, Theodosius. *Genetics and the Origin of Species*. New York: Columbia University Press, 1937.

——. *Mankind Evolving*. New Haven, CT: Yale University Press, 1962.

Fisher, Helen E. *The Sex Contract: The Evolution of Human Behavior*. New York: William Morrow, 1983.

Fleagle, John G. *Primate Adaptation and Evolution*. Cambridge, MA: Academic, 2013.

Foley, Robert. *Another Unique Species: Patterns in Human Evolutionary Ecology*. Harlow, UK: Pearson Education, 1987.

Fuhlrott, Johann Carl. "Menschliche Überreste aus einer Felsengrotte des Düsselthals. Ein Beitrag zur Frage über die Existenz fossiler Menschen." *Verhandlungen des Nationales Verein des Preusisches Rheinlandisches und Westfalens* 16 (1859): 131–53.

Gabounia, Léo, Marie-Antoinette de Lumley, Abesalom Vekua, David Lordkipanidze, and Henry de Lumley. "Découverte d'un nouvel hominidé à Dmanissi (Transcaucasie, Géorgie)." *Comptes Rendus Palevol* 1, no. 4 (2002): 243–53.

Gould, Stephen Jay. *Ontogeny and Phylogeny*. Cambridge, MA: Harvard University Press, 1977.

——. *Wonderful Life: The Burgess Shale and the Nature of History*. New York: Norton, 1990.

Green, Richard E., Johannes Krause, Adrian W. Briggs, Tomislav Maricic, Udo Stenzel, Martin Kircher, Nick Patterson, et al. "A Draft Sequence of the Neandertal Genome." *Science* 328, no. 5979 (2010): 710–22.

Hennig, Willi. *Phylogenetic Systematics*. Urbana: University of Illinois Press, 1966.

Higuchi, Russell, Barbara Bowman, Mary Freiberger, Oliver A. Ryder, and Allan C. Wilson. "DNA Sequences from the Quagga, an Extinct Member of the Horse Family." *Nature* 312, no. 5991 (1984): 282–84.

Howell, F. Clark. *Early Man*. New York: Time-Life, 1965.

Howells, William. *Getting Here: The Story of Human Evolution*. Washington DC: Compass Press, 1993.

——. *Mankind in the Making*. Garden City, NY: Doubleday, 1959.

Huxley, Julian. *Evolution. The Modern Synthesis*. London: George Alien & Unwin, 1942.

Huxley, Thomas H. *Evidence as to Man's Place in Nature*. London: Williams and Norgate, 1863.

Hrdy, Sarah Blaffer. "Infanticide as a Primate Reproductive Strategy." *American Scientist* 65, no. 1 (1977): 40–49.

Isaac, Glynn. *The Archaeology of Human Origins: Papers by Glynn Isaac*. Cambridge: Cambridge University Press, 1989.

Jacob, François. "Evolution and Tinkering." *Science* 196, no. 4295 (1977): 1161–66.

Jensen, Arthur R. *The g Factor: The Science of Mental Ability*. Westport, CT: Praeger, 1998.

Johanson, D. C., and Maurice Taieb. "Plio-Pleistocene Hominid Discoveries in Hadar, Ethiopia." *Nature* 260 (1976): 293–97.

Johanson, Donald, and Maitland Edey. *Lucy: The Beginnings of Humankind*. New York: Simon & Schuster, 1981.

Johanson, Donald C., and Blake Edgar. *From Lucy to Language*. New York: Simon & Schuster, 1996.

Jungers, William L. "Lucy's Limbs: Skeletal Allometry and Locomotion in *Australopithecus afarensis*." *Nature* 297, no. 5868 (1982): 676–78.

Kimbel, William H., Tim D. White, and Donald C. Johanson. "Implications of KNM-WT 17000 for the Evolution of 'Robust' *Australopithecus*." In *Evolutionary History of the "Robust" Australopithecines*, ed. Frederick E. Grine, 259–68. Piscataway, NJ: Transaction, 1988.

Kingdon, Jonathan. *East African Mammals: Part A, Carnivores. Part B, Large Mammals. Part C, Bovids. Part D, Bovids*, 4 vols. Chicago: University of Chicago Press, 1984.

Krings, Matthias, Anne Stone, Ralf W. Schmitz, Heike Krainitzki, Mark Stoneking, and Svante Pääbo. "Neandertal DNA Sequences and the Origin of Modern Humans." *Cell* 90 (1997): 19–30.

Laland, Kevin N. *Darwin's Unfinished Symphony*. Princeton, NJ: Princeton University Press, 2018.

Lamarck, Jean-Baptiste. *Philosophie Zoologique, Ou Exposition des Considérations Relatives à l'Histoire Naturelle des Animaux*. Paris: Musée d'Histoire Naturelle, 1809.

Landau, Misia. *Narratives of Human Evolution*. New Haven, CT: Yale University Press, 1993.

Goodall, Jane. *In the Shadow of Man*. London: Collins, 1971.

——. "Tool-Using and Aimed Throwing in a Community of Free-Living Chimpanzees." *Nature* 201, no. 4926 (1964): 1264–66.

Le Gros Clark, Wilfrid E. "Observations on the Anatomy of the Fossil Australopithecinae." *Journal of Anatomy* 81, part 3 (1947): 300–33.

Leakey, Louis S. B. *Adam's Ancestors. The Evolution of Man and His Culture*. New York: Harper Torchbooks, 1960.

——. "A New Fossil Skull from Olduvai." *Nature* 184, no. 4685 (1959): 491–93.

——. "Recent Discoveries at Olduvai Gorge." *Nature* 188 (1960): 1050–52.

Leakey, Louis S. B., Jack F. Evernden, and Garniss H. Curtis. "Age of Bed I, Olduvai Gorge, Tanganyika." *Nature* 191, no. 4787 (1961): 478–79.

Leakey, Louis S. B., Phillip V. Tobias, and John R. Napier. "A New Species of the Genus *Homo* from Olduvai Gorge." *Nature* 202 (1964): 7–9.

Leakey, Mary D., and Richard L. Hay. "Pliocene Footprints in the Laetoli Beds at Laetoli, Northern Tanzania." *Nature* 278 (1979): 317–23.

Leakey, Meave G., Craig S. Feibel, Ian McDougall, and Alan Walker. "New Four-Million-Year-Old Hominid Species from Kanapoi and Allia Bay, Kenya." *Nature* 376, no. 6541 (1995): 565–71.

Leakey, Richard E. F. "Evidence for an Advanced Plio-Pleistocene Hominid from East Rudolf, Kenya." *Nature* 242 (1973): 447–50.

Lewin, Roger. *Bones of Contention: Controversies in the Search for Human Origins.* Chicago: University of Chicago Press, 1997.

Lordkipanidze, David, Tea Jashashvili, Abesalom Vekua, Marcia S. Ponce de León, Christoph P. E. Zollikofer, G. Philip Rightmire, Herman Pontzer, et al. "Postcranial Evidence from Early *Homo* from Dmanisi, Georgia." *Nature* 449 (2007): 305–10.

Lordkipanidze, David, Marcia S. Ponce de León, Ann Margvelashvili, Yoel Rak, G. Philip Rightmire, Abesalom Vekua, and Christoph P. E. Zollikofer. "A Complete Skull from Dmanisi, Georgia, and the Evolutionary Biology of Early *Homo*." *Science* 342, no. 6156 (2013): 326–31.

Lovejoy, C. Owen. "The Origin of Man." *Science* 211, no. 4480 (1981): 341–50.

Lovejoy, C. Owen, Kingsbury G. Heiple, and Albert H. Burstein. "The Gait of *Australopithecus*." *American Journal of Physical Anthropology* 38, no. 3 (1973): 757–79.

MacAskill, William. *Doing Good Better: How Effective Altruism Can Help You Make a Difference.* New York: Random House, 2015.

Mayr, Ernst. *Animal Species and Evolution.* Cambridge, MA: Harvard University Press, 1963.

——. *Systematics and the Origin of Species.* New York: Columbia University Press, 1942.

Morwood, Mike J., Raden P. Soejono, Richard G. Roberts, Thomas Sutikna, Chris S. M. Turney, Kira E. Westaway, W. Jack Rink, et al. "Archaeology and Age of a New Hominin from Flores in Eastern Indonesia." *Nature* 431 (2004): 1087–91.

Napier, John. "Fossil Hand Bones from Olduvai Gorge." *Nature* 196 (1962): 409–11.

Nesse, Randolph M., and George C. Williams. *Why We Get Sick: The New Science of Darwinian Medicine.* New York: Vintage, 1994.

Nisbett, Richard E., and Lee Ross. *Human Inference: Strategies and Shortcomings of Social Judgment.* Englewood Cliffs, NJ: Prentice-Hall, 1980.

Ord, Toby. *The Precipice: Existential Risk and the Future of Humanity.* New York: Hachette, 2020.

Pääbo, Svante. "The Diverse Origins of the Human Gene Pool." *Nature Reviews Genetics* 16, no. 6 (2015): 313–14. [The research articles on the topic are numerous. This review is just a good starting point.]

Pinker, Steven. *The Blank Slate: The Modern Denial of Human Nature.* New York: Penguin, 2002.

Plomin, Robert, John C. DeFries, Valerie S. Knopik, and Jenae M. Neiderhiser. *Behavioral Genetics*, 7th ed. New York: Worth, 2016.

Propp, Vladimir. *Morphology of the Folktale*, trans. Laurence Scott. Austin: University of Texas Press, 1968.

Raff, Rudolf A. *The Shape of Life: Genes, Development, and the Evolution of Animal Form.* Chicago: University of Chicago Press, 1996.

Reich, David. *Who We Are and How We Got Here: Ancient DNA and the New Science of the Human Past.* New York: Pantheon, 2018.

Richards, David A., Simon H. Bottrell, Robert A. Cliff, Klaus Ströhle, and Peter J. Rowe. "U-Pb Dating of a Speleothem of Quaternary Age." *Geochimica et Cosmochimica Acta* 62, no. 23: 3683–88.

Robinson, John T. *Early Hominid Posture and Locomotion*. Chicago: University of Chicago Press, 1972.

Russo, Camila. *The Infinite Machine: How an Army of Crypto Hackers Is Building the Next Internet with Ethereum*. New York: Harper Business, 2020.

Sarich, Vincent M., and Allan C. Wilson. "Immunological Time Scale for Hominid Evolution." *Science* 158, no. 3805 (1967): 1200–1203.

Schaller, George. *The Mountain Gorilla: Ecology and Behavior*. Chicago: University of Chicago Press, 1963.

Senut, Brigitte, Martin Pickford, and Dudely Wessels. "Panafrican Distribution of Lower Miocene Hominoidea." *Comptes Rendus de l'Académie des Sciences, Paris* 325 (1997): 741–46.

Senut, Brigitte, Martin Pickford, Dominique Gommery, Pierre Mein, Kiptalam Cheboi, and Yves Coppens. "First Hominid from the Miocene (Lukeino Formation, Kenya)." *Comptes Rendus de l'Académie des Sciences, Paris* 332 (2001): 137–44.

Sibley, Charles G., and Jon E. Ahlquist. "The Phylogeny of the Hominoid Primates, as Indicated by DNA-DNA Hybridization." *Journal of Molecular Evolution* 20, no. 1 (1984): 2–15.

Simpson, George Gaylord. *Tempo and Mode in Evolution*. New York: Columbia University Press, 1944.

Slon, Viviane, Fabrizio Mafessoni, Benjamin Vernot, Cesare de Filippo, Steffi Grote, Bence Viola, Mateja Hajdinjak, et al. "The Genome of the Offspring of a Neanderthal Mother and a Denisovan Father." *Nature* 561 (2018): 113–16.

Stern Jr., Jack T., and Randall L. Susman. "The Locomotor Anatomy of *Australopithecus afarensis*." *American Journal of Physical Anthropology* 60 (1983): 279–317.

Thompson, D'Arcy W. *On Growth and Form*. Cambridge: Cambridge University Press, 1917.

Van Lawick-Goodall, Jane. "The Behaviour of Free-Living Chimpanzees in the Gombe Stream Reserve." *Animal Behaviour Monographs* 1, part 3 (1968): 161–311.

Vrba, Elisabeth S. "Mammals as a Key to Evolutionary Theory." *Journal of Mammalogy* 73, no. 1 (1992): 1–28.

Walker, Alan, Richard E. Leakey, John M. Harris, and Francis H. Brown. "2.5-Myr *Australopithecus boisei* from West of Lake Turkana, Kenya." *Nature* 322, no. 6079 (1986): 517–22.

Washburn, Sherwood L. "The New Physical Anthropology." *Transactions of the New York Academy of Sciences* 13, no. 7, series II (1951): 298–304.

——. "Tools and Human Evolution." *Scientific American* 203 (1960): 62–75.

Weidenreich, Franz. "The Skull of *Sinanthropus pekinensis*: A Comparative Study on a Primitive Hominid Skull." *Paleontologia Sinica, Series D* 10 (1943): 96–157.

Weiss, Kenneth M. "A Tooth, a Toe, and a Vertebra: The Genetic Dimensions of Complex Morphological Traits." *Evolutionary Anthropology: Issues, News, and Reviews* 2, no. 4 (1993): 121–34.

West-Eberhard, Mary Jane. *Developmental Plasticity and Evolution*. Oxford: Oxford University Press, 2003.

White, Tim D., Berhane Asfaw, Yonas Beyene, Yohannes Haile-Selassie, C. Owen Lovejoy, Gen Suwa, and Giday WoldeGabriel. "*Ardipithecus ramidus* and the Paleobiology of Early Hominids." *Science* 326, no. 5949 (2009): 64–86.

Wilson, Edward O. "Social Insects." *Science* 172, no. 3981 (1971): 406.

——. *Sociobiology: The New Synthesis*. Cambridge, MA: Harvard University Press, 1975.

Zahavi, Amotz, and Avishag Zahavi. *The Handicap Principle: A Missing Piece of Darwin's Puzzle*. Oxford: Oxford University Press, 1997.

NOTES

1. DAVID ALBA

1. David M. Alba et al., "Miocene Small-Bodied Ape from Eurasia Sheds Light on Hominoid Evolution," *Science* 350 (2015): aab2625.
2. Stephen Jay Gould, *Wonderful Life: The Burgess Shale and the Nature of History* (New York: Norton, 1990).
3. Roger Lewin, *Bones of Contention: Controversies in the Search for Human Origins* (Chicago: University of Chicago Press, 1997).
4. John G. Fleagle, *Primate Adaptation and Evolution* (Cambridge, MA: Academic, 2013).
5. Thomas H. Huxley, *Evidence as to Man's Place in Nature* (London: Williams and Norgate, 1863).
6. Stephen Jay Gould, "Nonoverlapping Magisteria," *Natural History* 106 (1997): 16–22.

2. PETER ANDREWS

1. Peter Andrews, *An Ape's View of Human Evolution* (Cambridge: Cambridge University Press, 2016).
2. Louis S. B. Leakey, *Adam's Ancestors. The Evolution of Man and His Culture* (New York: Harper Torchbooks, 1960).

3. DAVID BEGUN

1. David R. Begun, *The Real Planet of the Apes* (Princeton, NJ: Princeton University Press, 2015).
2. Jean-Baptiste Lamarck, *Philosophie Zoologique, Ou Exposition des Considérations Relatives à l'Histoire Naturelle des Animaux* (Paris: Musée d'Histoire Naturelle, 1809).
3. Laurent Goulven, "Lamarck: De la Philosophie du Continu à la Science du Discontinu," *Revue d'Histoire des Sciences* 28, no. 4 (1975): 327–60.

4. BRENDA BENEFIT

1. Brenda R. Benefit and Monte L. McCrossin, "Earliest Known Old World Monkey Skull," *Nature* 388 (1997): 368–71.
2. Richard E. F. Leakey, "Evidence for an Advanced Plio-Pleistocene Hominid from East Rudolf, Kenya," *Nature* 242 (1973): 447–50.

5. MICHAEL "MIKE" BENTON

1. Michael J. Benton, *Vertebrate Palaeontology* (Hoboken, NJ: Wiley, 2014).

6. MATT CARTMILL

1. Matt Cartmill and Fred H. Smith, *The Human Lineage*, 2nd ed. (Hoboken, NJ: Wiley, 2022).
2. Simon Conway Morris, *Life's Solution: Inevitable Humans in a Lonely Universe* (Cambridge: Cambridge University Press, 2003).
3. Kevin N. Laland, *Darwin's Unfinished Symphony* (Princeton, NJ: Princeton University Press, 2018).

7. YAOWALAK CHAIMANEE

1. Yaowalak Chaimanee et al., "A New Orang-Utan Relative from the Late Miocene of Thailand," *Nature* 427 (2004): 439–41.

8. GLENN CONROY

1. Glenn C. Conroy, Martin Pickford, Brigitte Senut, and Pierre Mein, "Diamonds in the Desert: The Discovery of *Otavipithecus nambiensis*," *Evolutionary Anthropology* 2, no. 2 (1993): 46–52.
2. Glenn C. Conroy and Herman Pontzer, *Reconstructing Human Origins: A Modern Synthesis* (New York: Norton, 2012).
3. Robert Ardrey, *African Genesis* (London: Collins, 1961).
4. Raymond A. Dart, *Adventures with the Missing Link* (New York: Harper, 1959). See also Raymond A. Dart, "*Australopithecus africanus*: The Man-Ape of South Africa," *Nature* 115 (1925): 195–99.
5. Edward Abbey, *The Monkey Wrench Gang* (Philadelphia: Lippincott, 1975).

9. SIMON CONWAY MORRIS

1. Stephen Jay Gould, *Wonderful Life: The Burgess Shale and the Nature of History* (New York: Norton, 1990).
2. Simon Conway Morris, *The Crucible of Creation* (Oxford: Oxford University Press, 1998).

3. Simon Conway Morris, *Life's Solution: Inevitable Humans in a Lonely Universe* (Cambridge: Cambridge University Press, 2003).

4. Simon Conway Morris, *The Runes of Evolution: How the Universe Became Self-Aware* (West Conshohocken: Templeton, 2015).

5. Simon Conway Morris, *From Extraterrestrials to Animal Minds: Six Myths of Evolution* (West Conshohocken: Templeton Foundation, 2022).

6. Jacques Jaubert et al., "Early Neanderthal Constructions Deep in Bruniquel Cave in Southwestern France," *Nature* 534 (2016): 111–14.

7. Eric Saidel, "Through the Looking Glass, and What We (Don't) Find There," *Biology & Philosophy* 31, no. 3 (2016): 335–52.

10. ERIC DELSON

1. F. Clark Howell, *Early Man* (New York: Time-Life, 1965).

11. MARC FURIÓ

1. Marc Furió and Pere Figuerola, *La Especie Humana: Los Caminos Para Evitar la Extinción* (Spain: RBA Coleccionables, S.A.U., 2017).

12. DAN GEBO

1. Daniel L. Gebo, *Primate Comparative Anatomy* (Baltimore, MD: John Hopkins University Press, 2014).

2. Raymond A. Dart, "*Australopithecus africanus*: The Man-Ape of South Africa," *Nature* 115 (1925): 195–99.

3. Robert Broom, "The South African Fossil Ape-Men. The Australopithecinae. Part I. The Occurrence and General Structure of the South African Ape-Men," *Transvaal Museum Memoirs* no. 2 (1946): 7–144.

13. JAY KELLEY

1. Charles Darwin, *On the Origin of Species. Or the Preservation of Favoured Races in the Struggle for Life* (London: John Murray, 1859).

14. YUTAKA KUNIMATSU

1. Yutaka Kunimatsu et al., "A New Late Miocene Great Ape from Kenya and Its Implications for the Origin of African Great Apes and Humans," *Proceedings of the National Academy of Science, U.S.A.* 104, no. 49 (2007): 19220–25.

2. William E. Swinton, *The Dinosaurs* (London: Natural History Museum, 1969).

3. Wilfrid E. Le Gros Clark, *History of the Primates* (London: British Museum, 1950).

4. James P. Hogan, *Inherit the Stars* (Riverdale: Baen, 1977).

5. Tim D. White et al., "*Ardipithecus ramidus* and the Paleobiology of Early Hominids," *Science* 326, no. 5949 (October 2009): 64–86.

15. LAURA MACLATCHY

1. Matt Cartmill, "Arboreal Adaptations and the Origin of the Order Primates," in *The Functional and Evolutionary Biology of Primates*, ed. Russell H. Tuttle (Chicago: Aldine-Atherton, 1972), 97–122.
2. Peter S. Rodman and Henry M. McHenry, "Bioenergetics and the Origin of Hominid Bipedalism," *American Journal of Physical Anthropology* 52, no. 1 (1980): 103–6.
3. C. Owen Lovejoy, "The Origin of Man," *Science* 211, no. 4480 (1981): 341–50.
4. For example, see William L. Hylander, "Canine Height and Jaw Gape in Catarrhines with Reference to Canine Reduction in Early Hominins," in *Human Paleontology and Prehistory*, ed. Assaf Marom and Erella Hovers (Cham, Switzerland: Springer, 2017), 71–93.

16. SALVADOR MOYÀ-SOLÀ

1. Salvador Moyà-Solà, Meike Köhler, David M Alba, Isaac Casanovas-Vilar, and Jordi Galindo, "*Pierolapithecus catalaunicus*, a New Middle Miocene Great Ape from Spain," *Science* 306 (2004): 1339–44.
2. Salvador Moyà-Solà and Meike Köhler, "A *Dryopithecus* Skeleton and the Origins of Great-Ape Locomotion," *Nature* 379, no. 6561 (1996): 156–59. At the time, the genus *Hispanopithecus* was collapsed into *Dryopithecus*. Not anymore.

18. MARTIN PICKFORD

1. Brigitte Senut, Martin Pickford, Dominique Gommery, Pierre Mein, Kiptalam Cheboi, and Yves Coppens, "First Hominid from the Miocene (Lukeino Formation, Kenya)," *Comptes Rendus de l'Académie des Sciences, Paris* 332 (2001): 137–44.

19. DAVID PILBEAM

1. David Pilbeam, "New Hominoid Skull Material from the Miocene of Pakistan," *Nature* 295, no. 5846 (1982): 232–34.
2. Elwyn L. Simons, "The Phyletic Position of *Ramapithecus*," *Postilla* 57 (1961): 1–9.
3. Robert Ardrey, *African Genesis* (London: Collins, 1961).

20. LESLIE AIELLO

1. Leslie Aiello and Christopher Dean, *An Introduction to Human Evolutionary Anatomy* (London: Academic, 1990).

2. Leslie C. Aiello and Peter Wheeler, "The Expensive-Tissue Hypothesis: The Brain and the Digestive System in Human and Primate Evolution," *Current Anthropology* 36, no. 2 (1995): 199–221.

21. BERHANE ASFAW

1. Berhane Asfaw et al., "The Earliest Acheulean from Konso-Gardula," *Nature* 360, no. 6406 (1991): 732–35.
2. Berhane Asfaw, Tim White, Owen Lovejoy, Bruce Latimer, Scott Simpson, and Gen Suwa, "*Australopithecus garhi*: A New Species of Early Hominid from Ethiopia," *Science* 284 (1999): 629–35.
3. Tim D. White, Gen Suwa, and Berhane Asfaw, "*Australopithecus ramidus*, a New Species of Early Hominid from Aramis, Ethiopia," *Nature* 371 (1994): 306–12.

22. ANNA "KAY" BEHRENSMEYER

1. Glynn Ll. Isaac and John W. K. Harris, "Sites Stratified Within the KBS Tuff: Reports," in *Koobi Fora Research Project Volume 5: Plio-Pleistocene Archaeology*, ed. Glynn Ll. Isaac, assisted by Barbara Isaac (Oxford: Clarendon, 1997), 71–114.
2. Mary D. Leakey and Richard L. Hay, "Pliocene Footprints in the Laetoli Beds at Laetoli, Northern Tanzania," *Nature* 278 (1979): 317–23.

23. RENÉ BOBE

1. Robert Foley, *Another Unique Species: Patterns in Human Evolutionary Ecology* (Harlow: Pearson Education, 1987).
2. Jonathan Kingdon, *East African Mammals: Part A, Carnivores. Part B, Large Mammals. Part C, Bovids. Part D, Bovids*, 4 vols. (Chicago: University of Chicago Press, 1984).

24. TIM BROMAGE

1. Yves Coppens, F. Clark Howell, Glynn Ll. Isaac, and Richard E. F. Leakey, eds., *Earliest Man and Environments in the Lake Rudolf Basin: Stratigraphy, Paleoecology, and Evolution* (Chicago: University of Chicago Press, 1976).

25. JEREMY "JERRY" DESILVA

1. Jeremy DeSilva, *First Steps: How Walking Upright Made Us Human* (New York: Harper, 2021).
2. Lee R. Berger et al., "*Australopithecus sediba*: A New Species of *Homo*-Like Australopith from South Africa," *Science* 328 (2010): 195–204.
3. Jeremy M. DeSilva et al., "The Lower Limb and Mechanics of Walking in *Australopithecus sediba*," *Science* 340, no. 6129 (April 2013): 1232999.

26. STEVE FROST

1. Elisabeth S. Vrba, "Mammals as a Key to Evolutionary Theory," *Journal of Mammalogy* 73, no. 1 (1992): 1–28.

27. YOHANNES HAILE-SELASSIE

1. Yohannes Haile-Selassie et al., "An Early *Australopithecus afarensis* Postcranium from Woranso-Mille, Ethiopia," *Proceedings of the National Academy of Sciences, U.S.A.* 107 (2010): 12121–26.
2. Yohannes Haile-Selassie, Beverly Z. Saylor, Alan Deino, Naomi E. Levin, Mulugeta Alene, and Bruce M. Latimer, "A New Hominin Foot from Ethiopia Shows Multiple Pliocene Bipedal Adaptations," *Nature* 483, no. 7391 (2012): 565–69.
3. Yohannes Haile-Selassie, Stephanie M. Melillo, Antonino Vazzana, Stefano Benazzi, and Timothy M. Ryan, "A 3.8-Million-Year-Old Hominin Cranium from Woranso-Mille, Ethiopia," *Nature* 573, no. 7773 (2019): 214–19.
4. Yohannes Haile-Selassie et al., "New Species from Ethiopia Further Expands Middle Pliocene Hominin Diversity," *Nature* 521, no. 7553 (2015): 483–88.
5. Yohannes Haile-Selassie, "Late Miocene Hominids from the Middle Awash, Ethiopia," *Nature* 412, no. 6843 (2001): 178–81.

28. ASHLEY HAMMOND

1. Tim D. White et al., "*Ardipithecus ramidus* and the Paleobiology of Early Hominids," *Science* 326, no. 5949 (2009): 64–86.

29. SONIA HARMAND

1. Christopher J. Lepre et al., "An Earlier Origin for the Acheulian," *Nature* 477, no. 7362 (2011): 82–85.
2. Sonia Harmand et al., "3.3-Million-Year-Old Stone Tools from Lomekwi 3, West Turkana, Kenya," *Nature* 521, no. 7552 (2015): 310–15.
3. Jane Goodall, "Tool-Using and Aimed Throwing in a Community of Free-Living Chimpanzees," *Nature* 201, no. 4926 (1964): 1264–66.

30. RALPH HOLLOWAY

1. Julian Huxley, *Evolution: The Modern Synthesis* (London: George Alien & Unwin, 1942).

31. KEVIN D. HUNT

1. Jack T. Stern Jr. and Randall L. Susman, "The Locomotor Anatomy of *Australopithecus afarensis*," *American Journal of Physical Anthropology* 60 (1983): 279–317.

2. Donald Johanson and Maitland Edey, *Lucy: The Beginnings of Humankind* (New York: Simon & Schuster, 1981).

3. Brian G. Richmond and William L. Jungers, "Size Variation and Sexual Dimorphism in *Australopithecus afarensis* and Living Hominoids," *Journal of Human Evolution* 29 (1995): 229–45; J. Michael Plavcan, Charles A. Lockwood, William H. Kimbel, Michael R. Lague, and Elizabeth H. Harmon, "Sexual Dimorphism in *Australopithecus afarensis* Revisited: How Strong Is the Case for a Human-Like Pattern of Dimorphism?," *Journal of Human Evolution* 48, no. 3 (2005): 313–20; and Adam D. Gordon, David J. Green, and Brian G. Richmond, "Strong Postcranial Size Dimorphism in *Australopithecus afarensis*: Results from Two New Resampling Methods for Multivariate Data Sets with Missing Data," *American Journal of Physical Anthropology* 135, no. 3 (2008): 311–28.

4. Kim Hill, "Hunting and Human Evolution," *Journal of Human Evolution* 11, no. 6 (1982): 521–44; and Robert A. Foley and Phyllis C. Lee, "Finite Social Space, Evolutionary Pathways, and Reconstructing Hominid Behavior," *Science* 243, no. 4893 (1989): 901–6.

5. Richard F. Kay, "Dental Evidence for the Diet of *Australopithecus*," *Annual Review of Anthropology* 14, no. 1 (1985): 315–41; and William L. Hylander, "Incisor Size and Diet in Anthropoids with Special Reference to Cercopithecidae," *Science* 189, no. 4208 (1975): 1095–98.

6. Yves Coppens, "East Side Story: The Origin of Humankind," *Scientific American* 270, no. 5 (1994): 62–69.

7. Stephen Jay Gould, "Nonoverlapping Magisteria," *Natural History* 106 (March 1997): 16–22.

32. WILLIAM "BILL" KIMBEL

1. Donald C. Johanson and Maurice Taieb, "Plio-Pleistocene Hominid Discoveries in Hadar, Ethiopia," *Nature* 260 (1976): 293–97.

2. Franz Weidenreich, "The Skull of *Sinanthropus pekinensis*: A Comparative Study on a Primitive Hominid Skull," *Paleontologia Sinica, Series D* 10 (1943): 96–157.

3. Clifford J. Jolly, "The Seed-Eaters: A New Model of Hominid Differentiation Based on a Baboon Analogy," *Man* 5, no. 1 (1970): 5–26.

4. C. Owen Lovejoy, "The Origin of Man," *Science* 211, no. 4480 (1981): 341–50.

33. FREDRICK "KYALO" MANTHI

1. Meave G. Leakey, Craig S. Feibel, Ian McDougall, and Alan Walker, "New Four-Million-Year-Old Hominid Species from Kanapoi and Allia Bay, Kenya," *Nature* 376, no. 6541 (1995): 565–71.

34. MARY MARZKE

1. Sherwood L. Washburn, "The New Physical Anthropology," *Transactions of the New York Academy of Sciences* 13, no. 7, series II (1951): 298–304.

35. EMMA MBUA

1. Emma Mbua et al., "Kantis: A New *Australopithecus* Site on the Shoulders of the Rift Valley Near Nairobi, Kenya," *Journal of Human Evolution* 94 (2016): 28–44; Tim D. White, "Evolutionary Implications of Pliocene Hominid Footprints," *Science* 208, no. 4440 (1980): 175–76; Donald C. Johanson et al., "Morphology of the Pliocene Partial Hominid Skeleton (A.L. 288–1) from the Hadar Formation, Ethiopia," *American Journal of Physical Anthropology* 57 (1982): 403–51; and Zeresenay Alemseged et al., "A Juvenile Early Hominin Skeleton from Dikika, Ethiopia," *Nature* 443 (2006): 296–301.
2. Meave G. Leakey et al., "New Hominin Genus from Eastern Africa Shows Diverse Middle Pliocene Lineages," *Nature* 410 (2001): 433–40.
3. Yohannes Haile-Selassie et al., "New Species from Ethiopia Further Expands Middle Pliocene Hominin Diversity," *Nature* 521, no. 7553 (2015): 483–88.
4. Michel Brunet, Alain Beauvilain, Yves Coppens, Emile Heintz, Aladji H. E. Moutaye, and David Pilbeam, "The First Australopithecine 2,500 Kilometres West of the Rift Valley (Chad)," *Nature* 378 (1995): 273–75.

36. ROBYN PICKERING

1. Donald C. Johanson and Blake Edgar, *From Lucy to Language* (New York: Simon & Schuster, 1996).
2. David A. Richards, Simon H. Bottrell, Robert A. Cliff, Klaus Ströhle, and Peter J. Rowe, "U-Pb Dating of a Speleothem of Quaternary Age," *Geochimica et Cosmochimica Acta* 62, no. 23 (1998): 3683–88.
3. Robyn Pickering et al., "U–Pb-Dated Flowstones Restrict South African Early Hominin Record to Dry Climate Phases," *Nature* 565, no. 7738 (2019): 226–29.

37. J. MICHAEL "MIKE" PLAVCAN

1. Raymond A. Dart, "*Australopithecus africanus*: The Man-Ape of South Africa," *Nature* 115 (1925): 195–99.

38. KAYE REED

1. Brian Villmoore et al., "Early *Homo* at 2.8 Ma from Ledi-Geraru, Afar, Ethiopia," *Science* 347 (2015): 1352–55.

39. BRIGITTE SENUT

1. Glenn C. Conroy, Martin Pickford, Brigitte Senut, John Van Couvering, and Pierre Mein, "*Otavipithecus namibiensis*, First Miocene Hominoid from Southern Africa," *Nature* 356, no. 6365 (1992): 144–48; and Glenn C. Conroy, Martin Pickford, Brigitte Senut, and Pierre Mein, "Diamonds in the Desert: The Discovery of *Otavipithecus namibiensis*," *Evolutionary Anthropology* 2, no. 2 (1993): 46–52.

2. Brigitte Senut, Martin Pickford, Dominique Gommery, Pierre Mein, Kiptalam Cheboi, and Yves Coppens, "First Hominid from the Miocene (Lukeino Formation, Kenya)," *Comptes Rendus de l'Académie des Sciences, Paris* 332 (2001): 137–44.

3. Donald C. Johanson and Maurice Taieb, "Plio-Pleistocene Hominid Discoveries in Hadar, Ethiopia," *Nature* 260 (1976): 293–97.

4. Brigitte Senut, Martin Pickford, and Dudely Wessels, "Panafrican Distribution of Lower Miocene Hominoidea," *Comptes Rendus de l'Académie des Sciences, Paris* 325 (1997): 741–46.

5. Aldous Huxley, *Brave New World* (London: Chatto & Windus, 1932).

40. RICHARD "RICH" SMITH

1. Theodosius Dobzhansky, *Genetics and the Origin of Species* (New York: Columbia University Press, 1937).

2. Ernest Mayr, *Systematics and the Origin of Species* (New York: Columbia University Press, 1942).

3. George Gaylord Simpson, *Tempo and Mode in Evolution* (New York: Columbia University Press, 1944).

4. Sherwood L. Washburn, "The New Physical Anthropology," *Transactions of the New York Academy of Sciences* 13, no. 7 series II (1951): 298–304.

5. Jonathan Scott Friedlaender, *The Solomon Islands Project: A Long-Term Study of Health, Human Biology, and Culture Change* (Oxford: Clarendon, 1987).

6. Pascal Boyer, *Religion Explained: The Evolutionary Origins of Religious Thought* (New York: Basic Books, 2001).

41. DAVID STRAIT

1. Alan Walker, Richard E. Leakey, John M. Harris, and Francis H. Brown, "2.5-Myr *Australopithecus boisei* from West of Lake Turkana, Kenya," *Nature* 322, no. 6079 (1986): 517–22.

2. William H. Kimbel, Tim D. White, and Donald C. Johanson, "Implications of KNM-WT 17000 for the Evolution of 'Robust' *Australopithecus*," in *Evolutionary History of the "Robust" Australopithecines*, ed. Frederick E. Grine (Piscataway: Transaction, 1988), 259–68.

42. RANDALL "RANDY" SUSMAN

1. Randall L. Susman, ed., *The Pygmy Chimpanzee: Evolutionary Biology and Behavior* (New York: Plenum, 1984).

2. Richard G. Klein, *The Human Career: Human Biological and Cultural Origins* (Chicago: University of Chicago Press, 2009).

3. George V. Lauder, "Form and Function: Structural Analysis in Evolutionary Morphology," *Paleobiology* 7, no. 4 (1981): 430–42.

4. George Schaller, *The Mountain Gorilla: Ecology and Behavior* (Chicago: Chicago University Press, 1963); and Jane van Lawick-Goodall, "The Behaviour of Free-Living

Chimpanzees in the Gombe Stream Reserve," *Animal Behaviour Monographs* 1 (1968): 161-IN12.

5. For example, see the works of Louis S. B. Leakey, "Recent Discoveries at Olduvai Gorge," *Nature* 188 (1960): 1050–52; John Napier, "Fossil Hand Bones from Olduvai Gorge," *Nature* 196 (1962): 409–11; Michael H. Day and John R. Napier, "Hominid Fossils from Bed I, Olduvai Gorge, Tanganyika: Fossil Foot Bones," *Nature* 201, no. 4923 (1964): 969–70; Louis S. B. Leakey, Phillip V. Tobias, and John R. Napier, "A New Species of the Genus *Homo* from Olduvai Gorge," *Nature* 202 (1964): 7–9; and John T. Robinson, *Early Hominid Posture and Locomotion* (Chicago: University of Chicago Press, 1972).

43. PETER UNGAR

1. Peter Ungar, *Evolution's Bite* (Princeton, NJ: Princeton University Press, 2017).

2. These are the two classics, Charles Darwin and Alfred Wallace, "On the Tendency of Species to Form Varieties; and on the Perpetuation of Varieties and Species by Natural Means of Selection," *Zoological Journal of the Linnean Society* 3, no. 9 (1858): 45–62; and Charles Darwin, *On the Origin of Species. Or the Preservation of Favoured Races in the Struggle for Life* (London: John Murray, 1859).

44. CAROL WARD

1. Richard D. Alexander, *How Did Humans Evolve? Reflections on the Uniquely Unique Species* (Ann Arbor: Museum of Zoology, University of Michigan, 1990).

45. TIM WHITE

1. Donald C. Johanson and Tim D. White, "A Systematic Assessment of Early African Hominids," *Science* 203 (1979): 321–30.

2. Berhane Asfaw, Tim White, Owen Lovejoy, Bruce Latimer, Scott Simpson, and Gen Suwa, "*Australopithecus garhi*: A New Species of Early Hominid from Ethiopia," *Science* 284 (1999): 629–35.

3. These are the original description and subsequent summary papers describing the "Ardi" partial skeleton. Tim D. White, Gen Suwa, and Berhane Asfaw, "*Australopithecus ramidus*, a New Species of Early Hominid from Aramis, Ethiopia," *Nature* 371 (1994): 306–12; and Tim D. White et al., "*Ardipithecus ramidus* and the Paleobiology of Early Hominids," *Science* 326, no. 5949 (2009): 64–86.

4. Charles Darwin, *On the Origin of Species. Or the Preservation of Favoured Races in the Struggle for Life* (London: John Murray, 1859).

46. BERNARD WOOD

1. Bernard Wood, *Human Evolution: A Very Short Introduction* (Oxford: Oxford University Press, 2019).

2. Bernard Wood, ed., *Wiley-Blackwell Encyclopedia of Human Evolution* (Chichester: Wiley-Blackwell, 2013).
3. Theodosius Dobzhansky, *Mankind Evolving* (New Haven, CT: Yale University Press, 1962).
4. William Howells, *Mankind in the Making* (Garden City, NY: Doubleday, 1959).
5. William Howells, *Getting Here: The Story of Human Evolution* (Washington DC: Compass, 1993).
6. Wilfrid E. Le Gros Clark, "Observations on the Anatomy of the Fossil Australopithecinae," *Journal of Anatomy* 81, part 3 (1947): 300–333.

47. EUDALD CARBONELL

1. Read the summary in Juan Luis Arsuaga et al., "Neandertal Roots: Cranial and Chronological Evidence from Sima De Los Huesos," *Science* 344, no. 6190 (2014): 1358–63.
2. José María Bermúdez de Castro, Juan Luis Arsuaga, Eudald Carbonell, Antonio Rosas, Ignacio Martínez, and Marina Mosquera, "A Hominid from the Lower Pleistocene of Atapuerca, Spain: Possible Ancestor to Neandertals and Modern Humans," *Science* 276 (1997): 1392–95.

48. MANUEL DOMÍNGUEZ-RODRIGO

1. Donald Johanson and Maitland Edey, *Lucy: The Beginnings of Humankind* (New York: Simon & Schuster, 1981).
2. Helen E. Fisher, *The Sex Contract: The Evolution of Human Behavior* (New York: William Morrow, 1983).
3. Glynn Isaac, *The Archaeology of Human Origins: Papers by Glynn Isaac* (Cambridge: Cambridge University Press, 1989).
4. John Yellen and Henry Harpending, "Hunter-Gatherer Populations and Archaeological Inference," *World Archaeology* 4, no. 2 (1972): 244–53.

49. DEAN FALK

1. Dean Falk, Christoph P. E. Zollikofer, Naoki Morimoto, and Marcia S. Ponce de León, "Metopic Suture of Taung (*Australopithecus africanus*) and Its Implications for Hominin Brain Evolution," *Proceedings of the National Academy of Sciences* 109, no. 22 (2012): 8467–70.
2. Dean Falk et al., "The Brain of LB1, *Homo floresiensis*," *Science* 308 (2005): 242–45.
3. Weiwei Men et al., "The Corpus Callosum of Albert Einstein's Brain: Another Clue to His High Intelligence?," *Brain* 137, no. 4 (2013): e268.
4. Dean Falk and Eve Penelope Schofield, *Geeks, Genes, and the Evolution of Asperger Syndrome* (Albuquerque: University of New Mexico Press, 2018).

51. YOUSUKE KAIFU

1. Peter Brown et al., "A New Small-Bodied Hominin from the Late Pleistocene of Flores, Indonesia," *Nature* 431 (2004): 1055–61.
2. Lee R. Berger et al., "*Homo naledi*, a New Species of the Genus *Homo* from the Dinaledi Chamber, South Africa," *eLife* 4 (2015): e09560.
3. Florent Détroit et al., "A New Species of *Homo* from the Late Pleistocene of the Philippines," *Nature* 568, no. 7751 (2019): 181–86.
4. Robert Foley, *Another Unique Species: Patterns in Human Evolutionary Ecology* (Harlow: Pearson Education, 1987).

52. RICHARD KLEIN

1. Richard G. Klein, *The Human Career: Human Biological and Cultural Origins* (Chicago: University of Chicago Press, 2009).
2. Rebecca L. Cann, Mark Stoneking, and Allan C. Wilson, "Mitochondrial DNA and Human Evolution," *Nature* 325 (1987): 31–36.
3. Matthias Krings, Anne Stone, Ralf W. Schmitz, Heike Krainitzki, Mark Stoneking, and Svante Pääbo, "Neandertal DNA Sequences and the Origin of Modern Humans," *Cell* 90 (1997): 19–30.
4. David Reich, *Who We Are and How We Got Here: Ancient DNA and the New Science of the Human Past* (New York: Pantheon, 2018).
5. Kathryn Cruz-Uribe et al., "Excavation of Buried Late Acheulean (Mid-Quaternary) Land Surfaces at Duinefontein 2, Western Cape Province, South Africa," *Journal of Archaeological Science* 30, no. 5 (2003): 559–75.
6. William R. Leonard and Marcia L. Robertson, "Comparative Primate Energetics and Hominid Evolution," *American Journal of Physical Anthropology* 102 (1997): 265–81; Herman Pontzer, David A. Raichlen, and Michael D. Sockol, "The Metabolic Cost of Walking in Humans, Chimpanzees, and Early Hominins," *Journal of Human Evolution* 56 (2009): 43–54; Peter S. Rodman and Henry M. McHenry, "Bioenergetics and the Origin of Hominid Bipedalism," *American Journal of Physical Anthropology* 52, no. 1 (1980): 103–6; and Michael D. Sockol, David A. Raichlen, and Herman Pontzer, "Chimpanzee Locomotor Energetics and the Origin of Human Bipedalism," *Proceedings of the National Academy of Sciences* 104, no. 30 (2007): 12265–69.
7. Thure E. Cerling et al., "Global Vegetation Change Through the Miocene/Pliocene Boundary," *Nature* 389, no. 6647 (1997): 153–58.
8. Phillip V. Tobias, "Foreword: Evolution, Encephalization, Environment," in *Human Brain Evolution. The Influence of Freshwater and Marine Food Resources*, ed. Stephen C. Cunnane and Kathlyn M. Stewart (Hoboken, NJ: Wiley-Blackwell, 2010).

53. CARLES LALUEZA-FOX

1. Iñigo Olalde et al., "Derived Immune and Ancestral Pigmentation Alleles in a 7,000-Year-Old Mesolithic European," *Nature* 507 (2014): 225–28.
2. F. Clark Howell, *Early Man* (New York: Time-Life, 1965).
3. Ruth E. Moore, *Evolution* (New York: Time-Life, 1962).

4. C. W. Ceram, *Gods, Graves, and Scholars: The Story of Archaeology* (New York: Knopf, 1951).
5. Donald Johanson and Maitland Edey, *Lucy: The Beginnings of Humankind* (New York: Simon & Schuster, 1981).

54. RICHARD LEAKEY

1. Frank Brown, John Harris, Richard Leakey, and Alan Walker, "Early *Homo erectus* Skeleton from West Lake Turkana, Kenya," *Nature* 316 (1985): 788–92; and Alan Walker and Richard Leakey, eds., *The Nariokotome* Homo erectus *Skeleton* (Cambridge, MA: Harvard University Press, 1993).
2. Henry Gray and Henry Vandyke Carter, *Anatomy Descriptive and Surgical* (London: John William Parker, 1858).
3. Richard E. F. Leakey, "New Cercopithecidae from the Chemeron Beds of Lake Baringo, Kenya," in *Fossil Vertebrates of Africa*, Vol. 1, ed. Louis S. B. Leakey (New York: Academic, 1969): 53–69.
4. Mary found the fossil, but Louis published the report: Louis S. B. Leakey, "A New Fossil Skull from Olduvai," *Nature* 184, no. 4685 (1959): 491–93.
5. Louis S. B. Leakey, Jack F. Evernden, and Garniss H. Curtis, "Age of Bed I, Olduvai Gorge, Tanganyika," *Nature* 191, no. 4787 (1961): 478–79.

55. DANIEL "DAN" LIEBERMAN

1. Daniel E. Lieberman, *The Evolution of the Human Head* (Cambridge, MA: Harvard University Press, 2011).
2. Daniel E. Lieberman, *The Story of the Human Body: Evolution, Health, and Disease* (New York: Vintage, 2014).
3. Daniel E. Lieberman, *Exercised: Why Something We Never Evolved to Do Is Healthy and Rewarding* (New York: Pantheon, 2021).
4. J. Rimas Vaišnys, Dan Lieberman, and David Pilbeam, "An Alternative Method of Estimating the Cranial Capacity of Olduvai Hominid 7," *American Journal of Physical Anthropology* 65, no. 1 (1984): 71–81.
5. Daniel E. Lieberman, David R. Pilbeam, and Bernard A. Wood, "A Probabilistic Approach to the Problem of Sexual Dimorphism in *Homo habilis*: A Comparison of KNM-ER 1870 and KNM-ER 1813," *Journal of Human Evolution* 17 (1988): 503–11.
6. Randolph M. Nesse and George C. Williams, *Why We Get Sick: The New Science of Darwinian Medicine* (New York: Vintage, 1994).

56. BIENVENIDO MARTÍNEZ-NAVARRO

1. Bienvenido Martínez-Navarro, *El Sapiens Asesino y el Ocaso de los Neandertales* (Córdoba: Editorial Almuzara, 2020).
2. Donald Johanson and Maitland Edey, *Lucy: The Beginnings of Humankind* (New York: Simon & Schuster, 1981).
3. Leslie C. Aiello and Peter Wheeler, "The Expensive-Tissue Hypothesis: The Brain and the Digestive System in Human and Primate Evolution," *Current Anthropology* 36, no. 2 (1995): 199–221.

57. BRIANA POBINER

1. Stephen Jay Gould, "Nonoverlapping Magisteria," *Natural History* 106, no. 2 (1997): 16–22.

59. MARY PRENDERGAST

1. The research articles on this topic are numerous, and this review is a good starting point. Svante Pääbo, "The Diverse Origins of the Human Gene Pool," *Nature Reviews Genetics* 16, no. 6 (2015): 313–14.
2. Viviane Slon et al., "The Genome of the Offspring of a Neanderthal Mother and a Denisovan Father," *Nature* 561 (August 2018): 113–16.

60. LORENZO ROOK

1. Roberto Macchiarelli et al., "The Late Early Pleistocene Human Remains from Buia, Danakil Depression, Eritrea," *Rivista Italiana di Paleontologia e Stratigrafia* 110 (2004): 133–44.
2. Léo Gabounia, Marie-Antoinette de Lumley, Abesalom Vekua, David Lordkipanidze, and Henry de Lumley, "Découverte d'un nouvel hominidé à Dmanissi (Transcaucasie, Géorgie)," *Comptes Rendus Palevol* 1, no. 4 (2002): 243–53.
3. For example, see David Lordkipanidze et al., "Postcranial Evidence from Early *Homo* from Dmanisi, Georgia," *Nature* 449 (October 2007): 305–10; and David Lordkipanidze et al., "A Complete Skull from Dmanisi, Georgia, and the Evolutionary Biology of Early *Homo*," *Science* 342, no. 6156 (2013): 326–31.

61. ANTONIO ROSAS

1. Antonio Rosas et al., "The Growth Pattern of Neandertals, Reconstructed from a Juvenile Skeleton from El Sidrón (Spain)," *Science* 357, no. 6357 (2017): 1282–87.

62. CHRIS RUFF

1. C. Owen Lovejoy, Kingsbury G. Heiple, and Albert H. Burstein, "The Gait of *Australopithecus*," *American Journal of Physical Anthropology* 38, no. 3 (1973): 757–79.
2. Sherwood L. Washburn, "Tools and Human Evolution," *Scientific American* 203 (1960): 62–75.
3. Loren Eiseley, *The Immense Journey* (New York: Random House, 1957).

63. JEFFREY "JEFF" SCHWARTZ

1. Jeffrey H. Schwartz, *The Red Ape* (Boston: Houghton Mifflin, 1987).

64. JOHN SHEA

1. John J. Shea, *Stone Tools in Human Evolution: Behavioral Differences Among Technological Primates* (Cambridge: Cambridge University Press, 2017).
2. John J. Shea, *Prehistoric Stone Tools of Eastern Africa: A Guide* (Cambridge: Cambridge University Press, 2020).
3. Misia Landau, *Narratives of Human Evolution* (New Haven, CT: Yale University Press, 1993).
4. Richard E. F. Leakey, "Early *Homo sapiens* Remains from the Omo River Region of South-West Ethiopia: Faunal Remains from the Omo Valley," *Nature* 222, no. 5199 (1969): 1132–34.
5. Ian McDougall, Francis H. Brown, and John G. Fleagle, "Sapropels and the Age of Hominins Omo I and II, Kibish, Ethiopia," *Journal of Human Evolution* 55, no. 3 (2008): 409–20.

65. TANYA SMITH

1. Tanya M. Smith, *The Tales Teeth Tell: Development, Evolution, Behavior* (Cambridge, MA: MIT Press, 2018).
2. Richard E. Green et al., "A Draft Sequence of the Neandertal Genome," *Science* 328, no. 5979 (2010): 710–22.

66. IAN TATTERSALL

1. Ian Tattersall and Robert DeSalle, *The Accidental* Homo sapiens: *Genetics, Behavior, and Free Will* (New York: Pegasus, 2019).
2. Ian Tattersall and Rob DeSalle, *A Natural History of Wine* (New Haven, CT: Yale University Press, 2015); and Rob DeSalle and Ian Tattersall, *A Natural History of Beer* (New Haven, CT: Yale University Press, 2019).

67. MATT TOCHERI

1. Peter Brown et al., "A New Small-Bodied Hominin from the Late Pleistocene of Flores, Indonesia," *Nature* 431, no. 7021 (2004): 1055–61; and Mike J. Morwood et al., "Archaeology and Age of a New Hominin from Flores in Eastern Indonesia," *Nature* 431, no. 7012 (2004): 1087–91.
2. Matthew W. Tocheri et al., "The Primitive Wrist of *Homo floresiensis* and Its Implications for Hominin Evolution," *Science* 317, no. 5845 (2007): 1743–45.

68. MILFORD WOLPOFF

1. Alan G. Thorne and Milford H. Wolpoff, "The Multiregional Evolution of Humans," *Scientific American* 266, no. 4 (1992): 76–83.

2. Milford H. Wolpoff, "'Telanthropus' and the Single Species Hypothesis," *American Anthropologist* 70, no. 3 (1968): 477–93.

3. Richard E. F. Leakey and Alan C. Walker, "*Australopithecus, Homo erectus* and the Single Species Hypothesis," *Nature* 261, no. 5561 (1976): 572–74.

69. CHRISTOPH ZOLLIKOFER

1. Johann Carl Fuhlrott, "Menschliche Überreste aus einer Felsengrotte des Düsselthals. Ein Beitrag zur Frage über die Existenz fossiler Menschen," *Verhandlungen des Nationales Verein des Preusisches Rheinlandisches und Westfalens* 16 (1859): 131–53.

70. SUSANA CARVALHO

1. Raymond A. Dart, "*Australopithecus africanus*: The Man-Ape of South Africa," *Nature* 115, no. 2884 (1925): 195–99.

2. Jane Goodall, "Tool-Using and Aimed Throwing in a Community of Free-Living Chimpanzees," *Nature* 201, no. 4926 (1964): 1264–66.

3. See, for example, Jonathan Kingdon, *East African Mammals: Part A, Carnivores. Part B, Large Mammals. Part C, Bovids. Part D, Bovids*, 4 vols. (Chicago: University of Chicago Press, 1984).

71. FRANS DE WAAL

1. Frans de Waal, *Chimpanzee Politics: Power and Sex Among Apes*, 25th anniversary ed. (Baltimore, MD: John Hopkins University Press, 2007).

2. Frans de Waal, *The Age of Empathy: Nature's Lessons for a Kinder Society* (New York: Three Rivers, 2009).

3. Frans de Waal, *Mama's Last Hug: Animal Emotions and What They Tell Us About Ourselves* (New York: Norton, 2019).

4. Frans de Waal, *Different: Gender Through the Eyes of a Primatologist* (New York: Norton, 2022).

5. Mary-Claire King and Allan C. Wilson, "Evolution at Two Levels in Humans and Chimpanzees," *Science* 188, no. 4184 (1975): 107–16.

73. KRISTEN HAWKES

1. Kristen Hawkes, James F. O'Connell, Nicholas G. Blurton Jones, Helen Alvarez, and Eric L. Charnov, "Grandmothering, Menopause, and the Evolution of Human Life Histories," *Proceedings of the National Academy of Sciences* 95, no. 3 (1998): 1336–39.

2. Nicholas Blurton Jones, *Demography and Evolutionary Ecology of Hadza Hunter-Gatherers* (Cambridge: Cambridge University Press, 2016).

74. LESLEA HLUSKO

1. Rudolf A. Raff, *The Shape of Life: Genes, Development, and the Evolution of Animal Form* (Chicago: University of Chicago Press, 1996).

75. SARAH HRDY

1. Sarah Blaffer Hrdy, *Mothers and Others: The Evolutionary Origins of Mutual Understanding* (Cambridge, MA: Harvard University Press, 2009).
2. Paul R. Ehrlich, *The Population Bomb* (New York: Ballantine, 1968).
3. See Sarah B. Hrdy, "Male-Male Competition and Infanticide Among the Langurs (*Presbytis entellus*) of Abu, Rajasthan," *Folia Primatologica* 22, no. 1 (1974): 19–58.
4. Sarah Blaffer Hrdy, "Myths, Monkeys and Motherhood: A Compromising Life," in *Leaders in Animal Behavior: The Second Generation*, ed. Lee Drickamer and Donald Dewsbury, 343–74 (Cambridge: Cambridge University Press, 2010).
5. Edward O. Wilson, "Social Insects," *Science* 172, no. 3981 (1971): 406.
6. Edward C. Wilson, *Sociobiology: The New Synthesis* (Cambridge, MA: Harvard University Press, 1975).
7. Charles Darwin, *Descent of Man and Selection in Relation to Sex* (London: John Murray, 1871).
8. Richard Dawkins, *The Selfish Gene* (Oxford: Oxford University Press, 1976).
9. Sarah Blaffer Hrdy, *Mother Nature: A History of Mothers, Infants, and Natural Selection* (New York: Pantheon, 1999).
10. Mary Jane West-Eberhard, *Developmental Plasticity and Evolution* (Oxford: Oxford University Press, 2003).
11. Karin Isler and Carel P. Van Schaik, "How Our Ancestors Broke Through the Gray Ceiling: Comparative Evidence for Cooperative Breeding in Early *Homo*," *Current Anthropology* 53, no. S6 (2012): S453–65.
12. J. Anderson Thomson, *Why We Believe in God(s)* (Charlottesville, VA: Pitchstone, 2011).

76. NINA JABLONSKI

1. Nina G. Jablonski, *Skin: A Natural History* (Berkeley: University of California Press, 2008).
2. Nina G. Jablonski, *Living Color: The Biological and Social Meaning of Skin Color* (Berkeley: University of California Press, 2014).
3. Sindiwe Magona, Nina G. Jablonski, and Lynn Fellman, *Skin We Are In* (South Africa: David Philip, 2018).
4. D'Arcy W. Thompson, *On Growth and Form* (Cambridge: Cambridge University Press, 1917).

77. CLIFFORD "CLIFF" JOLLY

1. Clifford J. Jolly, "The Seed-Eaters: A New Model of Hominid Differentiation Based on a Baboon Analogy," *Man* 5, no. 1 (1970): 5–26.

2. Charles Darwin, *On the Origin of Species. Or the Preservation of Favoured Races in the Struggle for Life* (London: John Murray, 1859).
3. Thomas H. Huxley, *Evidence as to Man's Place in Nature* (London: Williams and Norgate, 1863).

79. LEAH KRUBITZER

1. François Jacob, "Evolution and Tinkering," *Science* 196, no. 4295 (1977): 1161–66.
2. Leah Krubitzer, "The Organization of Neocortex in Mammals: Are Species Differences Really So Different?," *Trends in Neurosciences* 18, no. 9 (1995): 408–17.

80. SUSAN LARSON

1. Donald C. Johanson and Maurice Taieb, "Plio-Pleistocene Hominid Discoveries in Hadar, Ethiopia," *Nature* 260, no. 5549 (1976): 293–97.
2. For example, David Pilbeam, "New Hominoid Skull Material from the Miocene of Pakistan," *Nature* 295, no. 5846 (1982): 232–34; and David Pilbeam, Michael D. Rose, John C. Barry, and S. M. Ibrahim Shah, "New *Sivapithecus* Humeri from Pakistan and the Relationship of *Sivapithecus* and *Pongo*," *Nature* 348 (1990): 237–39.
3. Salvador Moyà-Solà, Meike Köhler, David M Alba, Isaac Casanovas-Vilar, and Jordi Galindo, "*Pierolapithecus catalaunicus*, a New Middle Miocene Great Ape from Spain," *Science* 306, no. 5700 (2004): 1339–44.
4. Salvador Moyà-Solà and Meike Köhler, "A *Dryopithecus* Skeleton and the Origins of Great-Ape Locomotion," *Nature* 379, no. 6561 (1996): 156–59.

81. ZARIN MACHANDA

1. It entailed a series of experiments that began in the 1960s. In the late 1980s, the first single-locus DNA sequencing studies followed, and in the 1990s multiple loci analyses finally resolved the phylogenetic relationships of African apes and humans.

82. TOMÀS MARQUÈS-BONET

1. Research articles on this topic are numerous; this review is a good starting point: Svante Pääbo, "The Diverse Origins of the Human Gene Pool," *Nature Reviews Genetics* 16, no. 6 (2015): 313–14.

83. ROBERT "BOB" MARTIN

1. Robert Martin, *How We Do It: The Evolution and Future of Human Reproduction* (New York: Basic Books, 2013).
2. Willi Hennig, *Phylogenetic Systematics* (Urbana: University of Illinois Press, 1966).

84. PRIYA MOORJANI

1. Priya Moorjani, Carlos Eduardo G. Amorim, Peter F. Arndt, and Molly Przeworski, "Variation in the Molecular Clock of Primates," *Proceedings of the National Academy of Sciences* 113, no. 38 (2016): 10607–12.
2. For example, Vincent M. Sarich and Allan C. Wilson, "Immunological Time Scale for Hominid Evolution," *Science* 158, no. 3805 (1967): 1200–1203.
3. For example, Russell Higuchi, Barbara Bowman, Mary Freiberger, Oliver A. Ryder, and Allan C. Wilson, "DNA Sequences from the Quagga, an Extinct Member of the Horse Family," *Nature* 312, no. 5991 (1984): 282–84.
4. Rebecca L. Cann, Mark Stoneking, and Allan C. Wilson, "Mitochondrial DNA and Human Evolution," *Nature* 325, no. 6099 (1987): 31–36.
5. Johannes Krause et al., "The Complete Mitochondrial DNA Genome of an Unknown Hominin from Southern Siberia," *Nature* 464, no. 7290 (2010): 894–97.
6. Viviane Slon et al., "The Genome of the Offspring of a Neanderthal Mother and a Denisovan Father," *Nature* 561, no. 7721 (2018): 113–16.

85. MARK PAGEL

1. Mark Pagel, *Wired for Culture: Origins of the Human Social Mind* (New York: Norton, 2012).

86. HERMAN PONTZER

1. Herman Pontzer, *Burn: New Research Blows the Lid Off How We Really Burn Calories, Lose Weight, and Stay Healthy* (New York: Avery, 2021).
2. Charles Darwin, *On the Origin of Species. Or the Preservation of Favoured Races in the Struggle for Life* (London: John Murray, 1859).
3. Leslie C. Aiello and Peter Wheeler, "The Expensive-Tissue Hypothesis: The Brain and the Digestive System in Human and Primate Evolution," *Current Anthropology* 36, no. 2 (1995): 199–221.

87. HOLGER PREUSCHOFT

1. Holger Preuschoft, *Understanding Body Shapes of Animals: Shapes as Mechanical Constructions and Systems Moving on Minimal Energy Level* (Cham: Springer, 2022).

88. JOAN RICHTSMEIER

1. Kenneth M. Weiss, "A Tooth, a Toe, and a Vertebra: The Genetic Dimensions of Complex Morphological Traits," *Evolutionary Anthropology: Issues, News, and Reviews* 2, no. 4 (1993): 121–34.

89. ROBERT SAPOLSKY

1. Robert M. Sapolsky, *Why Zebras Don't Get Ulcers: A Guide to Stress-Related Diseases, and Copying* (New York: Freeman, 1994).

2. Robert M. Sapolsky, *Behave: The Biology of Humans at Our Best and Worst* (New York: Penguin, 2017).

3. For example, Jane Goodall, "Tool-Using and Aimed Throwing in a Community of Free-Living Chimpanzees," *Nature* 201, no. 4926 (1964): 1264–66; Jane van Lawick-Goodall, "The Behaviour of Free-Living Chimpanzees in the Gombe Stream Reserve," *Animal Behaviour Monographs* 1, part 3 (1968): 161–311; and Jane Goodall, *In the Shadow of Man* (London: Collins, 1971).

4. See, for example, Sarah B. Hrdy, "Male-Male Competition and Infanticide Among the Langurs (*Presbytis entellus*) of Abu, Rajasthan," *Folia Primatologica* 22, no. 1 (1974): 19–58; and Sarah Blaffer Hrdy, "Infanticide as a Primate Reproductive Strategy," *American Scientist* 65, no. 1 (1977): 40–49.

90. CHET SHERWOOD

1. René Descartes, *Meditations on First Philosophy: In Which the Existence of God and the Immortality of the Soul Are Demonstrated*, trans. Elizabeth S. Haldane (Cambridge: Cambridge University Press, 1911).

91. CRAIG STANFORD

1. Robert Ardrey, *African Genesis* (London: Collins, 1961).

2. Donald C. Johanson and Maurice Taieb, "Plio-Pleistocene Hominid Discoveries in Hadar, Ethiopia," *Nature* 260, no. 5549 (1976): 293–97.

3. For example, see C. Owen Lovejoy, "The Origin of Man," *Science* 211, no. 4480 (1981): 341–50; and Donald C. Johanson et al., "Morphology of the Pliocene Partial Hominid Skeleton (A.L. 288-1) from the Hadar Formation, Ethiopia," *American Journal of Physical Anthropology* 57, no. 4 (1982): 403–51.

4. For example, Jack T. Stern Jr. and Randall L. Susman, "The Locomotor Anatomy of *Australopithecus afarensis*," *American Journal of Physical Anthropology* 60, no. 3 (1983): 279–317; and William L. Jungers, "Lucy's Limbs: Skeletal Allometry and Locomotion in *Australopithecus afarensis*," *Nature* 297, no. 5868 (1982): 676–78.

92. JACK STERN

1. John Fowles, *Daniel Martin* (London: Jonathan Cape, 1977), 156.

2. "Diversity in *Australopithecus*: Tracking the Earliest Bipeds," Fourth Annual Stony Brook Human Evolution Symposium, Convened by Richard Leakey (Turkana Basin Institute), September 25, 2007.

93. ANDREA B. TAYLOR

1. Donald C. Johanson and Maurice Taieb, "Plio-Pleistocene Hominid Discoveries in Hadar, Ethiopia," *Nature* 260, no. 5549 (1976): 293–97.

2. Stephen Jay Gould, *Ontogeny and Phylogeny* (Cambridge, MA: Harvard University Press, 1977).

3. Gavin Rylands de Beer, "Embryology and Evolution," *Journal of Philosophical Studies* 5, no. 19 (1930): 482–84.

94. KENNETH "KEN" WEISS

1. Kenneth M. Weiss and Anne V. Buchanan, *Genetics and the Logic of Evolution* (Hoboken, NJ: Wiley, 2004).

2. Kenneth M. Weiss and Anne V. Buchanan, *The Mermaid's Tale: Four Billion Years of Cooperation in the Making of Living Things* (Cambridge, MA: Harvard University Press, 2009).

3. William Howells, *Mankind in the Making* (Garden City, NY: Doubleday, 1959).

95. RICHARD WRANGHAM

1. Richard Wrangham and Dale Peterson, *Demonic Males: Apes and the Origins of Human Violence* (New York: Houghton Mifflin, 1997).

2. Richard Wrangham, *Catching Fire: How Cooking Made Us Human* (New York: Basic Books, 2009).

3. Richard Wrangham, *The Goodness Paradox: The Strange Relationship Between Virtue and Violence in Human Evolution* (New York: Pantheon, 2019).

4. Charles G. Sibley and Jon E. Ahlquist, "The Phylogeny of the Hominoid Primates, as Indicated by DNA-DNA Hybridization," *Journal of Molecular Evolution* 20, no. 1 (1984): 2–15.

96. DANIEL GILBERT

1. Daniel Gilbert, *Stumbling on Happiness* (New York: Knopf, 2006).

2. Richard E. Nisbett and Lee Ross, *Human Inference: Strategies and Shortcomings of Social Judgment* (Englewood Cliffs, NJ: Prentice-Hall, 1980).

97. HENRY T. "HANK" GREELY

1. Henry T. Greely, *The End of Sex: And the Future of Human Reproduction* (Cambridge, MA: Harvard University Press, 2016).

2. Henry T. Greely, *CRISPR People: The Science and Ethics of Editing Humans* (Cambridge, MA: MIT Press, 2021).

98. KEVIN KELLY

1. Kevin Kelly, *The Inevitable: Understanding the 12 Technological Forces That Will Shape Our Future* (New York: Viking, 2016).

2. Kevin Kelly, *Vanishing Asia* (Pacifica, CA: Cool Tools Lab, 2021).

99. DAVID KRAKAUER

1. Jean-Pierre Changeux, *Neuronal Man: The Biology of Mind* (New York: Pantheon, 1985).
2. Friedrich W. Nietzsche, *Human All Too Human: A Book for Free Spirits*, trans. A. Harvey (Chicago: Charles H. Kerr, 1908).
3. Richard Dawkins, *The Blind Watchmaker: Why the Evidence of Evolution Reveals a Universe Without Design* (New York: Norton, 1986).
4. Daniel C. Dennett, *Darwin's Dangerous Idea: Evolution and the Meaning of Life* (New York: Touchstone, 1995).

100. MISIA LANDAU

1. Misia Landau, *Narratives of Human Evolution* (New Haven, CT: Yale University Press, 1991).
2. Vladimir Propp, *Morphology of the Folktale*, trans. Laurence Scott (Austin: University of Texas Press, 1968).
3. William Howells, *Mankind So Far* (London: Sigma, 1948).
4. Hendrik Willem van Loon, *The Story of Mankind: World History* (New York: Boni and Liveright, 1921).
5. Thomas Henry Huxley, *Evolution and Ethics: And Other Essays* (London: Macmillan, 1894).

101. GEOFFREY MILLER

1. Geoffrey F. Miller, *The Mating Mind: How Sexual Choice Shaped the Evolution of Human Nature* (New York: Doubleday, 2000).
2. Geoffrey Miller, *Spent: Sex, Evolution, and Consumer Behavior* (New York: Viking, 2009).
3. Tucker Max and Geoffrey Miller, *Mate: Become the Man Women Want* (New York: Little, Brown, 2015).
4. Geoffrey Miller, *Virtue Signaling: Essays on Darwinian Politics & Free Speech* (Cambrian Moon, 2019).
5. This book is the first in the series: Iain M. Banks, *Consider Phlebas* (New York: Macmillan, 1987).
6. Charles Darwin, *Descent of Man and Selection in Relation to Sex* (London: John Murray, 1871).
7. Richard Dawkins, *The Extended Phenotype: The Long Reach of the Gene* (Oxford: Oxford University Press, 1982).
8. Amotz Zahavi and Avishag Zahavi, *The Handicap Principle: A Missing Piece of Darwin's Puzzle* (Oxford: Oxford University Press, 1997).
9. Steven Pinker, *The Blank Slate: The Modern Denial of Human Nature* (New York: Penguin, 2002).
10. Arthur R. Jensen, *The g Factor: The Science of Mental Ability* (Westport, CT: Praeger, 1998).
11. Robert Plomin, John C. DeFries, Valerie S. Knopik, and Jenae M. Neiderhiser, *Behavioral Genetics*, 7th ed. (New York: Worth, 2016).

12. William MacAskill, *Doing Good Better: How Effective Altruism Can Help You Make a Difference* (New York: Random House, 2015).

13. Toby Ord, *The Precipice: Existential Risk and the Future of Humanity* (New York: Hachette, 2020).

14. Camila Russo, *The Infinite Machine: How an Army of Crypto Hackers Is Building the Next Internet with Ethereum* (New York: Harper Business, 2020).

15. Robin Hanson, *The Age of Em: Work, Love and Life When Robots Rule the Earth* (Oxford: Oxford University Press, 2016).

16. William Gibson, *Neuromancer* (New York: Ace, 1984).

102. EUGENIE "GENIE" SCOTT

1. Eugenie C. Scott, *Evolution vs. Creationism: An Introduction* (Berkeley: University of California Press, 2009).

2. Eugenie C. Scott and Glenn Branch, *Not in Our Classrooms: Why Intelligent Design Is Wrong for Our Schools* (Boston: Beacon, 2006).

3. Ernst Mayr, *Animal Species and Evolution* (Cambridge, MA: Harvard University Press, 1963).

103. ANIL SETH

1. Anil Seth, *Being You: A New Science of Consciousness* (New York: Dutton, 2021).

2. Daniel C. Dennett, *Consciousness Explained* (New York: Back Bay, 1991).

INDEX

Page numbers in *italics* indicate figures.